Digital Movement

Palgrave Studies in Performance and Technology

Series Editors: **Susan Broadhurst** and **Josephine Machon**

Books included in this cutting-edge series centre on global and embodied approaches to performance and technology. As well as focusing on digital performance and art, books in the series also include the theoretical and historical context relevant to these practices. The series offers fresh artistic and theoretical perspectives on this exciting and growing area of contemporary performance practice, and includes contributors from a wide range of international locations working within this varied discipline. Titles in the series will include edited collections and monographs on issues including (but not limited to): identity and live art; intimacy and engagement with technology; biotechnology and artistic practices; technology, architecture and performance; performance, gender and technology; and space and performance.

Titles include:

Susan Broadhurst and Josephine Machon (*editors*)
SENSUALITIES/TEXTUALITIES AND TECHNOLOGIES
Writings of the Body in 21st Century Performance

Susan Broadhurst and Josephine Machon (*editors*)
IDENTITY, PERFORMANCE AND TECHNOLOGY
Practices of Empowerment, Embodiment and Technicity

Maria Chatzichristodoulou and Rachel Zerihan (*editors*)
INTIMACY ACROSS VISCERAL AND DIGITAL PERFORMANCE

Aneta Mancewicz
INTERMEDIAL SHAKESPEARES ON EUROPEAN STAGES

Kara Reilly (*editor*)
THEATRE, PERFORMANCE AND ANALOGUE TECHNOLOGY
Historical Interfaces and Intermedialities

Nicolás Salazar Sutil and Sita Popat
DIGITAL MOVEMENT
Essays in Motion Technology and Performance

Piotr Woycicki
POST-CINEMATIC THEATRE AND PERFORMANCE
Palgrave Studies in Performance and Technology

Palgrave Studies in Performance and Technology
Series Standing Order ISBN 978–0–230–29362–5 Hardback
978–0–230–29363–2 Paperback
(*outside North America only*)

You can receive future titles in this series as they are published by placing a standing order. Please contact your bookseller or, in case of difficulty, write to us at the address below with your name and address, the title of the series and the ISBN quoted above.

Customer Services Department, Macmillan Distribution Ltd, Houndmills, Basingstoke, Hampshire RG21 6XS, England

---

Digital Movement

Essays in Motion Technology and Performance

Edited by

Nicolás Salazar Sutil
University of Surrey, UK

Sita Popat
University of Leeds, UK

palgrave
macmillan

First published 2015 by
PALGRAVE MACMILLAN

Palgrave Macmillan in the UK is an imprint of Macmillan Publishers Limited, registered in England, company number 785998, of Houndmills, Basingstoke, Hampshire RG21 6XS.

Palgrave Macmillan in the US is a division of St Martin's Press LLC, 175 Fifth Avenue, New York, NY 10010.

Palgrave Macmillan is the global academic imprint of the above companies and has companies and representatives throughout the world.

Palgrave® and Macmillan® are registered trademarks in the United States, the United Kingdom, Europe and other countries.

ISBN 978–1–137–43040–3

This book is printed on paper suitable for recycling and made from fully managed and sustained forest sources. Logging, pulping and manufacturing processes are expected to conform to the environmental regulations of the country of origin.

A catalogue record for this book is available from the British Library.

A catalog record for this book is available from the Library of Congress.

Contents

Illustrations

Figures

Tables

Series Editors' Preface

Susan Broadhurst was invited to be Series Editor of *Palgrave Studies in Performance and Technology* in 2009 and she invited Josephine Machon to be co-editor soon afterwards. Performance and technological resources, combined in various forms, constitute an increasingly popular area of artistic practice. In a relatively short time a proliferation of new technological applications have infiltrated and irrevocably altered everyday life. The consequences of this might not be unproblematic, but the ambitions of performance practitioners have been extended by the availability of such resources. The remit of this important series is to acknowledge the progressive and diverse approaches to various performances and artworks employing technology in their practices. The series was launched in 2010 at the Digital Resources Arts and Humanities Conference, hosted by Broadhurst and held at Brunel University, London. Books included in this cutting-edge series centre on embodied approaches to performance and technology globally. As well as focusing on digital performance and art, books in the series also include the theoretical and historical context relevant to these practices. The series offers fresh artistic and theoretical perspectives on this exciting and growing area of contemporary performance and arts practice, and includes contributors working within this varied discipline from a wide range of international locations. Titles in the series will include edited collections and monographs on issues including (but not limited to) identity and live art; intimacy and engagement with technology; biotechnology and artistic practices; technology, architecture and performance; performance, gender and technology; and space and performance.

Susan Broadhurst
Josephine Machon

Acknowledgements

This book came about as a result of the event "Corporeal Computing", which was held at the University of Surrey in September 2013, and which gave rise to many of the contributions collected in this volume. We would like to thank those individuals and groups who sponsored and made this event possible, especially Mirela Dumic and Nigel Gilbert at the Institute of Advanced Studies, University of Surrey. The additional contributions of Brian Rotman, Johannes Birringer, Stamatia Portanova and Maaike Bleeker further endowed this collection with a richer and sharper focus.

We are exceedingly indebted to Laura Griffiths and Eleanor Gribbin, who were entrusted with the task of editing and revising many of the manuscripts, and whose conscientious and detailed work has greatly improved the quality of this publication. Thanks are also due to Susan Broadhurst, who read early drafts of this project and encouraged us to see the book through to the state in which it finds itself now.

We would like to thank Paula Kennedy, our publisher at Palgrave Macmillan, and Peter Cary, assistant editor, for their patience and support. And finally we are eternally grateful to our respective families for their encouragement and forbearance throughout this project.

Contributors

Johannes Birringer is a choreographer and media artist, and with Michèle Danjoux he co-directs the DAP-Lab (http://www.brunel.ac.uk/dap) at Brunel University, London, where he is a professor of performance technologies in the school of arts. As artistic director of the Houston-based AlienNation Co. (www.aliennationcompany.com), he has created numerous dance-theatre works, video installations and digital projects in collaboration with artists in Europe, the Americas, China, and Japan. His digital oratorio *Corpo, Carne e Espírito,* premiered in Brasil at the FIT Theatre Festival in 2008; the interactive dancework *Suna no Onna* was featured at festivals in London; and DAP-Lab's mixed-reality installation *Ukiyo [Moveable Worlds]* went on European tour in 2010. DAP-Lab's most recent production is a dance opera, *for the time being,* created as an homage to the 1913 Futurist Russian opera *Victory over the Sun*. He is the founding director of Interaktionslabor (http://interaktionslabor.de), a media lab located in an abandoned coal mine in southwest Germany and dedicated to long-term research into interactive systems and real-time processes. His books include: *Theatre, Theory, Postmodernism* (1989), *Media and Performance* (1998), *Performance on the Edge* (2000), *Performance, Technology and Science* (2009), and two edited volumes on *Dance and Cognition* (2005), and *Dance and Choreomania* (2011).

Maaike Bleeker is a professor and chair of theatre studies at Utrecht University. She studied art history, theatre studies and philosophy at the University of Amsterdam where she also completed her PhD on *Visuality in the Theatre* (2002). Previously, she lectured at the department of theatre studies of the University of Amsterdam, the Piet Zwart postgraduate programme in fine arts (Rotterdam), Media Gn: Centre for Emergent Media (Groningen), the School for New Dance Development (Amsterdam), the postgraduate programme Arts Performance Theatricality (Antwerp), and in the IPP performance and media studies summer school of the Johannes Gutenberg Universität, Mainz. Since 1991, she has also worked as a dramaturge for various theatre directors, choreographers and visual artists. She performed in several lecture performances, ran her own theatre company (Het Oranjehotel) and translated five plays that were performed by major Dutch theatre companies. She was an artist in residence at the Amsterdam School for the Arts (2006–7) and member of the jury of the Dutch National Theatre Festival TF (2007–8). Professor Bleeker is president of Performance Studies International. She is also a member of the International Advisory Board of Maska (Ljubljana) and of *Inflexions: A Journal of Research-Creation* (Montreal).

Jamie Bullock is director of Integra Lab, a research centre focused on enhancing music technology user experience through design-led research and development. He led the team that developed Integra Live, an interactive sound design environment with over 19,000 downloads worldwide. He is also the author of LibXtract, a C library for audio feature extraction used by interactive arts group Nocte and widely deployed in analysis applications such as Sonic Visualiser. He has published widely on topics related to music technology interfaces and interaction design, and has worked as a consultant to numerous companies, developing audio- and music-based applications for mobile and desktop platforms. Website: http://jamiebullock.com.

Tom Calvert is professor of computing science at Simon Fraser University, Vancouver and director of information technology and the interactive arts programme. Professor Calvert's other activities at Simon Fraser University have included serving as dean of interdisciplinary studies for eight years, vice-president for research and information systems for five years, and leader of the initiative to develop the new school of engineering science. He completed his PhD at Carnegie Mellon University in 1967, and then taught there for five years, serving as chair of the biomedical engineering programme. In 1972, Professor Calvert's interest in biomedical engineering led him to a position in kinesiology (and later computing science) at Simon Fraser University. As a result of working with choreographers (most notably with Merce Cunningham), he developed software to create human figure animation from dance notation. This led to the development in 1990 of Life Forms, a software system for choreographers and animators. Life Forms continues to be developed and marketed by Credo Interactive, Inc., a Vancouver company of which Professor Calvert is founder and board chair.

James Charlton is a New Zealand-based transdisciplinary artist whose recent work has developed out of an established sculptural practice. His work spans a spectrum of media that includes object-based sculpture, stereo-lithography, installation, robotics, and interactive screen-based and performance work. As senior lecturer and programme leader of postgraduate studies at Colab–AUT University's interdisciplinary unit, he lectures on post-material practice and interactive installation, topics that parallel his PhD research into digital materiality with Transart Institute (US/DE) and Plymouth University (UK). Recent publications include *Remembering a Post-Digital Future* (ARPA for Transmediale, DE, 2013), *Acts of Materiality* (Making Futures, Plymouth, UK, 2013), *Exhalant Gestures* (Corporeal Computing Conference, Surrey, UK, 2013). Recent practice includes *Three Action in 56 Bytes*, Berlin, DE, nCompass, (Berlin, DE, 2013), *nbreathMe*, (Manchester, UK, 2013). Selected works can be seen at https://www.idot.net.nz.

John Collomosse is a senior lecturer in the Centre for Vision Speech and Signal Processing (CVSSP) at the University of Surrey. John joined CVSSP in 2009, following four years lecturing at Bath where he also completed his PhD in *Computer Vision and Graphics* (2004). John has spent periods of time at IBM UK Labs, Vodafone R&D Munich, and HP Labs Bristol. John's research focuses on ways to add value and make sense of large, unstructured media collections – to visually search media collections, and present them in aesthetic and comprehensible ways. Recent projects spanning vision and graphics include sketch-based search of images/video; plagiarism detection in the arts; visual search of dance; structuring and presenting large visual media collections using artistic rendering; developing characters animation from 3D multiview capture data. John holds ~50 refereed publications, including oral presentations at ICCV, BMVC, and journal papers in IJCV, IEEE TVCG and TMM. He was general chair for CVMP 2014, BMVC 2012, NPAR 2010–11 (at SIGGRAPH), and is an AE for C&G and CGF. He regularly reviews for EPSRC and the major vision and graphics conferences.

Mark Coniglio is recognized as a pioneering force in the field of interactive performance. With Dawn Stoppiello, he co-founded the media intensive dance group Troika Ranch in 1994. As a result of this artistic practice, he created the software Isadora® – a software tool that provides deep interactive control over digital media. Isadora has become the tool of choice for thousands worldwide, including such notables as Francis Ford Coppola, the Wooster Group, and the Royal Shakespeare Company. Coniglio has received a "Bessie" award, a prize from Prix Ars Electronica, and the World Technology Award, which recognized his long-term legacy in art and technology.

Martine Époque is professor emeritus at University of Québec à Montréal (UQAM). She is an established choreographer and Dalcroze eurhythmics specialist. At Université de Montréal, where she taught from 1967 to 1974, she founded the famous contemporary dance company Groupe Nouvelle Aire (1968), which she directed until 1981. At UQAM, where she taught from 1980 to 2007, she created the dance department (1985) and established the Master's degree programme in dance. She has also been one of the main actors in the foundation of L'Agora de la danse in Montreal (1991). She has earned international recognition for her stage work, conferences and publications. She is the recipient of the Clifford E. Lee prize for choreography 1983 (Alberta) and the Prix du Québec Denise Pelletier (arts d'interprétation) 1994. A number of articles, book chapters and television broadcasts have focused on her career.

Rachel Fensham is professor of dance and theatre, and currently head of the school of culture and communication at the University of Melbourne.

Recent publications include essays on black modern dance (*DRJ* 2013); the ontology of the archive (Museum Terculanum 2013); the monograph *To Watch Theatre* (2009); and the co-edited volume *Dancing Naturally* (with Alexandra Carter, Palgrave 2011). In the wider context of the digital humanities, and as a dance scholar, Fensham is engaged with movement analysis, costume history and archival practices for which she has managed funded projects including Digital Dance Archives (www.dance-archives.ac.uk); iWeave; Pioneer Women; and MoVE (www.visualisingmotion in3d.blogspot.com) (with Nicolás Salazar Sutil). With Peter Boenisch (Kent) she is co-editor of the Palgrave series, *New World Choreographies*.

Lise Amy Hansen is an interaction design researcher with a special interest in artistic explorations of movement data, as well as the technologically mediated role of movement in health and rehabilitation work. She is currently a researcher at the Oslo School of Architecture and Design (AHO) and a designer. She holds an MA from the Royal College of Art, London and a BA (Hons) from Central Saint Martins (CSM), and has been a lecturer at CSM and AHO. Her research revolves around the role of movement as a material in digital interactions and how we may explore an understanding of movement through design research.

Wan-Gi Lee is a Korean cultural theorist specializing in cultural analysis of finance and economy, areas which he has recently explored in his doctoral thesis, *Narrative and Finance: Cultural Logic of Financial Capitalism*, at Goldsmiths, University of London. Lee worked for several years as facility team manager at Korean electronics giant LG (UK), before moving into a career in academia. His other major articles include "The Formation of the Economic Value and Subsumption Mechanism under Financial Capitalism: On the Cultural Problem of Distance in the Making of the Economic Subject", and "Value, Performance and Narrative: Cultural Politics of Narrative as the Meta-Frame of Value Transference in Financial Capitalism".

Tyler Martin works in computer graphics, and is currently a lead software developer at Bespoke, a San Francisco startup developing virtual fitting room technology. When not stuffing his face with all the delicious food of the city, he divides his time between studying character animation, body modelling, and the latest web development technologies. As an engineer he enjoys creating technologies that have a link to the physical senses, yet augment reality to enhance a product experience. While completing his Master's thesis at the University of California, Davis, he built a quadruped animation system which uses embodied 3D motion specification to ease the learning curve for novice animators.

Tychonas Michailidis is a composer and performer working with sensors and interactive technologies. His research interests focus around the role and meaning of sensors and haptics in performances. His work includes collaboration with dancers, video artists, and writers and he has performed in various venues and festivals across Europe and the UK. Currently he is a visiting lecturer at the School of Digital Media Technology at Birmingham City University. For more information visit www.tychonas.com.

Fernando Nabais is active in research, industry and education, while maintaining an independent artistic production. He has directed and managed projects in areas such as augmented reality, digital signage, immersive environments, natural user interaction and social robotics, among many others, and their application to several vertical market segments such as advertising and culture for many Fortune 500 companies and world-renowned museums and cultural events. Several of these projects and their outcomes have been covered by prestigious media such as *Wired* magazine, *MSNBC, The Economist*, among many others. He was the recipient of the Auggie Award for best augmented reality application at the Augmented Reality Event in 2010, San Francisco, USA. He has co-authored several software and hardware patents. Nabais holds an engineering degree in electronics and telecommunications, an MSc in communication sciences and is a PhD candidate on the Nova University of Lisbon/UTAustin digital media programme.

Sally Jane Norman is professor of performance technologies at the University of Sussex. She has led the development of the Attenborough Centre for creative arts, promoting cross-campus initiatives while teaching and supervising postgraduates. From 2015 she will be co-investigator of the Sussex Humanities Lab. Holder of dual nationality (Aotearoa-New Zealand/France), a Doctorat d'Etat and Doctorat de 3è cycle from the Institut d'études théâtrales (Paris III), her work spans art and technology, performance, scenography and sound. She has contributed to collective editions of the Performing Arts Laboratory of the French Centre National de la Recherche Scientifique, ensured commissioned reports for UNESCO and the French Ministry of Culture, and published a wide range of academic and cultural policy papers. Her work has been nurtured by practical experimentation on embodiment and digital technologies since a motion capture workshop she instigated in 1994 at the International Institute of Puppetry, Charleville-Mézières. Sally Jane is regularly engaged with Telefonica Foundation's Vida Art and Artificial Life competition (founding jury member in 1999). As artistic co-director of the Studio for Electro-Instrumental Music, Amsterdam (1998–2000), she co-organized STEIM's first Touch Festival with Michel Waisvisz and Joel Ryan. From 2004 to 2009, she was founding director of Culture Lab, Newcastle University's interdisciplinary digital hub, co-developing the ongoing Music & Machines research series with Bennett Hogg.

Michael Neff is an associate professor in computer science and cinema and digital media at the University of California, Davis, where he directs the Motion Lab, an interdisciplinary research effort in character animation and embodied input. He holds a PhD from the University of Toronto and is also a certified Laban movement analyst. His interests include character animation tools, especially modelling expressive movement, physics-based animation, gesture and applying performing arts knowledge to animation. He received an NSF CAREER Award, the Alain Fournier Award for his dissertation, a best paper award from Intelligent Virtual Agents and the Isadora Duncan Award for Visual Design. He also currently serves as co-director for the programme in cinema and digital media.

Doros Polydorou is currently programme leader of the digital media and creative cultures programme at the University of Hertfordshire. He is a creative practitioner combining skills in both art and technology. His focus is to investigate technological embodiment through the integration of technology with live stage performances. During the last few years, he has done extensive work on Game Engines, Arduino boards, wireless sensors, camera vision systems and motion capture systems. He has worked with dancers, with live music composers, sound engineers, costume designers, stage designers and visual artists. He has published extensively on a topic related with art and technology: robotics. His current research interests include wearable technology, holography and haptic feedback technologies. Website: http://www.techbodiment.com.

Sita Popat is professor of performance and technology at the University of Leeds. Her research interests lie at the intersections between the body, performance, and new media technologies, and she has danced and choreographed with humans, robots and digital sprites. She is the author of *Invisible Connections: Dance, Choreography and Internet Communities* (2007) and co-editor of *Performance Perspectives: A Critical Introduction* (Palgrave 2011). She is co-founder and associate editor of the *International Journal of Performance Arts and Digital Media* (Taylor & Francis). She sits on the board of trustees for DV8 Physical Theatre. In her spare time, Sita plays a gnome healer in World of Warcraft.

Stamatia Portanova is a part-time lecturer in cultural and media studies at the Università degli Studi di Napoli "L'Orientale" (Naples), where she is also a member of the research unit on technocultures. Her research focuses on philosophy, digital culture and aesthetics. She is the author of *Moving without a Body. Digital Philosophy and Choreographic Thoughts* (MIT Press, *Technologies of*

Lived Abstraction series), and of several articles published in journals such as *Body and Society, Computational Culture, Space and Culture, Angelaki, Inflexions* and *Fibreculture*. She is a member of the Senselab, a Montrèal-based interdisciplinary laboratory on research-creation focusing on intersections between philosophy, technology and art through the sensing body in motion.

Denis Poulin is an associate professor in the dance department at Université du Québec à Montréal. At Collège Montmorency in Laval, where he was dance creation teacher from 1985 to 2008, he founded the dance department and its dance programme (1985), and the 3D human movement animation programme (2007). Fascinated by onscreen dance, he has directed many films and videos for his choreographic works and those of Martine Époque. As the winner of the National Film Board of Canada's (NFB) first animation concourse, "Cinéaste recherché(e)" (1981), he produced his first fiction film (*Le Bouffe-petrole*) in 1981. His dance film *Beyond Curtains*, produced by the NFB (1978), won the Copa Enta Maratone dei Templi of the Salerno International Film Festival (1982). In 1999, he and Martine Époque founded the Laboratory for Research-creation in Technochoreography (LARTech, www.lartech.uqam.ca) where they create digital dance films and works for computer using particle animation and motion captured choreography. Their film *CODA*, the finale of *NoBody dance: the Rite of Spring* (2014), produced by the NFB, was part of the official selection of the Toronto International Film Festival (September, 2014) and is in national competition at the Festival du nouveau cinema de Montréal (October, 2014).

Brian Rotman is emeritus, humanities distinguished professor in the department of comparative studies, Ohio State University. His writings have appeared in various scholarly journals, as well as the *Guardian, London Review of Books*, and other publications. Published books include *Signifying Nothing: the Semiotics of Zero* (1991), *Ad Infinitum... the Ghost in Turing's Machine* (1993), *Mathematics as Sign: Writing, Imagining, Counting* (2000), and *Becoming Beside Ourselves: the Alphabet, Ghosts, and Distributed Human Being* (2008). He is the author of a number of stage plays and a radio drama.

Nicolás Salazar Sutil is a Chilean cultural theorist and performance practitioner. He currently works as lecturer in digital arts at the University of Surrey. He funded and directs C8, an artistic collaborative that focuses on the interaction between formal languages (mathematics and programming languages) and digital performance (http://salazarsutil.net). He is also the co-founder, with Rachel Fensham, of the research hub MOVE (Movement Visualization in e-Cultures). Between 2011 and 2014 he was director of performance studies international and chair of the independent scholar committee. Recent publications include *Motion and Representation: the Language of Human Movement* (MIT Press, forthcoming). He is the recipient of the Beca de Excelencia Académica, Universidad de Chile, and the Premio IX Instituto Chileno Norteamericano award, Santiago de Chile.

Thecla Schiphorst is associate director and associate professor in the school of interactive arts and technology at Simon Fraser University in Vancouver, Canada. Her background in dance and computing form the basis for her research in embodied interaction, focusing on movement knowledge representation, tangible and wearable technologies, media and digital art, and the aesthetics of interaction. She is a member of the original design team that developed Life Forms, the computer compositional tool for choreography, and collaborated with Merce Cunningham from 1990 to 2005, supporting his creation of new dance with the computer. She is the recipient of the 1998 PetroCanada Award in New Media. Her media art installations have been exhibited internationally in Europe, Canada, the United States and Asia in many venues including Ars Electronica, the Dutch Electronic Arts Festival (DEAF), Future Physical, Siggraph, the Wexner Centre for the Arts, the Canadian Cultural Centre in Paris, and the London ICA. Thecla Schiphorst leads the whisper[s] research group, an acronym for "wearable, handheld, intimate, sensory, personal, expressive, responsive systems", and is the director of *MovingStories*: Digital Tools for Movement, Meaning and Interaction, a SSHRC-funded international institutional research partnership. Visit: www.movingstories.caies.

Nicholas Toothman is a computer science PhD student at the University of California, Davis. After completing his BS in computer science at California State University, Bakersfield, he joined the character animation research group led by Michael Neff at UC Davis. His research focuses on developing novel ways to create high-quality animation with help from psychology, human-computer interaction, and graphics hardware. A passionate programmer and gamer, he is also a member of the UC Davis Modlab, a research team created to explore the growing number of intersections between sciences and humanities. When Nicholas isn't writing code or replacing dead video cards, he enjoys spending time with his family, playing instruments, and hoarding games he may never get the time to play.

Introduction

Nicolás Salazar Sutil and Sita Popat

Tools to move to think

Human movement is not only anatomical and biological. It is also defined by historical and material conditions of motor issuance – conditions of interpretation, of attempted representation and enactment. Thus, human movement is transformed by developing techniques and evolving technological tools that expand and refine our motile capabilities, and thus help us to understand ourselves through cultures of movement. But if tooled thinking is affecting the evolution of human movement, then how is our capacity to move, and to know through movement, characterized by the digital? What techniques and technologies define our culture of computerized movement today? Throughout this book, we have acknowledged that digital movement is an area of great significance, and that it demands plenty of attention, not least because movement (and likewise sound) are fields that enable the direct (unworded) communication between humans and machines. If we can now move in order to be recognized by motion and gesture sensing devices and systems, what novel communicational interplays and dialogues between human and machine can be established, and do we need to critically assess this kinetic intercreativity?

Martin Heidegger (1962 [2007]) argued that since human knowledge is instrumental before it can become transcendental, we can therefore insert ourselves in a world defined by instruments, where Things can be defined as either "present-to-hand" or "ready-to-hand". The example of the latter, famously, is Heidegger's hammer. Interestingly, Heidegger speaks of a "state of equipment" (114). Is he implying that equipment is in a condition of stasis, of standing, or that things that are ready-to-hand are not capable of being motivated and motorized by their own will? Because the hammer is an instrument devoid of autonomous agency we can pick it up, we can bang a nail, we can put the tool back where it belongs. Heidegger clarifies: "the hammering of the hammer is not a property of entities" (114), thus it is not possible for "Things" to possess their own character. When we use a computer, however, we do not necessarily know what actions or moves are being performed once the computer is switched on. Sometimes we do not even know what the motive behind automated computerization is, nor whose motives they belie, nor what the outcome of computerized actions might be.

Heidegger's hammer can be thrown up as a question: if we are no longer defined by instruments that are "ready-to-hand", but which are "already-handed" (in the sense that computerization involves an independent agency that is handed back to humans), where does technologization of movement land in a contemporary context? Admittedly, in a condition of possibilities defined by equipment whose kind is not static but kinetic. The digital computer can process information numerically using a two-base language – we all know that. Let us not forget, then, that the machine can also *walk* through data, and then produce a memory of these walks. The computer machine can be fed an input, from where it will walk, algorithmically (i.e. step by step), to an output. If a computer can process information thus, and if it can think and plan movement in abstract terms, then it can also perform movement outwardly at the level of a physical hardware or machinic/robotic movement performance. Unlike the hammer, computerized tools have the capacity to evolve independent character and agency. Computers can programme movement as thought – as software – and then perform it at the level of a command to move, closer to some physical hardware or technologized body. (That computers can *perform* movement is a point of enormous significance throughout this book – that is clear from our book title – especially since the performability of movement via technological materiality affords various levels of interpretation: i.e. performance as an artistic, technological or even economic/financial paradigm.) Drawing on Tim Ingold, one might say that because computers can think and perform movement, they have the means of being alive *to* our world, if only artificially. This highlights the power of autokinesis, of self-generated motion and self-generative creation, which distinguishes digital and smart technologies from, say, hammers. And so, unlike earlier histories of tooled movement and tooled thinking, we find ourselves transformed by a tool that can think and move outside our own human ontology, and which can force us to question, doubt and even debunk the subjectivity of being human, and the subjectivity of anthropocentric movement.

Movement returns to us from a technological agency as digital movement, provoking an undisclosable gap. The problem here is not that we see ourselves move on the screen as though we were some other person. This, for instance, is how Brian Massumi (2002) describes Ronald Reagan. The actor sees himself on the screen as others see him. Massumi brackets the terms "mirror-vision" and "movement-vision" to discriminate between two modalities of intersubjective sight. "To see oneself standing as others sec one is not the same as walking as others see one" (50). The sight of movement at the interface is not, as in Massumi's biopic, the sight of a human that sees another human. What is happening is that we are sensing ourselves through computers (optically or otherwise). We perceive ourselves not only as a different person, but also as a different being. Digital movement is more than just intersubjective (human watching human). It is, if you can bear the mouthful, interontological, or interspecies: human sensing machine (or animal, or chimera), which in turn is sensing the human back. There is no mirror gaze here, but an inter-sensorium that is profoundly transformative given the ontological de-centredness of the experience.

Movement on its own

This book was conceived with the following premise in mind: let us look at human movement as a modality of knowledge, as a way of knowing off-balance and in-between. And so if we recognize that we move in a world that is in turn moving around us, how in the face of so much information and transformation can we know *through* movement – how can we know-on-the-go? The question here is not what movement is per se, or what it means. The question is how and what we know *in* and *through* movement. Kinetic and mobile knowledge are perhaps less understood in the scholarly community, and this might set expectations on our book to mark a field as yet unmarked. Perhaps this book can start providing new directions toward this more kinetic way of thinking and knowing. After all, "it is by moving that we know", going back to Ingold, "and it is by moving, too, that we describe. [...] A being that moves, knows and describes must be observant. Being observant means being alive *to* the world" (2011: xii).

There is more to be said on kinetically sensed knowledge – i.e. on knowledge that flows directly from motor perception. Before we move on, however, let us assume movement can be an independent mode of sensation and let us decouple it from its neighbouring sense: vision. If there is more than one modality of seeing (thinking), then *to see through movement* is not the same as *to see from standstill*. These are categorically different ways of seeing (thinking). The gaze "at rest" leads to knowledge that "arrests" things within still perception. This object-tethered vision serves the production of analytical and dialectical discourse well, but it poorly serves this experience of being alive *to* the world that Ingold speaks of. And so if ceaseless movement involves having no point at rest, where does knowledge in motion start and where does it finish? How do we begin to think in motion, and how do we stop? If our ambition is to put together a book that is conceived as thoughts in motion, how do we extract concepts (or shall we say kinecepts) from kinetic experience, and how do we put these to rest in the domain of inscription representation, and discourse?

Start from nothing

Henri Bergson (1922) made a key distinction in his attack on dialectical intellectualism. He argued that we often visualize our world through snapshots and standpoints. Every time we do this we switch on what the French philosopher called the "cinematograph inside us" (323). Bergson attacked this "ordinary knowledge of a cinematographical kind", which seeks to recompose movement through an artificial string of stillnesses. He wrote: "Vain is the attempt to range such states beside each other on the ego supposed to sustain them: never can these solids strung upon a solid make up that duration that flows" (4). To counter this knowledge, Bergson argued that it was necessary to stop taking refuge in this fixed consciousness we have of ourselves. "Install yourself within change," he rallied (324). The result is a coming and going of the mind between the without and the within, in the midst of which, according to Bergson, is a point at equal distance from both. From this moving point, he continued, "we no longer perceive the one, and we do not yet perceive the other: it is there that the image of Nothing is formed" (295).

Motion-blurred perception turns clearly defined objects (some things) into nothingness, a formless blurb, which is the potential for kinetic knowledge. This nothing is in fact full of things. Movement, if we are true to Bergson's word, is a nothingness full of potentiality. It is a blur from where virtually any form may materialize. To start from nothing then means that we must start this book from no single position, from uncertain theoretical perspectives, from no single categorical conception of movement, no single disciple and no ordinary method. That is where this book starts: with that formlessness and that in-betweenness that will allow us to install ourselves within change, and to be able to know-on-the-go – to think things through as we become immersed in the motion of thought and body. We do not know where we are, except in a position of nothingness and middleness. This knowledge in motion may not be suitable for an academic discipline (i.e. movement studies) concerned with arresting movement within cemented areas of study, but it is suited to a "creative evolution", drawing again on Bergson, since this immersion in nothingness fuels the motivation for kinetic creativity and novelty. What we *do* know is that we are directed toward the entanglement of movement and technology, and that our motivation is to sense (think) this entanglement in a kinetic way – movement is our method.

The blurred gaze

If what we propose by way of method is to make use of sensed knowledge (e.g. observation in motion) then the question that follows must be this: what kind of thinking comes about as a result of seeing blurred worlds? One thing that happens when we see through movement is that things appear distorted. Boundaries blur when we budge. The saccadic strategy does not work and motion perception becomes problematic, not least because perception turns to fuzz. The same applies to thinking and conceptualizing in motion. We conceptualize things that are in two or more places at once, and so we are confused. How do we know phenomena, which, by motor sensation, become paradoxical and uncanny? What kind of secretion of thought flows from this? What kind of words, what terms, what concepts (or kinecepts) can we use to give expression to this lack of definition? To generate concepts in and through movement it is necessary to think in-between one thing and the other, and to open up room for the instability that prevails in this condition of irresolvable in-betweenness. It seems that our terms become hyphenated to make room inside more unstable concepts. Thus the blurred gaze cannot rest in fixed representations or terminological language either: *state*-ments do not account for *move*-ments. Alfred Jarry (who was Bergson's student let us not forget) spoke of the "syzygy of words" (1965: 245),[1] which can yield blurred-out terms like here-there, male-female, or abstract-concrete. Deleuze also became a mouthpiece for Bergson when he spoke of movement in relation to "centers of indetermination", where we find a "coincidence of subject and object" (1986 [1997]: 65). If what we seek is some faithful representation of movement in materialized thought (i.e. in language), then we must accept the possibility of a non-representability or a syzygy where words slip along strange hyphenations and other linguistic gestures.

Our contributors try out many different ways of naming this way of thinking off-balance and in-between. Norman speaks of a "malice" in the object when the visual field correlated by spatial situation allows material accoutrements to give life to immaterial abstractions. Birringer speaks of "dis-alignment", Popat sees an "undisclosable gap" and a resident uncanny, which Fensham echoes when she speaks of virtual costumes as "sensory doppelgangers". Portanova speaks of "nebulized" and "intensive vision", Charlton speaks of "non-representability" and the state between art object and subject. Lee mentions "non-performability" in the unstable relation between performed and datarized movement. Schiphorst and Calvert reminisce about their collaborations with Merce Cunningham and speak of "doubting" as a way of seeing movement differently. Rotman's position is likewise sandwiched between the "virtual and the actual", in the midst of which he finds a movement whose mathematical determination as thought is linked with its physicalization in the body. Bleeker subsequently relates this same threshold (drawing also on Brian Massumi) to "lived abstraction". Our understanding of how to think movement is thus characterized by being caught off-balance, squeezed between thought that is conceived via abstraction and thought that is conceived via physical means. Without ever being cut off dialectically, we accept that when moving and thinking via movement one has to dwell in a threshold, in the midst of which lie all manner of complementarisms – that creative opportunity found inside Bergsonian nothingness. And because of this condition of in-betweenness inherent to movement, what we achieve when we move is the linking of differences. We can cancel differences out and make them appear as though they were part of a polarized sameness. We move to establish communicational links between humans, or between animal species (see Nabais's chapter), or between humans and computers (see Neff et al.), or between human sensation and machine hapticity (see Polydorou et al.). We exploit the gap as a potential for emergent artistic practice (see Époque and Poulin) as well as critical intervention (see Coniglio).

Computer-aided movement

The computerization of movement is an issue that demands many trajectories and many lines of enquiry. Rodolfo Llinás's line (2002) is a fine one to get us re-started. For Llinás, a Colombian neuroscientist, brain activity is the evolutionary upshot of motorization. "That which we call thinking is the evolutionary internalization of movement" (35). As soon as living organisms began to move purposefully, Llinás argues, they evolved nerve and brain systems of great complexity. The tunicate or sea squirt is Llinás's most memorable example: this creature starts life as a tadpole-like larva, and then undergoes metamorphosis, becoming sessile and plantlike, at which point it digests its ganglion brain. In its vegetal form, the tunicate has no need for a brain. Of course, even though the human brain has evolved to process much more than motorization, motricity and mentality are nonetheless engaged, according to this thesis, in a relationship of co-dependence, co-determination and co-evolution.

It would be interesting to float this insight over to the domain of artificial intelligence. After all, the human brain is a machine that has evolved as far from the sea squirt as digital computers have evolved from brainless tools. For mathematician John von Neumann, the so-called father of the modern computer, computers and human brains – or what he called "the two automata" – are comparable since the nervous system has "a prima facie digital character" (1958: 91). This is because nerve pulses can be viewed as two-valued markers. For von Neumann, the language of the brain and the language of mathematics are similar in this regard: "when we talk mathematics, we may be discussing a secondary language, built on the primary language used by the central nervous system" (82). In other words, mathematics and computer languages are layers of representation built over the natural digital language of the human brain. This brings us straight onto the in-between identity of all things digital: in the narrow sense, "digital" refers to a transit-state between discrete body part (finger), and discrete abstract value (0–9), or to the transit between physical signal (electricity on/off), and the representation of the physical signal as binary digits 0 and 1. Like movement, digital is neither here nor there. To speak of digital movement is thus to merge two ontological conditions (physical and abstract), and to reside in-between: digitalized human movement is caught between mathematics and mechanics, between computer languages and mechanics. Together, hardware and software can motorize digital things in ways other than our own neurologically disposed movement. Knowledge through movement, when faced with an agency of control that is both independent and exceedingly powerful at the processing level, leverages a politics of control that is inextricable from any serious debate on computer-assisted movement (see Birringer, in this volume). This is why the computer tool has radically shaken the entire history of tooled knowledge that has come before us, and this is why digital movement has the potential to shape every aspect of the human kinetic experience (see Salazar Sutil, in this volume). Digital movement opens up a political contestation that sees two agencies meet from opposite directions: we control the machine, but the machine can control us back.

Creativity and transformation

"An intelligent being bears within himself the means to transform his own nature," wrote Bergson (1922: 159). Let us replace the term "intelligent" with the term "smart" to provoke a slight but profound change of inflection. The significance of digital movement now becomes clearer: if we can move outside ourselves, through smart systems that move for us and through which we are moved, then digital movement is a kind of trance. On loan from Heidegger again, we can use the term "ek-stasis" (1962 [2007]: 377) to speak of a standing out of the self, not only in the temporal sense implied by Heidegger, but also in a kinetic sense (i.e. an "ex-kinesis"). Along with Drew Leder (1990), we might also argue that digital movement is an artificial "ex-corporation", an actualization of bodily movement outside the physical body, which can then be physically re-incorporated and re-embodied. Thus transformed, we return to the present body from its own digital absence. Whether our physical bodies are cut off from this digital ecstasy or not, and whether or not the passage from live to digital movement is continuous, we cannot tell. Are we being disconnected from our own live and living movement, are we withdrawing from some natural way of being to the world, or are we learning how to move anew every time we sense ourselves from the outside, from the bodiless ecstasies of the digital? These are questions that irk and divide our authors.

Having failed to agree on a definitive answer to these questions, is it assuming too much to say that computerized knowledge, which all our authors agree is profoundly transformative, is perhaps more transformational than any other historical form of tooled knowledge? André Leroi

Gourhan went as far as to say (1993) that the only belief system comparable to our own contemporary worldview, at least in the extraordinary permeability of our thinking, takes us back to the Upper Paleolithic, when humans saw and moved halfway between human and animal, and when the ontological distinction between human and natural environment was blurred. If so, some might argue that digital technology has brought on "future atavisms" (Salazar Sutil 2015), for instance through a recovery of the language of gesture (see Rotman, Charlton and Portanova, in this volume), which can express thought in a permeable way (after all, gestural communication can permeate the space between human and machine). Back then and right now gesture performs a modality of *mobile thinking*. Like the human-animal chimeras of Paleolithic culture, we now find ourselves re-enacting an atavism according to which we see and move halfway between human and machine.

Based on these themes, it makes sense to have asked our contributors to help us reveal some of the imponderables of digital movement by an integration of two common factors: creativity and transformation. First, we asked our contributors to consider the interaction between the inventive theory and creative practice of digital movement. Second, and following on from the above, we probed the connections between movement architectures of both a soft and a hard kind. In this way, we wanted to find connections between movement inside *digital thought* (as represented in mathematics, computation, software), vis-à-vis movement inside *digital embodiment* (i.e. devices, robots, doubles, avatars). Third, we asked our contributors to help us reveal how digital movement can be materialized within the ambit of performance, in the broad sense of this term. How is digital movement issued in technological, choreographic, theatrical or even economic performance? Fourth and finally, *Digital Movement* is intended to address the way in which an integrated knowledge of technologized movement can lead to an intervention: whether artistic, critical or political, even, so that creative movement may be used as an instrument to provoke a transformation of the kinetic self.

Kinetopoiesis

Elsewhere I put forward the term "kinetopoiesis" (Salazar Sutil 2015), to speak of the possibilities of making and doing things out of movement. Kinetopoiesis is the potential to build *from* movement, and to do things *through* movement. One final question our authors were faced with is this: what exactly can you make or create with movement? And across and between what kinds of creative disciplines? What creative opportunities does human movement afford, and to what extent do digital tools enhance these creative and decision-making possibilities?

One thing we build kinetically is the virtual, whose structures are as close to thought as they are to the body – perhaps they reside in that midpoint, that point at nothingness that Bergson spoke of. Unlike the structures of a more conventional architecture, kinetopoietic artifacts can exist in a world that is transient and immaterial like the movements of a dancer or the gestures of a mathematical diagram. Eleanor Metheny (1968) spoke of "kinestructs": the mental structures of movement that help build systems of kinesymbolic representation across a number of disciplines (in her case, sports and dance). In devoting this volume to the creative way in which kinestruction can be used to build, to structure, to compose new artifacts of the kinetopoietic imagination, we must not forget that, in its virtual appearance and in its virtual architecture, purposeful movement is a creative potential that exists before any formal language or any material expression. Before any actual representation or physicalization there is an inner motion-vision shared not only by those concerned with "aesthetic movement" (see Popat, this volume). The kinetopoietic imagination – and the making of cultural artifacts of structured movement – stems from an inner vision that can be shared by mathematicians (see chapters 1, 2, 3 and 4), sculptors (5), architects (7), choreographers (4, 6, 7, 12 and 13), theatre practitioners (1, 14 and Conclusion), musicians (11, 12, 13), computer scientists (10 and 14), roboticists (9 and 13). This is not, of course, an exhaustive list.

What concerns us most generally about the kinetopoietic imagination is the possibility of building and sharing creative and transformational knowledge through movement, across disciplines. This brings us back to where we started. Knowledge of a kinetopoietic kind demands not only an embodiment and a situation of the thinking body in space, but also a recognition of our living body related to its world, and to other occupants of this world.

Trajectivities

Paul Virilio (1997) has lamented that the question of the path is hardly ever addressed explicitly, adding: "One hears of objectivity and subjectivity, certainly, but never trajectivity." Then he hammers the nail on the head: "Between the subjective and the objective it seems we have no room for the *trajective*, that being of movement from here to there, from one to the other, without which we will never achieve a profound understanding of [...the] techniques and modalities of displacement" (1997: 24). This forces us to speak not only in terms of this book's trajectory, but also its trajectivity. As we run down the description of chapters, we find an ongoing desire amongst our authors to come to terms with this "being of movement", this "profound understanding of techniques of displacement", of this construction and articulation of a knowledge off-balance and in-between.

In the opening chapter Sally Jane Norman moves us through time, displacing the contemporary digital avant-garde (and its concern for computational formalisms) through consideration of a forerunner in the historical avant-garde (and its concern for mathematical formalisms). She contextualizes

Oskar Schlemmer's kinaesthetic art and its aesthetic underpinnings, highlighting the contemporary resonance of those historical endeavours to fuse embodied and abstract sensibilities. Schlemmer's *Triadic Ballet* (1922) deploys a mathematics of the body in kinaesthetic space, featuring the diverse corporeal behaviours induced by the geometries and materials of heavy architectural costuming. The creative tensions inherent in these earlier endeavours, Norman argues, resonate with those encountered by today's artists who seek to affirm the irreducibility of the corporeal presence whilst simultaneously sublimating it through digital processing.

A further historical frame is provided by Nicolás Salazar Sutil's reflection on the kinetic economy (or "kineconomy") of post-revolutionary Russia. Salazar Sutil focuses on a Soviet-sponsored programme that sought to design everyday movement according to a political agenda concerned with the industrialization of labour and the modernization of people's lives. Alexei Gastev's Central Institute of Labour (CIT) is the chosen case study for an examination of how technology and mathematical studies of movement provided the Soviet establishment with a scientific discourse (an adaptation of Taylor's scientific management), according to which a social re-engineering of the Russian peasantry could take place. Salazar Sutil's argument is that the laboratories of biomechanical science that sought to discover how movement is controlled by the brain and central nervous system were in fact controlled by a central political system. Thus, behind movement there is an intelligence, and behind intelligence lay a political intelligentsia. The author argues that this analogy between motor and state control via movement is an antecedent of contemporary histories of movement, in the sense that our own way of moving in the digital era is profoundly affected and informed by the digital technologies we control (or which control us). As such, our everyday practice of movement – our culture of movement – is indirectly influenced by a business intelligence driving production of new technology and the commoditization of human mobility, which deeply politicizes the relationship between movement and technology.

Part II is entitled "Gesture and the Digital Object". In his essay "Mathematical Movement: Gesture", Brian Rotman muses over a sense of movement – of gesture in particular – that can be baked inside certain mathematical languages and, by extension, inside the technologized mathematics that underpins digital computation. Gesture inducts us into speech, we are told, and it also prescribes many art objects where gesture is deployed as a creative visuo-kinetic resource. Rotman identifies gesture as a midwife to bring mathematics (and by extension digital computer languages) into contact with actual bodies, and indeed, with the business of staged performance. Drawing on an experimental choreography entitled *Ordinal 5*, conceived by Rotman and presented in London's Tate Modern in 2011, the body's disciplined mobilities are examined in a trajectory from frozen forms to formalized abstractions, and back. Rotman thus produces a choreographic proof of two concepts that link mathematical and corporeal motion from opposite directions: the first from the material body through the diagram to abstract, immaterial object, the second from diagrammatically abstracted objects to bodies performing in space.

Stamatia Portanova provides examples of both directions of Rotman's conceptual trajectories in Chapter 4, tracing Michael Jackson's *Thriller* and Psy's *Gangnam Style* dances on their pathway from specific corporeal manifestations through abstract mass mediations and back/onwards to move (other) bodies in space and time. She defines the *virtual choreographic object* as what is created when digital applications contribute to the generation and replication of movement ideas across people and cultures. Such contagious choreographies circulate, become adopted and adapted, continuously mutating as they are separated from their original contexts and meanings and employed, both consciously and serendipitously, to different ends. Portanova applies mathematical topology to contemporary cultural and social dynamics in order to trace the trajectories of these infinitely repeated dances. She describes digitalization as a form of movement codifier, an amplifier of reproduction and imitation, facilitating choreographic contamination between human bodies, and between the human and the movement experience.

James Charlton builds on the preceding discussions by arguing that gesture poses a fundamental sculptural concern as well. Object and subject are brought together, according to Charlton, when a sculptural practice inclusive of digital media can reposition the sculptural artifact as residue, or as "after-fact" (as opposed to "arti-fact"). Rather than seeing the digital as a system of representation, Charlton suggests digital actions should also be located in a dynamic non-representational space directly between artist and material, thus eliminating the necessity of any representational mediation by digital media. If gesture opens up new domains of sculptural creativity, it does so once we agree that sculptural materiality is an unstable moving point caught between ontological agencies. And if the communicational potential of gesture is said to be intersubjective, it is because it might engage the digital whilst maintaining our analogue existence in the body. This chapter finishes with a reflexive look at sculptural movement via Charlton's own practice as a sculptor and digital arts practitioner, especially in relation to his works *Leaking Into Ourselves* and *32-Bit Catch*, both of which provide Charlton with ample opportunities to reflect upon the ontology of being in movement, and being inside digitally mediated movement.

Part III is entitled "Movement Inscription/ Movement Performance", and it deals with the duplicity of movement as an alphabeticized, coded, or datarized form of inscription versus movement as a form of doing (a performance). The debate gets started with the essay "Movement and 21st Century Literacy", where Maaike Bleeker opens up a theoretical dialogue between Brian Rotman and Mark Hansen concerning the co-evolution of humans and technology from the point of view of an emergent corporeal literacy in 21st century technoculture. Once again, Bleeker identifies in moving bodies a capacity for abstraction. The continuous passage from live movement to formally scripted movement becomes a medium for the communication and mediation between humans and machines not only at the level of direct (i.e. non-worded) representation, but also enactment. This corporeal literacy has the potential to transform our sense of subjectivity and our construction of selfhood, and it is for this reason that it should be examined critically. The emergence of a post-alphabetic and enactive literacy performed through gesturo-haptic action requires not only technical, computational or kinetic know-how but also a critical awareness of how the intertwining of lived and abstract movement defines what emerges from our interactions with digital media, and how our own doing (our artistic and everyday performance of movement) is implicated within the affordances of these media.

A critique of the relationship between live and codified movement is also the subject of Lise Amy Hansen's contribution. Where Bleeker sees the production of 21st century literacy in relation to motion capture tools specifically, Hansen considers aesthetic and poetic practices found in other emergent technologies that are becoming ubiquitous not only in digital dance contexts, but also in everyday life. As in the preceding chapter, the medium of coded movement, or what this author calls "movement scripts", assumes the role of a reflexive thinking tool. The representation of movement also offers opportunities for subversion, according to Hansen. When representing movement in written or graphic form, and when materializing the coded body, we can deliberately leave our marks in the wrong place. Hansen then urges us to re-examine the strategies for categorization and interpretation in the visualization and application of movement through the *Sync* tool (developed by Hansen in collaboration with Hellicar and Lewis), to understand the role of movement and its potential for meaning-making, thus informing both a critical view and the skillsets with which to understand and create representable and performable movement.

Building on the previous two chapters, Wan-Gi Lee shows how the datarization of live movement can also inform practices of commercial signification and valorization, which directly inform creative strategies for the description (as opposed to inscription) of movement within the ambit of marketing and the development of state-of-the-art technology. Focusing on motion-controlled smart TVs, Lee uncovers a system of "movement performativity", according to which multinational technology firms like Samsung or LG can objectify a user's kinetic and cognitive behaviour, so as to design commercially successful products accordingly. Drawing on Georg Simmel's classic study of value, Lee follows the circulation of reified movement from its physical issuance, to processes of segmentation, to codification, and finally to valorization – a narrative that is construed by technology firms and the expert business intelligence they support.

Lee expands on the current politics of motion control by arguing that "subjection" (the making of technology-consuming subjects) is deeply engrained within our everyday relationship with high-tech commodities. This is of particular concern in the case of devices like the smart TV, which can sense our everyday movement in the intimate space of our living rooms, and which can register our private kinetic lives as data that firms will subsequently analyse, interpret and narrativize for commercial self-interest.

Moving on to Part IV, which deals broadly with the notions of technology, materiality and materials, Sita Popat turns to "aesthetic movement", or movement within an art context in general (dance, sculpture, performance, installation art, etc.), in order to find means of critically examining Heidegger's revelations surrounding the essence of technology. Reflecting Heidegger, Popat grounds her essay on an understanding of technology in relation to the notion of technê (which is also found inside art), which involves knowledge of a craft and the gathering of materials together to reveal an object (or artwork) from those materials. The knowledge that stems from this materializing process and this "bringing forth" opens up another trajective, in this case the "undisclosable gap" between the known and the unknown. Such a gap, which simultaneously opens up the rift between subject and object, gives leeway for a discussion of Freud's notion of the uncanny as a critique of Heidegger's position. Popat claims that movement in art does not have the power to reveal essential truth, in a Heideggerian sense, but that the aesthetic conception of movement pries open a position of in-betweenness (known-unknown, subject-object) where the uncanny resides, and as a result of which we find ourselves off-balance in relation to the world. It is because technology leads to our withdrawal from the world – a condition from which we in turn withdraw ourselves – that movement art can offer technology a salvation. It is from within this gap that the uncanny finds a way of linking movement art with technology, for instance through robot design and robotic art. Reflecting on her own experience in the development of the SpiderCrab robot (created in 2007/8 by a team of researchers at the University of Leeds and engineers at Shadow Robots Ltd), Popat concludes that the making of humanlike robots is perhaps as much about the fundamental revealing of essences that we do *not* want to acknowledge as it is about the confusion between technological and human essence.

In Chapter 7, Fensham and Collomosse's iWeave project also refers to the critical position of uncanniness in its exploration of digital archiving processes for dance and dance costumes. iWeave undertook the digitization of material objects in the form of four original costumes from the Madge Atkinson dance collection dating back to the 1920s, and immaterial objects including four reconstructions of original dances based on photographic footage. Computer-animated versions of the dresses and dances were rendered to support a Kinect interface, where users could try on the historical costumes virtually, as well as interact with choreographic reconstructions.

Working with dancers who understood the tensile properties of clothing and these earlier expressive dance forms became vital to understanding the flow, shape and pattern of what might constitute "digital kinesthesia". The project was a success from an archival perspective, since the dress could be seen dancing again. The uncanny, shadowy being of the costume avatar evoked a level of kinaesthetic empathy as she invited the viewer to watch her movement and the digital re-animation of the delicate costumes that she wore. Yet the transmission of this content was stilted by reductive notions of how the "natural" or the "touchless" and their correlation to a user interface should be designed around a gaming paradigm.

Wearable materials also provide an entry-point for Johannes Birringer, whose concern lies with what he calls a "gestural materiality" typically found inside the worn *dispositif* or sensor-based wearable technology systems used in a number of digital dance and digital music practices. In addition to reintroducing gesture to the discussion, and its historicity within the technologically inflected performing arts of the historical avant-garde, Birringer finds the need to reflect on the issue of control, or control via gesture, in contemporary avant-garde practice. Birringer then takes a detour – he finds the concept of "dis-alignment" on route to a debate on wearable space or the mutual embedding of body and space through sensor technology. Destabilizing proprioception and intra-actional behaviour, technologized movement and wearable technologies can become new nervous systems, which afford innovative modalities of control. Drawing on two related case studies – Robert Rauschenberg's *Open Score* from *9 Evenings: Theatre and Engineering* (1966) and Chunky Move's *Mortal Engine* (2008/2012) – Birringer wonders whether, in the context of wearable *dispositifs* in digital performance, it is the performer who controls the instrument or whether the relationship player-instrument has become more unstable. After all, and this is the hard-boiled question Birringer leaves echoing behind him: who is controlling who, or indeed what is controlling whom?

Doros Polydorou, Tychonas Michailidis and Jamie Bullock also grapple with issues of human control in interactive performance in Chapter 12, but they take a different perspective. They refer to two performance pieces, *Ukiyo [Moveable Worlds]* and *Whisper[s]*, to illustrate their point. In *Ukiyo [Moveable Worlds]*, the performer is immersed in an interactive environment and her movement triggers particular visual imagery, yet the system is unable to provide feedback to the performer about what it is doing or when it has completed specific actions or elements of the scene. Thus the purportedly interactive environment operates in the performer's experience as a one-way channel of communication from human to technology. In contrast to this, they describe the *Whisper[s]* project, which uses haptic sensation to provide "kinaesthetic and proprioceptive intelligence" to the performer/participants about the proximity and activity of others around them. Their analysis leads these authors to propose a system of sensory feedback as a potential way to close the loop and create a "creative and corporeal link" between performer and technology, and between the performers themselves. Drawing on a theoretical framework grounded on Rotman's notion of "gesturo-haptic writing", and whilst acknowledging that "vibrotactile feedback" might not be appropriate in every case, they provide three scenarios to show how this approach might function in practice and to describe the benefits that it offers.

In Chapter 13, Fernando Nabais reflects on human fascination with creating life and its representations. He presents a historical lineage of robotic performances and artworks, starting in the 1960s and culminating with his own work in the present day. Ontological questions have been critical throughout this brief history, at the most fundamental level of questioning what a robot is. The expectations of the audiences for such works are as important to these questions as the technical and creative possibilities. The increasing independence of artificially intelligent robots leads to questions about the robot performer as an entity. Louis Philippe Demers refers to the machines as "it selves" media (2010), arguing that robotic art today is more concerned with the behaviour of the robot than its representational form. Nabais refers to his performances with flying robots or drones, where the robot does not attempt to represent the physical form of an animal, but its autonomy and its relationship with the human performer provides both performer and audience with a strong sense of its agency as a performer in its own right.

Nicholas Toothman, Tyler Martin and Michael Neff discuss in Chapter 14 the importance of embodiment and movement in the design of computer tools for creative activities. These authors argue that spontaneity and embodiment are key components of creativity, and so tools for creative purposes need to support improvisation and gesture in order to enhance their effectiveness. Toothman, Martin and Neff consider digital animation and drawing in particular, making the case that tools for these activities need to be real-time and interactive in order to maximize the imagination and creative potential of the animator. They describe experiments in which they asked participants to use tracked Wiimotes to undertake a series of computer-based drawing, editing and animation tasks. Parameters were varied across the tasks, including the scale of the Wiimote spatial tracking so that at times the participants had to walk across the space in order to use the edges of the projected screen on which they were working. The authors analysed the results of these experiments and discovered that enhancing the scale of gestural input and the level of spontaneity led to more creative responses from the participants and also to higher levels of task engagement or "flow", even though increased physical effort might have been expected to result in greater fatigue over longer tasks. The authors propose the need for further research in this area in order to build on their preliminary findings.

Thecla Schiphorst and Tom Calvert's chapter also tackles the question of mobile thinking from the point of view of emergent creative practice. As part of the original team behind the development of the LifeForms/DanceForms choreographic software, Schiphorst and Calvert have two decades of experience of working with digital movement representation. They provide an overview of the history of software for movement composition and notation purposes, tracing the lineage and placing their unique contribution into that context. Their work is situated at the confluence between technological and embodied epistemologies of practice, and they discuss the importance of interdisciplinary teams to the development of meaningful outcomes. For them, the question is not only how forms of technology impact upon choreographic processes but also how choreography as lived bodily movement informs technological design. Their historical perspective is brought right up to date with an insight into some of the latest developments of the DanceForms software in their current *MovingStories* project. *MovingStories* exploits the particular capacities of mobile computing (e.g. tablets, smartphones) which incorporate motion-sensors such as accelerometers and gyroscopes. Such sensors enable the capture of expressive movement parameters directly into the machine held in the hand of the dancer. The aim of this ongoing research is to enable dancers and choreographers to share motion data as meaningful, descriptive, rich movement language through digital means, and thus also to offer processes for mapping movement expressivity to inform methodological strategies for developing social and cultural knowledge.

In contrast to some extent, Martine Époque and Denis Poulin take issue with the possibilities motion capture technology and particle animation software can offer choreographers. Reflecting on their own digital choreographic practice, they contend that "infochoreography", or the choreographing of bodies through data, summons a "dance without bodies" or "nobody dance", which radically transforms the creative and aesthetic practice of movement-making within the discipline of digital dance. Speaking of a dancing body released from the "apparent sarcophagus" of the body, Époque and Poulin contend that MoCap can liberate the choreographic imagination to invisible, pure movement. However, the technological media raises a great many imponderables, once again highlighting the difficulties of seeing technologized movement through a single perspective. What ensues is a new creative possibility in the movement arts, where the performer is forced to accept that the delivery of the movement is not entirely within human control, inasmuch as it is part of a distributed process of aesthetic interpretation that is handed over to a digital agency.

Whilst the preceding chapters offer enthusiastic testaments for new and exciting approaches to digital movement theory and practice, Mark Coniglio concludes the book with a sobering dramaturgical reflection based on twenty years of experience as an artist in the field. He explains that "new" media or technology is never neutral in a performance, and the combination of "old" and "new" media always creates what he calls a dramaturgical problem.

Sensor-wearing dancers are no longer just *performing* on stage – they *are* "a technically extended body" with influence beyond their natural and physical sphere. Coniglio notes that sometimes the artist, the press or the producing house feels that it is sufficient to advertise the work simply as including technology, which only exacerbates the issue. The problem, he explains, is that "new" technology is not yet part of the performance vernacular and so it does not have a set of dramaturgical norms as recourse for artist and audience. Coniglio proposes two ways to consider technology in the making of performance – the *digital reflection* (in which the technology is treated as protagonist) and the *digital intervention* (where the technology becomes antagonist). He is largely dissuaded of the creative usefulness of digital reflection, although he cites some exceptions to his disappointment. In contrast he describes Troika Ranch's *Loopdiver*, in which he and Dawn Stoppiello used digital intervention techniques to present their dancers with impossible movement scenarios that challenged their very physicality to find solutions. This endeavour inspired innovative and sometimes surprising movement solutions that were triggered directly by the technology. If digital reflection is to create interesting work then Coniglio calls for new systems that are more sensitive to human gesture and that have a greater sense of context on which to draw, so that they too can provide input to the creative process that can surprise and innovate. One day, he hopes, we will have a sensing machine that can intervene at the same time as it reflects. Coniglio believes that this will herald the start of a new age for digital performance practice.

Note

1. Syzygy refers to the movement and alignment of planets, two of which are at the opposite antipodes of the third. The term was used in Hermetic philosophy to refer to an antinomy, a coming together of opposites (human-divine, male-female, subject-object). Jarry used the term to refer to a concept, central to pataphysics, according to which the poetic and noetic coincide.

Part I
Histories

1

Oskar Schlemmer's Programmatic Gesture Research

Sally Jane Norman

Oskar Schlemmer's place in a collection of essays on digital movement and motion technology may seem curious: his work is largely anchored in the nineteen-twenties, whereas the computerised processing of gesture – instrumental, communicative, or aesthetic – seemingly anticipated by his work commenced only decades after his death. The present chapter attempts to contextualise Schlemmer's unique kinaesthetic art and its aesthetic underpinnings in order to bridge this interval and highlight the contemporary resonance of his endeavours. The "programmatic" adjective of the title designates performative qualities of the codified gestures that are deployed in keeping with Schlemmer's formal analyses of corporeal dynamics and behaviours, and that may be subject to calculated modulations by virtue of worn and borne materials. It also more broadly designates the systematic, methodical nature and frameworks of Schlemmer's gesture research per se. While the notion of gesture as employed here refers essentially to the wilfully expressive movement of the live performer, it is useful in the context of this interdisciplinary study to recall Noland's wider definition of "gesture as the nodal point where culture (the imposition of bodily techniques), neurobiology (the given mechanics of a human sensorimotor apparatus), and embodied experience (the kinesthetic experience specific to an individual body) overlap and inform one another" (2009a: 8).

Schlemmer employed a vast array of media: graphics, lithography, painting, mural, relief and sculpture informed by architectonics and design; performing arts spanning what today is called "object theatre"; sets and costumes for avant-garde and mainstream opera, ballet, and theatre; and choreographic experiments built around systematic investigation of materials, colours, shapes. He was also a musician, dancer and performer, pedagogue and writer, whose short life (1888–1943) was marked by massive socio-cultural and technological transformations and by two world wars. A major figure in Bauhaus history (albeit on a mostly minor wage), Schlemmer taught at Weimar under Gropius from 1921, part-time as head of the mural painting workshop, then combining teaching with a role as founding director of the theatre workshop at Dessau from autumn 1925. His departure in late 1929 was motivated by disagreement with the director Hannes Meyer. The Breslau Academy he joined was shut down by the Nazis shortly after he took up his post. In 1930, the paintings and reliefs Schlemmer had created for the 1923 Weimar Bauhaus Exhibition were destroyed as "exercises devoid of artistic value" (Schlemmer 1972 [1958]: 274). His Stuttgart Kunstverein exhibition planned for March 1933 was closed down by the SA. His letter of protest to Goebbels in April 1933 denounced the mockery to which modern art was subjected. He was provisionally suspended then banned from teaching by National Socialist decree. In 1937 his work featured in the Munich exhibition of "degenerate art". His last forays as a painter included tarpaulins to camouflage strategic military buildings, on which he was employed by a Stuttgart company after the outbreak of the war.

The story of Schlemmer's tenacity as an artist, humanist, and deeply engaged person of his times is recounted elsewhere.[1] The stubbornly systematic thrust of his investigations defies simple chronological accounts. Schlemmer's creations and writings constitute a complex set of resources bound by persistent thematic threads that form a mesh open to diverse analytical approaches. Just as his creations emerge from processes iteratively interwoven throughout his active decades, so too do his published writings, diary entries and correspondence testify to his stubborn return to irreducible, avowedly unresolvable interrogations. The present chapter moves opportunistically across this mesh, looking for premonitory strands that might inform our current reflection on motion technology and performance.

The human form in space: Schlemmer's bodily calculus

From his marquetry apprenticeship aged fifteen to dreams of his future painting in a penultimate diary entry, Schlemmer was enthralled by the artistic exploration and composition of different types and qualities of space (Diserens 1999: 308). His exploration is haunted by human forms – from the instantly recognisable to the tenuously discernible, from the individual portrait to the group – rendered with a prolific visual arts vocabulary. These forms dynamically negotiate their place in the two or three dimensions of drawing and painting, relief and sculpture, and in the temporal dimensions of staged and architecturally sited actions, as familiar entities resurging ghost-like and acceding to new registers of presence across different supports. The human form is no mere figurative motif but a world-totality that is in turn part of a larger totality – the cosmos – which is construed as offering a universal system of references. Schlemmer's vision of human existence as integral and answerable to a greater order, demanding awareness and tuning to different scales of being, is explicitly formulated in his artistic production and teaching, culminating in his 1928–9 courses entitled *Der Mensch* (*Man*), which embraced drawing, measurement and proportion, natural sciences, philosophy and psychology (Schlemmer 1971 [1928]: 23).

Energies that are encapsulated by and radiate beyond an artistically crafted material-cum-symbolic nucleus are evoked throughout Schlemmer's writings. He describes his pictures as "tablets which burst out of their frames and ally themselves with the wall, thus becoming part of a large surface, a larger space than themselves, thus actually becoming part of an ideal architecture. Compressed in them, reduced to a miniature scale, is what should furnish the laws and form of their surroundings. In this sense: Tablets of the Law" (Schlemmer 1972 [1958]: 73–4, 1919 diary entry). His choreographic creations can equally be construed as kernels of energy subject to general laws governing the abstract and symbolic organisation of the body and space, as well as those pertaining to the physiological and psychological workings of the individual organism, as described in *Mensch und Kunstfigur* (*Man and Art Figure*).

> The laws of cubical space are the invisible network of planimetric and stereometric relationships. This mathematic corresponds to the inherent mathematic of the human body and creates its balance by means of movements, which by their very nature are determined *mechanically and rationally*. (...) The laws of organic man, on the other hand, reside in the invisible functions of his inner self: heartbeat, circulation, respiration, the activities of the brain and nervous system. (...) Cubical-abstract space is then only the horizontal and vertical framework for this flow. These movements are determined *organically and emotionally*. (Schlemmer 1925: 23–5)

It is the dancer who bridges and synthesises these sets of laws: "Invisibly involved with all these laws is Man as Dancer (Tänzermensch). He obeys the law of the body as well as the law of space; he follows his sense of himself as well as his sense of embracing space" (Schlemmer 1925: 25).[2] From 1912, when Schlemmer returned to the Stuttgart Fine Art Academy as a master student and teaching assistant, he began to reflect on and experiment with dance, engaging in daily exercise and gymnastic routines. He sought to heighten his understanding of the workings of bodily space and time, and find ways to objectify and universalise intimately experienced physiological space through reference to external constraints and laws. A later diary entry emphatically concedes that his discovery of dance from the standpoint of a painter and sculptor allowed him to appreciate "its essential element, movement, all the more because the expressive range of the former two arts is restricted to the static, the rigid, to "movement captured in a fixed moment" (Schlemmer 1972 [1958]: 283, 1931 diary entry).

By using the Dessau theatre workshop as a largely fixed apparatus, Schlemmer was able to reproducibly test and refine interactions of bodies, materials and colour, sounds and space. The confines of the small, cubic stage allowed systematic experimentation: though the walls could be and occasionally were removed to open the performance space onto the adjacent canteen and main hall, Schlemmer chose to limit variables by working mostly within its conventional proscenium format.[3] Like a conventional computational set-up, bounded by a machine and formally pre-encoded parameters, the Dessau stage was an essentially fixed *dispositif* within which different kinds of gestural sequences and corporeal interventions in space could be freely and repeatably tested. Their artistically bounded nature made these relations all the more aesthetically observable and valuable; in the controllable, scaled stage space, the body functioned as a stronghold of kinetic logic demanding to be variably disclosed as a function of its environment and accessories. At the same time, the alias of live performance ruled out prescriptive closure: Schlemmer valued the performers' idiosyncratic, ultimately arbitrary input to rigorously devised creations, insisting on the essential "fluid human element" that demarcated his works from the "mechanical ballet" tendencies and utopias that inspired many of his contemporaries (Schlemmer 1972 [1958]). Like artists today working to develop new kinds of scores and improvisational structures, the Bauhaus dancer was engaged in a crucially open system, as "a functionary charged with an operation to complete, a task to perform. In one sense, the artist's body might be conceived as a processing machine, acting out algorithms, performing code. Yet, human inflection ocurs within even the most banal task, the most prescriptive operation. Even the simplest sign is subject to interpretation, its meaning unstable, rarely fixed" (Cocker 2013: 72).

Mathematician and theatre historian Irina Kaldrack (2011) suggests that Schlemmer's melding of mathematical, experiential and affective principles gave rise to a kind of bodily calculus or performative programming. Kaldrack's emphasis on calculus – the mathematical study of change – and on the kinaesthetic and synthetic drivers of his work suggests connections between Schlemmer's choreographic experiments and motion computation: insofar as a programme is a declared plan of action, these experiments can be construed as precursors for live coding practices in all their generative vigour, i.e. as actions that construct and reconstruct themselves in the course of their execution.[4] The text entitled *Tänzerische Mathematik* (*Mathematics of the Dance*) literally provides the ground for such a programmatic approach (Schlemmer 1976 [1926]: 118–19). It posits the dancing body as a vital link between planar floor geometry and its stereometric spatial extensions. Through an unfurling of corporeal mechanics, the dancer plays out an embodied mathematics, vivifying space through motion accentuated by linear objects like bars or stilts, or accessories like spheres, cones and cylinders that generate reactive behaviours and alternative gestural programmes, materialising new spatiotemporal relations and topologies.

The live art figure and the malice of the object

Schlemmer sought to transcend the constraints of physical materials and laws by ascribing symbolic and poetic functions to colour, density, and texture, and fusing metaphysical and physical forces: "I shall always aspire to and defend a form of abstraction, which strives to create symbolic elements, thus necessarily surpassing the merely formal or formulaic (...). But at the same time I believe in the necessity of (or the necessary connection with) the human figure, seen as the measure of all things, as the link which makes understanding possible" (Schlemmer 1972 [1958]: 379, 1940 diary entry). This "necessary connection" is demonstrated by the *Abstract Figure* (1921–3), a white plaster sculpture whose geometric facets create an anthropomorphic form. The upright figure poised on its metal rod axis offers changing profiles to viewers whose circumnavigation endows it with movement by proxy: it "cannot be grasped in any given moment; rather it reveals itself in a temporal succession of vantage points and views. Since a piece of sculpture does not yield a total impression from one angle, the spectator is obliged to move, and only by walking around and adding up his impressions does he eventually grasp the sculpture" (Schlemmer 1972 [1958]: 148, 1924 diary entry).

Thematic consistency of Schlemmer's works across different media is attained through his rigorous conception of their specific affordances: "Painting should remain what it is, perfect itself within its own limits, just like music, architecture, the machine, technology, and science. I firmly believe that the laws of painting have not changed now and never will" (Schlemmer 1972 [1958]: 118, 1922 letter to Otto Meyer). It is because of his faith in the purportedly inalterable premises of drawing, painting, relief, sculpture, dance, theatre, that Schlemmer was able to move readily between them, skilfully transposing and redeploying figures and concepts. Thus, for example, the receding depths of his 1923 *Tischgesellschaft* (*Company at Table*) painting announce spatial qualities developed in the 1929 *Maskenchor* (*Mask Choir*) performance, where the masks of seated actors emerge from ghostly blackness. *Metalltanz* and *Glastanz* (*Metal and Glass Dances*, 1929) draw on visions of these materials grounded in sculptural and relief-related explorations.[5] Yet such associations notwithstanding, to seek strict counterparts across the media Schlemmer employs is to diminish his work's complexity and the influence of contextual factors – employment demands, collaborator availability, access to materials, commissioning constraints and costs, etc. Moreover, it risks undermining the uncategorisable events – pageants, spectacular photographic stagings, thematic parties – which so decisively fed into Bauhaus theatre research (e.g. *Metalltanz*, created for the 1929 Metallisches Fest [Metal Party]).

Schlemmer's discovery of dance freed him from the depictions of frozen exemplarity demanded by fixed media, famously described in Lessing's *Laocoon* as tied to subjects that best evoke what precedes and succeeds a given event. On the contrary, live, embodied exploration of the environment demands physical deployment and duration. At the helm of the Dessau workshop from autumn 1925, Schlemmer attempted to synthesise formal stage parameters with human morphologies and actions, exploring kinetic behaviours, corporeal typologies and trajectories, accessories and costumes, the properties of different materials, and interplay of sonic and visual actions. This research revolved around essentialist visions of the human figure in space. A 1926 photograph, *Figur im Raum mit Bodengeometrie* (*Figure in Space with Floor Geometry*), shows Werner Siedhoff in a white body stocking standing centre stage at the intersection of a painted white cross demarcating the orthogonal ground axes; the floor is also marked with black diagonal traces and an encompassing circle. Stereometry is made manifest by four white strings stretched diagonally from floor to ceiling, intersecting above the actor's head to designate spatiality of the stage box. A multiple exposure photograph from the same period by Lux Feininger, entitled *Raumlineatur mit Figur* (*Space Lines with Figure*), features Siedhoff standing on a black pedestal about a metre high, placed just outside the black circle at its farthest centre rear point. Caught in a web of diagonal traces, the superimposed legs and arms of successive poses produce a dense, chronophotographic-type motion capture image.[6] Schlemmer noted that marking techniques like these bring space to life, revealing coordinates that enhance awareness of previously unnoticed vectors and structures, and inducing kinds of movement that differ from those in indeterminate space. The performer thus "shrouded in magic" (*vom Nimbus des Magischen umwittert*) (Schlemmer 1971 [1928]) recurs in a diary entry corroborating the notion of the stage as a separate world: "One should have a deep respect and deference for any action performed by the human body, especially on the stage, that special realm of life and illusion, that second reality in which everything is surrounded with the nimbus of magic... " (Schlemmer 1972 [1958]: 243, 1929 diary entry).[7]

The Dessau stage provided a means to extend visual arts-informed reflection and vocabulary to live action, furthering investigation of "the ornament that forms between the body and the outer world" (Schlemmer 1972 [1958]: 32, 1915 diary entry) via a "Theatre of Types".[8] The Ba-Bü (Bauhaus-Bühne) programme[9] featured experiments where neutrality conferred by leotard-type garments, padded to conceal sexual and morphological differences, enhanced explorations of corporeal motion and gestural experiments within the known stage parameters. Performers were anonymised by masks made of papier mâché or plasterform coated with metallic gold, silver, or copper paper.[10] Schemata defined in *Mensch und Kunstfigur* served as starting points to devise motion typologies: the *ambulant architecture* figure transposed the laws of formally construed space to give cubical form to the head, torso, arms, and legs; the *marionette*'s egg-shaped head, vase-shaped torso, club-shaped arms and legs, and ball-shaped joints related functional laws of the body to space; the *technical organism* incarnated laws of corporeal motion to indicate potential rotation, direction, and spatial intersections; the *dematerialisation* figure's symbolic and metaphysical values were conveyed by the spread hand's starshape infinity sign formed by folded arms, and cross-shape of backbone and shoulders (Schlemmer 1925: 26–7). The programmatic deployment, modulation, and combination of this set of static postures – i.e. stilled or latent gestures – thus served to seed new choreographic sequences.

Efforts to engender a dramaturgical vocabulary from strictly formal premises were epitomised by various versions of the *Raumtanz, Gestentanz, Formentanz* (*Dances of Space, Gestures, and Forms*), each involving three performers, one of whom was often Schlemmer. Movements for *Gestentanz* were colour coded and accompanied by specific sounds, the yellow dancer's rapid, saccadic movements contrasting with the red dancer's heavier, right-angled trajectories, and the blue dancer's slow, dragging or crawling movements. "Theatre of Types" interpreter appearances were tailored by simple variations in apparel as in the *Raumtanz,* where padded garments bore variously positioned, dissymmetrical panels of different colours (e.g. covering a leg or torso). Adornments added to masks further individualised the characters – e.g. crude moustaches and spectacles worn in the *Gestentanz,* which gave them a comical pomposity. For full dramatic effect, such modulations required the players to express their identities and interactions through theatrical stylisation, employing exaggerated gestural and vocal techniques.

In contrast to these endogenous principles of differentiation dependent on essentially dramatic technique, Schlemmer also explored the exogenous forces of carried artefacts to shape distinctive dynamic modes of conduct. He explains how the manipulation of forms and objects gives rise to new sets of movements: a rod, bar, sphere or club each "has its own specific law, which dictates how it can be handled" (Schlemmer 1929: 13). The actor who wears, turns, balances, builds with or throws different kinds of objects transforms both the immediate visual field (*optische Bild*) and the correlative spatial situation – i.e. the projected environment to which we witnesses can relate as equally embodied and phenomenologically entrained beings. By investigating how material accoutrements give life to and consolidate links between the human actor and his/her surroundings, Schlemmer creatively exploited what he called the "malice" (*Tücke*) of the object (Schlemmer 1929: 15). Its performative qualities may be grounded in geometric affordances: *Stäbentanz* and *Reifentanz* (*Stick and Hoop Dances*) were choreographed to optimise the visual and kinaesthetic transfigurations wrought by these expertly worn and wielded forms. Alternatively, actions could be derived from specific attributes of thematically treated matter, as in the hallucinatory personifications of metal and glass in the dances of the same name.[11] Whether they have as prime mover the formal constraints of a dominant shape or material, or gross dramatic stereotypes, these staged behaviours all manifest the same programmatic qualities, dramatically spelling out their "source code". Such exploration of schematised gestural repertories in the responsive, temporally and spatially malleable confines of the Dessau theatre triggered startling affective responses in both performers and observers – kinaesthesia being elicited by one's own feelings as a moving body, as well as by one's feelings as observer of another moving body.

Bodily transformations: *Das Triadische Ballett/The Triadic Ballet*

The history of theatre is equated by Schlemmer (1925: 25) with the history of transfiguration of the human body, "made possible by the costume; the disguise. Costume and mask emphasize the body's identity or they change it; they express its nature or they are purposely misleading about it; they stress its conformity to organic or mechanical laws or they invalidate this conformity." The historical tradition of theatrical dance, masked and spectacularly costumed as at the time of Vestris, is an art form seen as a starting point for the renewal of theatrical performance, freed from subservience to the literary word, tone, or narrative gesture (Schlemmer 1972 [1958]: 127, 1922 diary entry). Schlemmer attributed his discovery of live art as a creative medium relevant for his own practice to his preparation of a first version of *The Triadic Ballet* for a charity performance by his regiment in 1916. Granted a furlough to supervise rehearsals, he engaged Stuttgart-based professional dancer Albert Burger with whom he had studied painting under Adolf Hoelzel (Burger and his partner Elsa Hötzel Burger were inspired by Jaques-Dalcroze and Appia, whose work they witnessed in Hellerau in 1912). His later correspondence states that the ballet "brought out an aspect of me which usually remains invisible; I felt transformed and struck people as being very different. The new artistic medium was a much more direct one; the human body" (Schlemmer 1972 [1958]: 50, 1918 letter to Otto Meyer). Re-engaged on rehearsals with the Burgers in 1919, he noted: "What we are doing now, rehearsing the dances, makes it thoroughly apparent that I am the key to them. What suits my body is determined by my kinesthetic sense and cannot be adequately reproduced by other bodies" (Schlemmer 1972 [1958]: 70, 1919 diary entry). He admired the Burgers' technique and saw his own role as ideally that of a director (sensing that as a performer he lacked critical distance), while realising that artistic investigation centred on the human figure required first-hand experience of the "dynamics of the body. Movement. Dance. Kinesthetic sense. Man in his relationship to the world about him" (Schlemmer 1972 [1958]: 133, 1922 diary entry).

The Triadic Ballet haunted Schlemmer for decades, evolving from a 1912 dream scenario and early trial sequences to the full Stuttgart creation in 1922, a version scored for player piano by Paul Hindemith for the 1926 Donaueschingen Music Festival, a performance at the Paris International Dance Competition in 1932, and a set of sketches from 1936 testifying to unrealised plans to revive the work. Its three tonal sequences, traversing a yellow comedy-burlesque opening, a pink festive-ceremonial sequence, and a mystical-fantastical black finale, recall the artist's earlier notes regarding a dance-dramaturgy wrought by chromatic transformations: music, demon and angel protagonists are conjured up by metamorphoses of the stage universe, its hues and timbres ultimately dissolving into a silver-white atmosphere.[12] Departing from this expressionist scenario, *The Triadic Ballet* "flirts with hilarity without falling into grotesqueness, brushes against conventions without running through baseness, and ultimately aspires to dematerialize bodies without purging itself through an occultist direction".[13] It is triadic by virtue of its "three dancers and the three parts of its symphonic architectonic composition and the fusion of the dance, the costumes, and the music. The special characteristics of the ballet are the costumes which are of a colored, three-dimensional design, the human figure which is in an environment of basic mathematical shapes, and the corresponding movements of that figure in space" (Schlemmer 1976 [1926]: 119).

Eighteen costumes donned for the twelve successive dances, crafted by Schlemmer's brother Carl, employed a range of materials and configurations to steer dynamic gestural behaviours evocative of specific characters. Like paintings Schlemmer saw as compressed spatial laws that transcend their framed confines, the *Triadic Ballet* costumes were considered as programmatic entities, bearing within themselves their own temporally spatialised finality or kinetic destiny, through design that enforced the expression – or extrusion – of specific choreographic forms. For example, the awkwardly encased diver's bouncing steps were accentuated by pompom-weighted fringing around his torso, giving him a drolly graceless allure. The Spiral Dancer executed spinning-top fouettés across the stage whose spectacular diagonals were dictated by her dark lacquered skirt and spiral headpiece with contrasting edging. Shields slicing the steel-masked twin disk dancers vertically down their centres reinforced their profiles as time-pieces, ventral and dorsal portions of the shields showing contrasting colours during slow 180° sentinel rotations. Metallic ball torsos and hats of the GoldBall twins were offset by vibrating steel string leggings as they jumped with joined feet, while circular motions of the Wire Dancer's spherically bound wire skirt and headdress formed a shimmering moiré pattern. The eerily dissymmetrical abstract apparition that loomed out of darkness brandishing its curious appendages combined archaic totemic and premonitory cyborgian qualities for a brief, breathtaking finale.

Despite limited rehearsal time – the 1922 creation was the first opportunity to test costumes which so interfered with movement that they had to be completely revised (Schlemmer 1972 [1958]: 129, 1922 diary entry) – *The Triadic Ballet*'s plastic and choreographic composition displayed remarkable cohesion.[14] Like computer graphics entities whose dynamics are correlated with ascribed masses, shapes, and constituent materials, or determined by forward and inverse kinematics, the dancers' movements were dictated by the morphological constraints of their very ostensible apparel. Yet at the same time, beyond systematic deployment of physically encoded behaviours, *The Triadic Ballet* effectively exploited symbolic and metaphysical associations triggered by the rich array of costume finishes (soft padding, metal, lacquer, etc.), playing on figurative possibilities opened up by the novel allure of these materials.[15] Last but not least, *The Triadic Ballet* demonstrates deep theatrical intelligence in its integration of conventions spanning buffoonery and whimsy, solemn regalia, the grotesque, and the disturbingly enigmatic.

Motion mappings

The dramaturgically productive relations Schlemmer orchestrated between live human agents and environments apprehended as their spatially, temporally, and materially active counterparts are best documented by his own creations and critical writings. His programming of interactions between bodies and their staged surroundings, denuded and starkly geometrised or playfully inflected, makes him an inspiring precursor for artists today working in performance and motion. For all these achievements, Schlemmer's slight profile, notably compared with Laban's, with whom he coorganised the 1929 Magdeburg Dance Congress,[16] seems paradoxical if not ironic given their very different destinies.[17] While the international dance community has made good use of Laban's kinesphere, dynamosphere, and notational concepts for its technological advances (e.g. William Forsythe's *Improvisation Technologies: A Tool for the Analytical Dance Eye*, 1999), Schlemmer's singular insights into the traces and cross-media renderings of human motion have still to be fully gauged.

Analogies can be readily drawn between the grids and wireframe features of computer-generated imagery and Schlemmer's graphical expression of potential relations between a human figure and its surroundings, notably in the *Figur und Raumlineatur/Figure and Space-Delineation* and *Egozentrische Raumlineatur/Egocentric Space-Delineation* ink drawings of 1924 (see illustrations in Schlemmer 1925: 23–4), which respectively make the human form subservient to an external geometric web, or empower it to generate its own organic environment (Norman 2001). Such analogies are all the more easily substantiated given Schlemmer's atypical mastery of multiple, variably dimensioned media: few graphics developers of wireframe figures have actually physically used wire, let alone with the expertise Schlemmer acquired at the Bauhaus metal workshop. Similar connections can be surmised between contemporary computer graphics libraries of body types, and the basic corporeal schemata described in *Man and Art Figure*.

Schlemmer's overlooked insights into ways of mapping movement include his speculation about a plastic, putty-like medium which, traversed by a moving body, would harden to serve as an imprint and record of the motion: "If one were to imagine a space filled with a soft, pliable substance in which the figures of the sequences of the dancer's movements were to harden as a negative form, then this would demonstrate the direct relationship of the geometry of the plane to the stereometry of space" (Schlemmer 1976 [1926]: 118). This vision of an exacerbated spatial tactility is developed in a lecture:

> Let's imagine a human body moving in space, and let's imagine the space as though it were filled with a soft mass that would harden once the movement had been executed. The movements of the body (torsions, thrusts, etc) would thus stay, in this mass which has become solid, like plastic shapes generated by the body. For example, if I were to move my arm or leg parallel to the axis of the body, this would create a disk shape. If I were to rotate my arm or my leg, this would produce a cone or funnel shape. So this kind of gestural sectioning of space might produce spinning tops, volutes and coiling, whirling, spiralling shapes, that would look like technical organisms. (Schlemmer 1929: 19)

Schlemmer thus anticipated procedures that have recently become feasible through computational renderings of movement, including that used for Forsythe's *Synchronous Objects* interdisciplinary project to facilitate visualisation of the choreographic structure of *One Flat Thing, reproduced* (2000).[18] The project's "3DAlignment Forms" section presents a shifting landscape of three-dimensional surfaces that indicate relationships between traces of the dancers' motions and explores changing viewpoints over time, producing forms uncannily aligned with Schlemmer's prescient vision of ninety years earlier.

Possibilities of mapping and recording the graphic traces of motion have fired notational strategies for centuries, as exemplified by Balthazar de Beaujoyeux's documents for the *Ballet comique de la Reine* (1581), which sequentially display the dancers' elaborate floor patterns. This technique, where ground plans are projected as though from the elevated viewing galleries of dedicated performance spaces (e.g. the Louvre royal reception hall used by de Beaujoyeux), is evoked in Schlemmer's 1927 *Stage Elements* lecture presentation of a floor plan diagram (possibly for *Dance of Gestures II*, where swirling and linear geometries animate the square surface), and his accompanying statement that performances in which choreography plays an important role are known to work best when viewed from above (Schlemmer 1929: 27). Graphic materialisation of motion is described as akin to trails that might be deposited by the dancers' soles. An elaborate set of diagrams of coloured, aerially viewed traces of *Triadic Ballet* sequences for a planned 1936 reconstitution further evidences Schlemmer's interests in notation, and in possibilities of adopting different and changing viewpoints, in keeping with the affordances of recent technological devices.

Schlemmer (1972 [1958]: 141, 1923 diary entry) recognised the potential of emerging technologies for artistic expression, just as he believed in the conjoined poetic and scientific thrust of new imaging techniques and strategies.[19] He was marked by his World War I experience in a military surveying unit, where "aerial photographs provide a hitherto unknown view of the earth's surfaces and forms from above, combined with the enormous tonal effects of photography (...) [T]his technique can derive new, untried stimuli from the visible world. I can think of nothing more fantastic and stimulating to the imagination" (Schlemmer 1972 [1958]: 26, 1916 diary entry). Fusing the camera eye with the superhuman perspectives of a Junkers-J1 airplane for photographic and map-making duties, Schlemmer learned to technically tune to different scales and vantage points: "Could there be anything more marvelous for my purposes than maps? Aerial photographs? After all, aren't they aspects of nature, seen from a long-neglected perspective? (...) I get to see nature from a completely new point of view, and from such an elevation that the earth's surface often appears not to have any" (Schlemmer 1972 [1958]: 43, 1917 letter to Otto Meyer).

The challenges of aerial perspectives would resurge in a very different context twenty-three years later: having endured the political destruction of some of his key works as degenerate, and been banned from teaching, Schlemmer was employed at the end of 1939 by a Stuttgart industrial firm to paint gigantic canvases directly onto seven barracks on the windswept Giebelstadt plateau. Each building measuring two-hundred-and-seventy feet wide, fifty feet high, and seventy-five feet deep, as well as many outbuildings, needed to be covered with anti-aircraft camouflage. Under harsh climatic conditions, restricted by frost until late morning, then by the afternoon chill, and with insufficient scaffolding for an acutely urgent task, Schlemmer executed his last "gigantic paintings" for an aerial viewpoint – this time with no trace of the human form.

Conclusions

Schlemmer's artistic stance anticipates phenomenological approaches to embodied motion (Trimingham 2011), and construals of the profoundly co-implicative relations between externally triggered and spontaneously performed behaviour, between organism and medium.[20] The codified, programmatic characteristics of his kinesic and kinaesthetic inventions appear eminently amenable to technological processing. Yet these systematic attributes are constantly tempered and entangled by wilfully elusive, symbolic qualities, and by a playful and ritual love of performance and the risky openness of live action.

As much as through its mediation and reinvention of the human form, it is Schlemmer's quest to creatively fuse embodied and abstract sensibilities that makes him profoundly contemporary. His work stands as a forerunner for postmodern "directors and performance artists who concentrate on the shifting and permeable visual, aural and haptic qualities of the performance space rather than notions of semiotic signing, narrative and mimesis" (Trimingham 2004: 128). Similarly, and to borrow terms from recent research into responsive media, his exploration of the aesthetic specificities of performative gesture was focused on its unique ability to generate "*temporal texture*, with multiple dynamics and multiple dimensions" (Wei 2002: 465). Schlemmer opened up artistic terrain by staging obdurate material forms that were bursting with symbolic associations, and seemingly intuitive expressive gestures and behaviours that were nonetheless governed by implacable external laws. This endeavour, and its inherent creative tensions, resonate with those encountered by artists seeking today to affirm the irreducibility of the very corporeal presence they are sublimating by digital processing.

Notes

1. Biographical information is provided in Maur (1982), Wingler (1976), Michaud (1978), Lehman and Richardson (1986), Scheper (1988), Diserens (1999), Trimingham (2011).
2. See also 1919 diary entry: "Let the world belong to the dancer, as Nietzsche would say" (77).
3. *The Triadic Ballet* was designed and tested to be performed on a stage visible from all sides, and Schlemmer used outdoor and requisitioned sites for some theatrical events (lantern and kite parades, performer photographs posed against the Dessau façade, thematic parties, etc.), but he chiefly exploited Dessau's stage in its frontal format (cf. Troy 1986: 141).
4. Thecla Schiphorst helpfully voiced this possibility during Corporeal Computing conference discussions (University of Surrey, September 2013).
5. Schlemmer (1925: 32) sees glass and metal as emblematic of future creations for the artist whose plans "founder on the impossibility of materialization. In the final analysis this is unimportant to him. His idea has been demonstrated, and its realization is a question of time, material, and technology. This realization will come with the construction of the new theatre of glass, metal, and the inventions of tomorrow."
6. Feininger's technique belongs to a chronophotographic tradition developed throughout the past century, forming a historical basis for much motion capture imagery. Other uses of multiple exposure in the Bauhaus theatre archives include overlaid shots of dancer Daisy Spies wearing the *Rundrock* (round skirt) for the 1926 *Triadic Ballet* reconstitution, making her pirouetting circular skirt a concrete pivot for her three smiling heads. A rich photographic iconography is reproduced in Scheper (1988).
7. The stage's magical force features in Schlemmer's *Bühnenelemente* lecture (1929). See also Richardson (1986).
8. Inspired by the imminent opening of Dessau's theatre, Schlemmer wrote in December 1925 to Otto Meyer: "My plan is to begin here with a theatrical ABC; I want to create a theater [*sic*] of types" (1972 [1958]: 185).
9. Affectionately abbreviated, as noted by ex-student Albert Flocon (1987: 52).
10. T. Lux Feininger, the photographer who assisted in the mask-making process, thus described their fabrication in response to Debra McCall's queries during reconstitutions of the dances (cf. Moynihan and Odom 1984: 55).

11. For further descriptions and references, see Norman (2001).
12. The notes correspond to a ballet project developed with Burger called *The Courtship*, where Schlemmer was to play the demon with Burger and Hötzel as lovers whose dramatic experience would progress from "the ecstatic and the demonic (...) to purity and form" (cf. Troy 1986: 128).
13. Oskar Schlemmer, playbill for the opening night of *The Triadic Ballet*, Württembergisches Landestheater, Stuttgart, 30 September 1922. Reproduced in the exhibition *Oskar Schlemmer*, Museo Nacional Centro de Arte Reina Sofia, Madrid, 1996–7.
14. It continues to do so, as evidenced by Debra McCall's reconstitutions (cf. Moynihan and Odom 1984). Gerhard Bohner's recreation, commissioned by the Berlin Academy of Arts for the 1977 *Tendenzen der Zwanziger Jahre* exhibition, has toured internationally ever since. Reconstitutions have also attracted criticism from participants in and witnesses of the original creations: Albert Flocon deplores the academism and idle reappropriation of what he calls "reconstitution treachery" (1987: 132–5).
15. "Possibilities are extraordinary in light of today's technological advancements: precision machinery, scientific apparatus of glass and metal, the artificial limbs developed by surgery, the fantastic costumes of the deep sea diver and the modern soldier, and so forth... *Consequently, potentialities of constructive configuration are extraordinary on the metaphysical side as well*" (Schlemmer 1925: 29, emphasis in the original).
16. The Congress was organised in the context of the Deutsche Theater-Ausstellung in Magdeburg by Rudolf Laban, Hanns Niedecken-Gebhard, and Oskar Schlemmer; its three hundred attendees were dominated by Laban and his followers, who sought – in vain – to unite all German dance organisations and institutions within a single federation (cf. Toepfer: 1997).
17. Schlemmer's letter to propaganda minister Joseph Goebbels in April 1933, to vehemently protest against the growing political condemnation of modern art, contrasts with Laban's leadership under Goebbels of dance initiatives dedicated to the *Volk* and *Führer*, before he moved to Dartington Hall, England, in 1937.
18. Co-directed by William Forsythe with Maria Palazzi and Norah Zuniga Shaw at Ohio State University, the project can be consulted at synchronousobjects.osu.edu (date accessed 14 July 2014).
19. Schlemmer cautions in this same text against artists who seek out new possibilities for the sake of their novelty, as a supposed source of new ideas.
20. "(...) living beings and their environments stand in relation to each other through *mutual specification or codetermination*" (Varela, Thompson and Rosch 1991: 198, emphasis in the original).

2

Intelligence behind Movement: Laboratories of Biomechanics and the Making of Movement Utopia

Nicolás Salazar Sutil

> *Movement is reality... because movement is life. But life has as its main element labor; consequently, the main element of reality is labor and the truest sign of reality is activity.*
>
> Nikolai Chernyshevsky, *What Is to Be Done?* (1989 [1863]: 181)

How human motion was engineered in the early 20th century through state-sponsored laboratories of industrial work movement is the topic of this essay. In this context, we find the emergence of an industrial modernity defined by how well (i.e. how productively) people move in the workplace, particularly in terms of the ergonomic and physiological means of achieving maximum performance with minimum effort. The subject of this essay is the making of a Soviet economic utopia during the 1920s in the course of which the Bolshevik state championed the engineering of a society of good movement (what I call "movement utopia"), particularly in terms of the normalization of everyday movement according to factory work training. Is it too much to assume that this history of the modernization of human movement (which is incidentally parallelled in the United States) should be looked at as a direct antecedent of Soviet (and American) cybernetic research, from where a historical sense of digital movement stems?

The point of this historical investigation, at least in the context of this collection, is to identify how people's intimate lives can be designed and engineered through everyday movement, and how movement constitutes a prime technique for the construction and control of selfhood outside the sphere of discursive ideology. What is true of Soviet Russia is also true of high-tech capitalist forms of production. Movement offered Soviet social engineers exactly what it has to offer technology firms or product designers and engineers in a capitalist context: a means of penetrating people's lives at the corporeal or sensory level and constructing identities from within – and not by means of ideology alone (i.e. through an imposed system of ideas and ideals). There is also a kind of "motorology" at work here, an imposed system of gestures, moves and motor functions that define how we interact with other humans and machines, and how our bodily and psychic selves are normalized at an embodied and kinaesthetic deep-level. Digital movement, as we will see presently, is yet another historical guise of a process that has been defining modernity through an alignment of technologized economic production with a re-engineered human motor. Over the course of this history, human movement has been persistently transformed to function in cooperation with the machine motor. In a digital capitalist context, the upshot of this motorology is the commoditization and fetishization of everyday movement performance. Kinetic identities, and by extension our own psychic selfhoods, can be defined by the moves, gestures, and locomotional habits that are corrected and normalized by our increasingly technologized lives, in the context of which the operability and performativity of digital technology has become entangled with the issuance of prescriptive physical movements. If some movements can be considered to be "good" for machine interaction and device operability and, more broadly, if some movements are deemed good as far as work and productivity are concerned, it is because the value of the moving body resides in its biological capacity to generate an economy *of* movements, and an economy *through* movement – what I call "kineconomy".

Intelligence behind movement

My argument is that whereas behind physical movement lies a motor intelligence, behind intelligence there is an intelligentsia. Put otherwise, human movement is programmed by the central nervous system and brain, which I see as an analogy of the way the central organs of the Bolshevik state could programme the everyday movement and rhythms of Soviet life. In the contemporary context, the engineering of a digital kineconomy has been redefined by mobile, network, sensor and haptic systems that rely on a controlling intelligence (as well as corporate intelligentsia) capable of normalizing and de-differentiating the way we move through a world of ubiquitous data representation and enactment. One of the reasons why the Soviet experiment is so significant as a historical antecedent is because it represents the first overt political and cultural attempt to study human movement as a form of intelligence – quite distinct from the automatic and unintelligent actions of reflex system research prevalent in pre-Revolutionary Russia and advanced by Ivan Pavlov's work in involuntary motor activity. The Soviet experiment of the 1920s sought to connect the science and scientific management behind programmed intelligence and voluntary movement to a political programme and a motorology that would help engineer human conduct, habit, identity and political selfhood through corrected movement and socio-politically determined motor control.

Three key issues will make up the overall framework of this investigation. First, I will explore the notion of the laboratory of movement in the Soviet Taylorist context, where the design and engineering of the utopian man-machine was tested and developed. My case study is the Moscow-based Central Institute of Labour (*Tsentralnyi Institut Truda*, variously abbreviated in English as TsIT or CIT). Second, I will turn my attention to the role played by motion technology in CIT's research agenda. I will argue that the scientific study of movement was grounded on the development of a new motion capture apparatus capable of observing, recording and breaking down motor actions to fine-grained detail. Technology also laid the foundations for mathematical and computational methods for the description of live motion, which provide the basis of a contemporary method for the formal representation and enactment of digital movement. Third, I will argue that the application of scientifically and technologically enhanced movement at CIT was promoted not only for high performance at the factory workshop, but also for a utopian artistic vision that was staged by the Projectionist Theatre resident at CIT, and which was led by Ukrainian vanguardist Solomon Nikritin. My case study, then, supports the historical emergence of a new culture of kinesis, defined by the integration of industrial, technological and sci-artistic knowledge to transform everyday movement. Finally, as we learn to move electronically and as we begin to define what it means to move digitally, a look back at the biomechanical history of corporeal movement must provoke a critical intervention and a caveat: the utopia of technologized movement in its contemporary historical context is fraught with many problems and shortcomings. The promise of a new culture of technologized living, and the utopian discourse digital technologies can afford, hinge upon the condition of inherent blindness of those who create new technology, and many of those who use it.

Soviet visions of future movement

In his book *The Human Motor: Energy, Fatigue and the Origins of Modernity*, Anson Rabinbach (1990) wrote that a unique vision of modernity emerged in industrial Europe that saw the working body as an exemplar of a universal process by means of which energy is converted into mechanical work. The shift from universal mechanics to the language of *Kraft* or mechanical labour required the elimination of the spiritual understanding of labour preserved in agrarian society, in lieu of a work ethic determined by abstract and quantitative relations. The argument tags along Georg Simmel's (1902) famous critique of the quantitative logic of industrial capitalism, which interrogates the function of numeration and quantification as sociological phenomena, particularly in terms of the growth of the technologized metropolis and the collateral effect of metropolitanism on the mental life. If we consider a strictly Simmelian view here, and inasmuch as group integration is hindered in large cities by the unavoidable differentiation of persons within it, a very large number of people can constitute a unity only with decisive division of labour, and not merely on the obvious grounds of economic technique suggested by Marx. Simmel argued that work had to be divided also at a more practical level. Thus one could argue, drawing on Simmel, that division of labour also involves the breakdown of *physical work* into component units, for instance through the identification of key actions and motions that make up an optimal way of achieving mechanical work, and which can be formally prescribed through training and physical indoctrination to improve kineconomic performance.

The Soviet utopia of the 1920s is a good example of how the quantitative rationality described by Simmel could be developed into a political programme of social engineering that would affect not only the mental life, but also the physical life of the new machine man. Having rejected Taylorism as an example of capitalist exploitation, Lenin was finally swayed by scientific management when he understood that the ultimate success of his Soviet Republic rested on the modernization of Russian agrarian society – a radical transformation that would involve electrification, industrialization, and the social re-engineering of the Russian peasantry. Lenin understood that the problem was not so much the Americanization of socialism, but the sudden and massive juxtaposition of two cultural frames – the village and the city. Richard Stites (1989) writes that the very language of peasants' gesture – "particularly the broad sweep of the arm (suggesting and no doubt derived from the wielding of the scythe) – seemed to reject the angular world of the industrial city" (163). Faced with the problem of an untrained and sluggish agrarian workforce, Lenin (1919) wrote that the Taylor system provided a number of most valuable scientific attainments in the analysis of mechanical motions during work, particularly in terms of "dismissing superfluous and useless motions, whilst determining the most correct methods of work, [and] the best systems of accounting and control" (25). It is clear that the ambition to engineer a modern eukinesis – i.e. a correct way of moving and gesturing – came from the very top of the Bolshevik intelligentsia, and not only from the premier of the Soviet Republic. Leon Trotsky wrote:

> Man at last will begin to harmonize himself... He will make it his business to achieve beauty by giving the movement of his own limbs the utmost precision, purposefulness, and economy of his work, his walk, and his play. He will try to master first the semiconscious and then the conscious processes in his own organism, such as breathing, the circulation of the blood, digestion, reproduction... and he will try to subordinate them to the control of reason and will. The human species, the coagulated

> *Homo sapiens*, will once more enter into a state of radical transformation, and, in his own hands, will become an object of artificial selection [...] Communist man will become immeasurably stronger, wiser and subtler; his body will become more harmonized, his movements more rhythmic... (2005: 206)

The language employed by Lenin and Trotsky is strikingly visionary and utopian. These political leaders were not just talking about movement, control and transformation in a purely ideological way. These are terms they wanted to apply to bodily training and somatic practices in order to conduct a sweeping reform of the Russian economy and Russian society at the deepest political and motorological level: i.e. at the level of how the new Soviet citizen ought to move, and how that movement ought to be harmonized and trained at a household level so as to achieve maximum productivity through human-machine integration. One who fully embraced these exhortations was the well-known Proletcult poet and critic Alexei Kapitonovich Gastev. In 1919, Gastev published a manifesto entitled *Tendencies of Proletarian Culture*, where he spelled out an industrialist and scientific vision in which machines were seen as extensions of the human body. Gastev would gain his reputation as the "Russian Taylor", also becoming one of the biggest national sponsors of the state science (and art) of biomechanics. Integration of biomechanics and Soviet Taylorism would lead, according to Gastev, to the transformation of Soviet society "in all the everyday and public life of people, regulating consumption and rationing family economy, even the movement of pedestrians on the streets" (1921).

NOT at work

Back in the late 1910s and early 1920s, the most significant system for the industrialization of workers' movement was scientific management, as proposed by Frederick W. Taylor, Henry Ford and Frank B. Gilbreth (in his time-motion studies). Although these methods had been developed in an overtly capitalist context, Taylorism was adopted and rebranded in Soviet Russia, where it became known as Scientific Organization of Labour (*Nauchnaya Organizatsiya Truda*, or NOT). In its Russian variation, scientific management evolved into a state-sponsored programme for the training of factory workers following a method that would eliminate superfluous and disorganized movement on the factory workshop and assembly line. Physical movements would have had to be corrected at the factory in order to be time-effective, following a one best method approach.

"Mechanisation," wrote Gastev in 1919, "will impart to proletarian psychology a striking anonymity, favoring the classification of an individual proletarian unit as A, B, or C, or 325, 0.075, 0, and so on" (in Bailes

1977: 378). The idea that numeration and quantification had become determining factors in the construction of a modern social lifeworld, going back to Simmel, is hereby fulfilled at an embodied level. Taylor himself had commented that when scientific management was being tested at the Bethlehem Steel Factory in Pennsylvania, it was necessary to bring to a halt the random handling of individuals and replace it with a detailed system that involved the numeration of workers (2005: 63). In his classic study of Soviet Taylorism, Kendall Bailes (1977) pointed out that the technical process of growing uniformity through quantitative and metrological tools would permeate every aspect of the worker's existence: "even his intimate life, including his aesthetic, intellectual and sexual values" (378). Consequently, the new Soviet man would no longer move according to personal qualities or subjective kinetic identity, but according to a mechanized collectivism, in whose domain movement is "devoid of expression, of a soul, of lyricism, of emotion, and is measured not by a shout or a smile but by a pressure gauge or a speed gauge" (379). What drove this depersonalization of the individual, or rather, the "dividual" – to borrow the term from Deleuze (1992) – was the need to raise productivity through a technologization of the means of production and labour. This in turn led to the re-engineering of the human body so that mechanical labour could be aligned to and coordinated with machinic production, which is why the European science of biomechanics proved so useful. "The cog in the wheel" is a commonly used metaphor for the new Soviet man which, according to Slava Gerovitch (2007), testifies to the fact that the passive individual was subsumed under the collective. This expression implied the machine-like operation of the party and state apparatus controlling social life (137). The image sums up my idea of an intelligentsia behind an intelligence, and a system of political control behind the issuance of motor control.

Chronization

It is worth pointing out that in order to change the way people move, Gastev first had to change the organization of a social sense of rhythm and time. Drawing directly on Taylor's time study methods and Gilbreth's time-motion studies, Gastev claimed that it was necessary to develop a stopwatch-clicking mentality in Russia to understand rhythm and the physical life as having outward manifestations in rationalized movement. Thus, the goal of the Soviet utopian enterprise according to Gastev ought to be the creation of a "plastic culture" that would harmonize the rhythm of the anatomical body and the machine. Where formerly workforce production was carried out in relation to an agrarian sense of temporality (seasonal time), or a religious sense of time (Church calendar), the Taylorist revolution in Russia also relied (like in the United States) on the objectification of time, or the "chronization" of social life. Thus, a time-ineffective labourer might yield one unit of production over a prescribed length of time. Synchronization of work would lead to the production of three units in the same amount of time.[1] This principle of time organization would be extended to everyday life, so that the body of the new Soviet man would now be forced to internalize the temporality of the city and the factory at a somatic level. Gastev wrote in his essay *Time* (1923), that subjective notions of timekeeping and time-motion would be eliminated, and that new methods for objective time-setting would be introduced instead.[2]

Gastev's vision was not free of controversy. Indeed, the émigré novelist and literary critic Yevgeny Zamyatin put forward a scathing parody of Gastev's project in his science-fiction classic *We* (1993 [1921]). The novel's narrator – referred to as D-503 – depicts a future society governed by a totalitarian state that forces its citizens to walk in even ranks and in metrical step to the beat of a state march. Zamyatin imagined Gastev's dystopia as a total chronization of life, such that people would have to plan their daily activities on a so-called "Table of Hours". Sleeping times, work times, sex hours, and even the number of bites performed during eating hours (fifty) – were all prescribed in Zamyatin's One State. Zamyatin wrote: "Like one body with a million hands, at one and the same second according to the Table, we lift the spoon to our lips. At one and the same second we leave for a stroll and go to the hall to do the Taylor exercises, and then to bed" (1993: 13). Zamyatin added: "I watched the [workers] how they would bend over, straighten up, turn around, all in accordance with Taylor, smoothly and quickly, keeping in time, like the levers of a single immense machine... It was the sublimest, the most moving beauty, harmony, music..." (81). It is clear that Zamyatin was foreseeing how Trotsky's utopian vision might turn into a dystopian nightmare.

Gastev's institute for the machinization of man

In 1921 Gastev sought financial support from Lenin to create the Central Institute for the Scientific Organization of Labour and the Mechanization of Man (CIT). The principal aim of this Moscow-based research hub was to train the new factory worker in the understanding of units of industrial motion, i.e. the stroke and the thrust, by developing a regular work rhythm aimed at full motor automation. After practice in the tempo and sureness of the hammer and chisel stroke, Gastev wanted his cadets at CIT not only to be ambidextrous, but also to apply their knowledge of biomechanical movement to generate efficiency in all kinds of fitting and machine work. Gastev gave particular attention to the psychophysiological behaviour of workers, which explains why he created a labour clinic at CIT responsible for studying cardiovascular and respiratory rhythms. The clinic was also responsible for overseeing the study of fatigue, energy cost, quantity and quality of work movement performance, as well as reorganization of diet.

Likewise, Gastev's institute also focused on how to design optimum housing conditions, clothing, transport and leisure activity to improve industrial movement performance. Indeed, over the coming years, and up until its closure by Stalin in 1938, CIT continued to develop a method known as "system (method) CIT", designed to standardize and streamline skilled workers based on determination of labour units, as well as segmentation and rationalization of key movements performed by the factory worker. The method could be taught to training instructors, factory administrators, military personnel and labourers working across the major Soviet industries (i.e. textile, construction, engineering, coal mining, metallurgy, auto and transport). Having visited the institute in the early 1920s, Austrian journalist and cultural historian René Fülöp-Miller (1927) wrote that anyone entering the front door as a normal living man would issue from the back door – after passing through countless laboratories – a completely perfected machine. He added:

> On entering the building, you find here a number of investigators engaged in fixing the general maximum output capacity of the human organism, and there, in the psychological-technical laboratory, other people, who are trying to ascertain how much energy is used in every movement, and how this movement can be made in the most economical way. The balance of energy is fixed as exact as possible, and efforts are made carefully to ascertain the optimum periods both of work and rest. The exact psychological working cadence has already been discovered; the effects of various physiological and psychological stimuli are exactly studied, and all micro and macro-motisms are determined with utmost exactness. Precision in the investigation of the energy of the organism here celebrates rousing triumphs. (1927: 210)

The main scientific area of research developed at CIT involved the study of biomechanics. Gastev was inspired by the early modern pioneers in the field (especially Giovanni Alfonso Borelli), as well as the landmark laboratory studies carried out in the late 19th century by Étienne-Jules Marey in France and Wilhelm Braune in Germany, both of whom were well known at CIT.[3] The study of industrial biomechanics was coupled with an application of artistic systems for movement analysis such as those proposed by François Delsarte and Émile-Jaques Dalcroze, whose methods served as platforms for the development of an industrial approach to movement analysis at CIT. One example of this is found in the work of theatre critic and practitioner Ippolit Sokolov, who became closely associated with Gastev's institute. In an essay published in 1921, Sokolov spelled out the need to develop a system he called "Industrial Delsarte"[4] at the heart of which was the development of a Delsarte-inspired system known as "industrial gymnastics". But whereas Delsarte (1887) had popularized the idea that the body could produce harmonious movement through a trinitary system underpinned by a belief in the spiritual essence of movement (involving the Good, the True and the Beautiful), Sokolov embraced the notion of good movement in its laicized connotation, and so as to serve the interests of the Soviet intelligentsia. And even though Delsarte's system would gain huge popularity also within a spiritual trend in the European movement arts of the 1920s, for instance with the development of the Delsarte-inspired idea of "eukinetics" put forward by Rudolf Laban, industrial gymnastics evolved into an application of good movement to Taylor's "one best way" philosophy, i.e. maximization of productivity in the minimum amount of time and with minimum effort. The approach championed by "system (method) CIT" can be described, with hindsight, as a kind of industrial eukinetic, whose effects would be to drive kineconomic performance, and lead the utopian transformation of agrarian Russia into a modernized, industrialized and technologized superpower.

Bernstein's lab

The person responsible for leading CIT's Laboratory of Biomechanics was the talented young physiologist Nikolai Aleksandrovich Bernstein. Bernstein would go on to become a leading figure in Soviet physiological cybernetics, as well as a world pioneer in the field of motor control. Bernstein (1983) himself acknowledged that the creation of a modern laboratory of movement was not a novelty. Unlike, say, the model of the *Station Physiologique* established by Marey in 1882, the Laboratory of Biomechanics at CIT was a state-sponsored lab – being thus mandated by political interest. And so whilst Bernstein was given the responsibility to carry out specialized work to advance European biomechanical science, his direct instruction would have been to apply knowledge to the Soviet context and according to the party's main agenda.

Bernstein took the laboratory in a direction that opposed the dominant physiological school, which at the time followed Ivan Pavlov's doctrine of conditional reflexes. Clearly, Bernstein was not interested in involuntary and conditioned movement. He wanted to study a dimension of movement Pavlov had utterly failed to recognize in his work: namely, programmed human motor activity. On the other hand, Bernstein's intention was to understand how the brain controls bodily movement, and *not* how to develop effective means of social and political control. The laws of industrial work movement could be formulated, according to Gastev, in terms of two determinations: an up, and a down. By identifying the shortest movement curve between the point of highest elevation of a hammer and its place of landing, Gastev believed CIT scientists could yield the most rational trajectory of a hammering motion, and thus help identify the most economical (i.e. the most cost effective) way of moving. Not only would automation of these good movements lead to an avoidance of fatigue and improved motor performance; the political ambition of Gastev's industrial movement training was to improve the productivity of factory work in general, in order to improve the Soviet economy (or its kineconomy). As Bernstein's work showed, this seemed like a crude simplification of a much more complex neurophysical process involving the neurophysical programming of motor actions. Bernstein's experiments showed that the same labour task was performed differently – with varying tensions of different muscles – at various times. Muscular movements were constructed, so to speak, each time the task was performed. Bernstein concluded that reflex theory could not adequately explain muscular movements. He understood movement in terms of a mathematical distinction between human locomotor mobility, which has no boundaries and is characterized by total freedom of movement, and what he called "deformational mobility" (1999), which refers to how anatomically bound parts of the body can move with respect to each other in a way that is limited by degrees of connectivity. Thus, the scientific knowledge that was being generated within CIT was not necessarily contained within reflex theory, and neither was it dumbed down by oversimplifications imposed by political agenda. Human movement was being discovered and re-invented in terms of a complex programme that the brain carries out programmatically and which the body must perform according to mathematically describable operations involving kinematic chains.

Kymocyclography and the mathematization of motion

Wilhelm Fischer and Otto Braune were the first to apply mathematical and computational mechanics to the study of live anatomical human movement in the 1890s. Insofar as their method focused on a mathematical description of live movement (this is of course before the invention of the electronic computer), and insofar as their samples were obtained using the very first modern optical motion capture system, their work can be credited as a monumental achievement in the history of movement science and technology. Bernstein's ambition was to lead the next generation of physiological studies through an improved mathematical approach and an enhanced technological medium used to support this framework. Indeed, Bernstein's position in the history of movement science is equally significant, not least because he drove the Soviet laboratory away from the biomechanical paradigms of the 1920s, onto the cybernetic paradigm of the 1950s and 60s, and thus saw the transition from combustion machine to computer machine, from biomechanical to robotic and cyborg human movement. At the heart of this drive for a new science of movement was the language of "technologized mathematics" (Rotman 2008), which is where the language of digital movement would inevitably emerge from.

Before the hammering action of the cadet could be described as "movement geometry" (Bernstein 1999 [1929]), however, Bernstein had to develop the technological means of capturing live movement, and representing the motion in a graphic way. To this effect, Fischer and Braune's chronocyclograph was used as the basis to develop a new optical motion capture system at CIT's Laboratory of Photocinematography (led by Nikolai Tikhonov). This in turn led to the development in 1927 of a novel motion capture apparatus devised for the more exact investigation of normal and pathological motions. Tikhonov and Bernstein captured CIT cadets using photographic sequences of images of selected lines or points on the body, similar to Marey's famous method of chronophotographic geometry. The drawback, in Marey's case, was that chronophotographic shots were perfectly readable only for locomotor movements where the object moved continuously along the visual field of the camera over the course of a single trajectory. Bernstein wrote: "For small, complex, and repetitive movements whose trajectories always come back to the starting point, these methods cannot be used" (1999: 127). Since these were precisely the kinds of movements Gastev wanted to investigate, Bernstein and Tikhonov had to find the means of recording the curving trajectories produced by moving limbs over repeatable cycles of localized movement (i.e. chiselling, hammering, rifle handling). To achieve a tridimensional representation of these movement forms, Bernstein also suggested the method of mirror-recording, which led to a tridimensional capture and stereoscopic vision of movement without the need for multi-camera calibration.

Unlike the Geissler tubes used by Braune and Fischer, CIT cadets were dressed in black suits connected to electrical light bulbs attached to the joints of the organ under investigation. CIT's motion capture set-up further combined the advantages of the cyclographic apparatus with suitable recording techniques (time-lapse), which enabled the registration of live movement in terms of motion curves. One advantage of seeing motion represented thus is that it could be mathematically analysed in terms of sine waves. The extraction of these curves of movement from the photographic print and their formal representation as sinusoids became known as kymocyclography (literally, the writing of wave cycles).[5] What Bernstein's kymocyclograms showed, in effect, was that the variation between movement trajectories and their complexity decreases significantly, even in richly innervated systems, under the action of human movement automation. As such, motion capture technology was intended to produce a kind of writing – what Peter Weibel (1996) calls a writing over the seeing, or "opseography", in the sense that one could read (mathematically) the curves that stereoscopic time-lapse photography yielded when capturing reflective body markers in studio conditions. This opseographic inscription generated multidimensional datasets, which, in their formalized representation, called for more numerical, tabulated and computational ways of representing movement. One important consequence of this mathematization of the knowledge of human movement, and the scientific discourse it began to support, was that scientific rationality began to be politicized to the extent that state-sponsored policies of full bodily automation and social re-engineering in Soviet Russia were justified according to a pseudo-mathematical ideology (or motorology). In other words, Soviet workers were told that in order to produce the best trajectory for the hammering motion in the assembly line, the worker had to follow the quantified and exactly measured movement curves produced by mathematicians and scientists working in Bernstein's laboratory, whose calculations would yield the most automatic, mechanized and cost-effective movement performance possible; a mathematically proven kineconomy of industrial movement.

No doubt the terminology Bernstein was beginning to develop at CIT would have chimed in with Gastev's political agenda. "Control", "central system", "motor program" – these terms could be readily interpreted as the conceptual building blocks of a political discourse. However, the two projects began diverging as the political climate in Soviet Russia changed with the rise of Stalinism. Indeed, the destiny of these two projects – and these two men – were to run very different courses. Whilst Gastev was killed during the Great Purge and CIT was completely disbanded, Bernstein went on to win the Stalin Prize for science in 1948. The tension between scientific intelligence and the political intelligentsia is of course a much broader question, and one that deserves more than a passing reference. What this relationship showcases, however, also bears directly on my underlying thesis. Where science objectified and instrumentalized the body and its physical intelligence, the political establishment in turn instrumentalized scientific knowledge for its own self-serving purposes. Hence, behind movement lies a motor intelligence, and behind this intelligence lies an intelligentsia.

It will be evident from this that a corollary statement is in order. Bernstein's capacity to see (think) the neurological process behind movement hinges on the optical power of his motion capture system. Whereas Braune and Fischer had described their motion capture technology as a "new microscopy of movement" (1987: 116), the technological instrument and accompanying method developed at the Laboratory of Biomechanics was both a microscope, an X-ray, and a proto-computer, in the sense that the apparatus could not only look into the fine-grained details of a micro-movement, but also reveal movement in its formalized manner as a mathematical wave or curve, or indeed as computable data. In other words, the technology allowed for a translation of the language of physicalized kinematics to its abstract mathematical description.

As scientists pushed toward a cybernetic era characterized by greater levels of information processing and mathematical and computational knowledge, and as new technology began to yield vast levels of movement data, so it became apparent that the challenge lay not only in refining physical intelligence, but also the mental intelligence and the technological means of processing vast data crops. Lev Manovich (1995) points out that there is a passage from Taylorism to cognitive science and cybernetics, which can be read "as one of the most important signs of the shift from industrial to post-industrial society", adding: "The point is not whether corporeal labour was indeed universally displaced by mental labour. What is important is that the obsession with the rationalization of corporeal work (Taylorism, European science of work, psychotechnics) disappeared, displaced by a new obsession with the rationalization of the mind" (1995: n.p.). It is important to stress that there is a historical confluence from one history to the other, in the sense that the degree of technologization and the mathematical/computational determination of live movement, which stemmed from research institutes like CIT, fed the need for new technologies of information processing and for a computer-aided science of movement. And it is here, with the emergence of a computer machine, that the history of a biomechanical utopia ends, and a contemporary eukinetic utopia – the utopia of the digital – begins.

Postscript: Nikritin's cartograms of utopia

In 1923, and partly in response to the unwelcome hijacking of the term "biomechanics" by Vsevolod Meyerhold and Nikolai Foregger,[6] Gastev awarded an artistic residency to the Projection Theatre at CIT headquarters in Moscow. Evidently, Gastev's intention was to promote further sci-artistic integration within his institute. More specifically, Gastev wanted the work of his resident artists to focus on "organizing a performance of work methods on stage", and on "creating an objective theater of normalized labour" (in Pchelkina 2013: 158). Gastev was granted the title of honorary member of the Projection Theatre, and members of the Projection Theatre were in turn allowed to work in close collaboration with CIT scientists, including Bernstein and Tikhonov.

Led by Ukrainian avant-garde artist Solomon Nikritin,[7] projectionism was an avant-garde movement whose principal intention was to "reflect the urge to rush ahead into the future" (Smirnov 2013: 13). Projectionism was thus committed to the conceptualization of utopian futures through sci-artistic integration, particularly through a cross-traffic between human actors and technological agents. Nikritin's principal idea was that the essence of the work of art resides in its method, not its material output. He wrote that the achievement of projectionism was "the intellectual work of the masses concentrated in one discipline – in projectionist expression of organizational classification and methodology". Nikritin added: "The art of projectionism is... the algebra of organizational science" (15).

It is also worth pointing out that around this time Nikritin entered into direct correspondence with higher members of the Soviet intelligentsia, including Trotsky himself. Trotsky (2005) had written some years earlier that through social construction "all the arts – literature, drama, painting, music, and architecture will lend to a process of beautiful form", adding: "The shell in which the cultural construction of communist man will be enclosed, will develop all the vital elements of contemporary art to the highest point, so Man will become immeasurably stronger...and his body will become more harmonized, his movements more rhythmic" (207). In order to materialize this vision, Nikritin's contribution at CIT involved much more than devising propaganda theatre in workers' clubs and businesses, as Gastev had originally intended. Nikritin also constructed CIT's so-called "perspective boxes", which helped calculate the local coordinates of moving subjects – similar to the lattice backgrounds used by Étienne-Jules Marey and Eadweard Muybridge in their chronophotographic studies. The Projection Theatre also experimented in the use of analytical scores that would serve as graphic records for the formal representation and composition of sound, rhythm, movement and emotional excitement. More ambitiously, Nikritin developed an educational programme for actors in terms of a system of self-growth, based on a practical training method known as "harmonious biomechanics". Liubov Pchelkina (2013) has explained that Nikritin's actor training system involved technical and expressive resources of the bodily organism, so that the performer could master their bodily movements, vocal apparatus and emotions, and so that this complete performer could in turn adapt to scenic and acoustic space. Like Meyerhold's laboratory, Nikritin's training method was grounded on an application of Gastev's biomechanics, and it included work on analytical gymnastics, breathing technique and physiological organization, as well as bodily positioning (158).

In his desire to model the movement of workers, dancers and pilots – the three great archetypes of the Soviet eukinetic utopia – Nikritin imagined a synthesis of the mathematico- scientific knowledge produced by Bernstein, the technologies of movement observation developed by Tikhonov, and the industry enterprise led by Gastev, in order to see where this synthesis might land in the near future. To this effect, and whilst still resident at Gastev's Central Institute of Labour, Nikritin produced three graphic representations he called the "cartograms of the program" (1924). Nikritin's triptych shows the stages through which society should pass – including periods of preparation, revolution and organization – before it congeals into a better future. In *Cartogram 2* (Figure 2.1), the artist specifies the stages of progress of development of this new society, and he indicates the main initial conditions related to the point of beginning. Among them are: "clear distinctness of responsibility, personal interest, general planned work."[8] Thus, a utopian sense of human creativity could be projected from a present techno-scientific and artistic creative context onto its as yet unimagined future.

What Nikritin envisaged was not the state mathematics parodied by Zamyatin, nor the dystopia that was to grip the Soviet Union during the

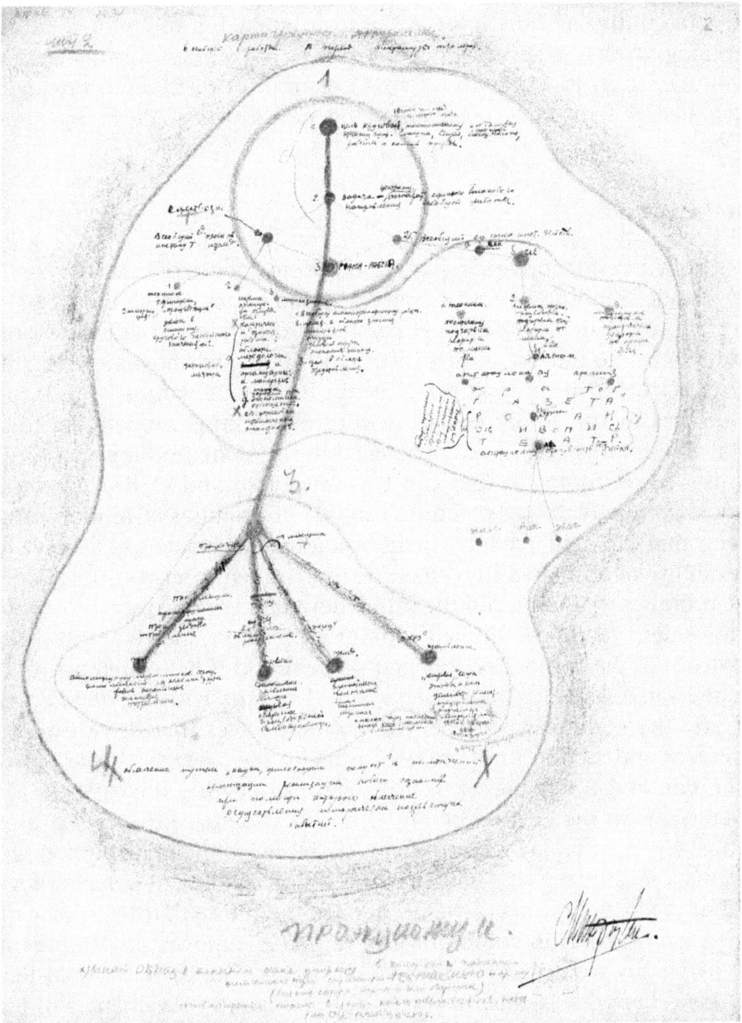

Figure 2.1 Solomon Nikritin's Cartogram of the Program No2 (1924). Courtesy of © State Tretyakov Gallery

Stalinist purge. Nikritin saw creativity branching into ever more complex networks made up of multitudinous nodes and distributive links, where the overall structure is devoid of any centralized control. According to Pchelkina (2013), the cartogram of the programme can be interpreted as an algorithm. A clear foresight into our own digital era is found in these pseudo-algorithms, which predict the emergence of a classless and authority-less society made up of evolved citizens. It is also a society whose movement is not prescriptive and disciplined, but distributed, self-disciplined and self-organizing. In this society we find the growth of consciousness issued forth organically, so as to make up a living organism of distributed intelligence (much like the Internet), performed by new citizens (or, if you prefer, netizens).

Conclusions

If this essay has not convincingly traced the emergence of our own contemporary understanding of technologized motion back to the Soviet biomechanical project (let us not forget parallel efforts were made in the United States), then I do not know what other history digital movement might be traced back to. What I do know, however, is that the failure of technological utopias in the Soviet case is a word of warning. Underlying today's digital kineconomy are questionable promises of free online mobility, or promises of technologies that can be transported and carried everywhere to boost productivity and communicability, or promises of motion capture systems that can map live movement onto digital characters to achieve total believability of animated movement for subsequent user consumption. All these promises that underpin the future development of contemporary digital movement cannot be studied without a critical perspective, and without knowledge of the corporate intelligentsia behind this new intelligence.

In the same way that movement typically starts with an *invisible* mark made by the corporeal body, so the political forces that drive intelligent movement and cash in on it are not visible to the average technology consumer. But whilst new digital technologies can develop the capacity to see movement with the greatest clarity, they can also efface the apparatuses of political control that lay behind them, enabling a blinded technologization of human movement. If I say that the vision behind new technology is blind, it is not only because we do not see the future that technology will fashion for us. Nikritin could not foresee that the movement utopia imagined by Gastev would end in Stalin's massacres. The point I would like to end this essay with is this, however: technologized movement withdraws us from our own inner vision of movement. The promises offered by technologized movement can alienate us from our own connection with the neurophysical and intellectual source of movement, allowing the machine to think the movement and to control it for us. This withdrawal from movement, or from the intelligence of movement, means that we are turning our backs on the internalized processes behind movement. Intelligence is being effaced, considering that not too many people know how computers think and how they programme good movement, nor do we necessarily know how our mobile gadgets move with us, nor how we are moved by them. We are bound to become less intelligent or at least more superficial in our understanding of physical intelligence if we let ourselves be driven by a world where everything moves automatically for us. Intelligence is outsourced – thought becomes the responsibility of the machine, not the user. The digital movement utopia with all its promises of free accessibility and mobility, with its promises of total motion recognition and motile interactivity, can lead to a dependence on total machine automation and the emergence of a society which is not, as Deleuze (1992) predicted, a society of control, but a future society out-of-our control.

Notes

1. In 1931, a public bricklaying competition was staged in Moscow as part of Gastev's "labour championships". Three different approaches were tested: that of an unskilled and untrained worker, that of a worker versed in Gilbreth technique (developed by Frank B. Gilbreth using time-motion and motion study), and the "system (method) CIT". The untrained bricklayer laid 327 bricks in an hour; whilst Gilbreth technique yielded 452 bricks in the same time. Using "system (method) CIT" the winning bricklayer laid 907 bricks. See Albert Kravchenko (2010).
2. Examples of this included the use of personnel sheets, personal report cards, record maintenance work orders for trains, trams and all vehicles, production schedules, chronometric cards, time control devices and time clocks in factories, alarm clocks, chronometers, metronomes and horns. See A. K. Gastev, *How to Work: Selected Writings by A. K. Gastev* (online).
3. Gastev's idea to develop a state science was inspired by the European science of biomechanics, in particular the book *The Movement of Animals* by Italian mathematician Giovanni Alfonso Borelli. Gastev was also inspired by French scientist Julien La Mettrie, whose book *Man-Machine* (1747) was fortuitously translated to Russian in 1911. Marey's work was also well known, as was the classic work on the biomechanics of the gait produced in Germany by Braune and Fischer in 1895 (*The Human Gait*). Bernstein (1983) quotes the Weber brothers, Marey, Muybridge and Fischer and Braune as pioneers in the field.
4. Sokolov was briefly associated with CIT, and with Gastev's project for the creation of a Taylorist theatre during the 1920s, before moving on to become a relatively well-known film critic. Whilst at CIT, Sokolov produced a number of essays on the subject of industrial movement including his essay *Industrialization of Gesture* (1921), where he spelled out his ideas for the taylorization of gesture and the creation of a stage reflexology, or what this author called "scenic-technical reflexology". Also in 1921 he published *Industrial and Rhythmic Gymnastics*, where he proposed a fusion of Delsarte technique and industrial movement to create a "gymnastics of labour". The work was intended to offer a rejection of the laboratory theatres of Meyerhold and Foregger, which Sokolov deemed a mere aestheticization of Gastev's scientific developments. On the subject of Industrial Delsarte, see John E. Bowlt (1996).
5. Bernstein described his motion capture system thus: "The frequency of exposure can be raised to 600 per sec., time being measured by means of the author's Sirenen-Nonius device with an accuracy of [plus or minus]4.10–5 sec. The method has already been applied to the study of bodily work, piano-playing, adiadocho-kinesis, tremors, arm-tonus reaction" (Bernstein 1927).
6. It is well known that Meyerhold and Foregger developed experimental laboratory theatre models based on CIT, and on the scientific narratives of biomechanics

and reflexology. Gastev dismissed Meyerhold's utilization of CIT's methods, however, and the application of biomechanics to Meyerhold's laboratory theatres (RSFSR). Crossovers between CIT and Meyerhold's theatre laboratories were frequent, particularly in the case of the Studio for Movement Registration at CIT, where Meyerhold's student Ilia Shlepianov was employed. Similarly to Meyerhold's programme for an alphabet of stage movements as part of his method for stage biomechanics, the Studio for Movement Registration at CIT developed techniques for the graphic and photographic recording of live movement, and its transcription into a language of movement and a "semantic cadence" (Misler 1991).

7. Gastev was impressed by Nikritin's avant-garde stage design, particularly the plotless "rhythmo-dynamic" compositions exemplified in stage productions like *The Tragedy of AOU* (1922) and *Pressing and Impact* (1923). These works featured a mixed media design that included a noise orchestra, a gymnastics apparatus, and mobile scenery. Special projectors were also included as well as large screens to depict "virtual characters, which could be extracted from the film projection, and which appeared to interact with actors on the stage" (Pchelkina 2013: 158). Gastev himself was no stranger to the multimedia studio – indeed, some of the laboratories at CIT would feature a number of interactive technologies, particularly within the Sensorics and Psychotechnics labs, which were furnished with simulation apparatuses for the training of car drivers and plane pilots (Smirnov 2013).
8. In the margin notes of *Cartogram 2*, Nikritin also writes: "simplification of realization of the historically predefined events is possible by means of sciences, philosophies of arts and the political organization and realization of new consciousness." This translation of Nikritin's *Cartogram* is by Liubov Pchelkina and Andrey Smirnov. I am grateful for their help in helping me access some of Nikritin's material, and for permission to reprint Nikritin's sketch.

Part II

Gesture and the Digital Object

3

Mathematical Movement: Gesture

Brian Rotman

> *Mathematics is infected by the ancient idea of being a science of quantity – which is doubly false, being neither a science nor more necessarily concerned with quantity than anything else. It is a practice and comparable to dance.*
>
> Paul Valery (qtd in Krauthausen 2010: 245)

> *In this sense* [about a blindfold, solo dance piece], *the thought of movement coincides with a movement of thought. Rather than being abstractly separated and antecedent to the moving body, thought coincides with movement in the critical moment of its appearance.*
>
> Stamatia Portanova, "Infinity in a Step" (available online)

How might movement and in particular gesture, a basic physiological capacity of all animate bodies, be connected to the inanimate – invisible, impalpable – objects of mathematics? What is the link between the movement of abstract mathematical thought and physical movements of the human body? Between virtual and actual entities? Certainly, gesture is deeply embedded in human culture, prior to, along with, and inside spoken language, as a means, a constitutive principle, of thought; and, as we shall see, the same is true for mathematical language.

This should not be surprising: before all else, there is gesture's role in the production of "the human". Gesture is the ur-medium of hominization, a crucial route and operative means of it: the proffered breast, the smile, facial "expressions of emotion", tones of voice, the innumerable gesturo-haptic movements of nurture, control, and communion (cradling, pointing, eye contact, turn-taking, and so on) that scaffold the production of a psyche. Gesture, in short, serves to induct the infant into speech and initiate the psychic mirroring of the other that allows a human subject to come into being. As a result, it is a source of pre-linguistic psychic affect, employed with powerful effect in rituals, ceremonies of social cohesion and collective enunciations, not least the many kinds of disciplined body practices – gestures of submission and piety – prescribed for their adherents by religions. And, of course, art forms such as dance, music and song deploy figured visuo-kinetic and sonic movements of the body – precise gestures – as their semiotic and ontological resource; a single move of a dancing body, for example, is an assemblage of choreographed disciplined movements – gestures – at many levels: from neuro-somatic and proprioceptive actions to those which relate the individual body to an ensemble of gesturing bodies. Theatrical art is predicated on the invention and staging of gesture: classical rules of pose, posture, delivery, gait for actors; Bertolt Brecht's post-Aristotelian concept of theatrical *gestus* – gesture signifying at the level of the entire play; Antonin Artaud's theatre of cruelty where "gesture… instead of serving as a decoration, an accompaniment of thought, instead causes its movement" (1958: 39); contemporary forms of dance- and movement-theatre in which gesture is a means of narration.

Less directly, for Gilles Deleuze the cinema is before all else a "cinema of bodies", an art form within which what he calls the "movement-image", a spatio-temporal entity that is to be understood corporeally, precisely as a gesture, is the irreducible element of filmic thought. According to Giorgio Agamben "to gesture" initiates the ethical. "What characterizes gesture," he insists, "is nothing being produced or acted, but rather something is being endured. The gesture, in other words, opens the sphere of *ethos*" (2000: 57). For Ludwig Wittgenstein, buildings, as soon as they exceed the purely instrumental, cannot but express ideas, display attitudes, make *gestures* in relation to their inhabitants and surroundings. "Architecture," he writes, "is a gesture, [it] expresses a thought, it makes one want to respond with a gesture" (1980: 22e). From a phenomenological perspective, Maurice Merleau-Ponty insists, one is "led to recognize a gestural or existential significance to speech" which is "the subject's taking up of a position in the world of his meanings" (1962: 193). Similarly: "Wittgenstein's finest insight," Antonio Negri tells us, "is precisely that language is a gestural form: to speak is to gesture [*gestire*], that is, to manage the body [*gestire il corpo*]; he taught us that language is all there within the body" (Negri 2004: 164).

Can mathematics be understood in terms such as these? In what sense could an abstract symbolic language be "all there within the body"? How might gestures be an irreducible element of mathematical thought? It will be evident from this that a certain disclaimer is in order. Unlike the interpretation of digital movement pursued in the other chapters of this volume – movement of information (transmission of data), digitized movement (motion capture), movement of bodies (human-computer interface) – I examine movement – understood as gesture – inside mathematical thought; the movement or motion, in other words, baked in or formalized and no longer visible within the static structures of mathematics. These include but are by no means confined to the algorithms and functions that constitute the mathematical infrastructure of the digital.

Mathematics is an operational practice. Mathematicians speak constantly of performing or executing or constructing or carrying out something such as a calculation, a geometrical construction, a computation, an algebraic operation, a mapping, an algorithm. They engage in adding, multiplying, differentiating, integrating, verifying, averaging, triangulating, along with numerous other operative practices, including the key actions of defining object-concepts and proving (showing, demonstrating, validating) a theorem. The objects of all this activity, the mathematical entities being acted upon, as well as the procedures and narratives organizing them, are immaterial; the actions occur not in actual space but in the virtual space of thought, instantaneously, at "infinite speed" as it is said. Where are the bodies of the mathematicians doing the thinking here? What's their connection, if any, to the carrying out of these virtual actions? Or, before that, how were bodies involved in inventing or creating the entities in the first place? For the most part, these bodies sit, scribble symbols and diagrams, stare into the distance, go walk about, and sketch demonstrations on blackboards. It is difficult to make any meaningful link between the nature of these bodily activities and the conceptual activity or origin of their thoughts. It would seem, as many would have it, that mathematical activity is in fact incorporeal; creating and practising it, and certainly its content, has nothing to do with the body. Seen thus, mathematics is entirely the work of *res cogitans*, a species of pure, disembodied thought, a vast thinking machine that discovers absolute truths about invisible, unembodied objects.

But this long, widely held view of mathematics, though natural when grappling with these invisibles – when thinking and imagining them somehow "out there" – is mistaken. On the contrary, mathematics' links to physical movement, though mediated, are undeniable: each of the primary object-concepts of the mathematical universe – number, relation, set, function, operation, variable, line, point, space, equation – can be seen to emerge from a small set of pre-conceptual physical actions, disciplined bodily movements or gestures, that, with their subsequent transformations, make up the corporeal wherewithal of mathematical thought. In other words, contemporary mathematics, though habitually understood in terms of static disembodied object-concepts, is constructed in/by a language whose basic conceptual vocabulary is rooted in gestural movement–schemata of the body. Chief among these are the gestures of pairing two things together; combining two things to make a third; replacing one thing by another; pointing at a thing; showing, exhibiting or manifesting a thing; displacing or extending the body in its space; making/altering a mark; and the meta-gesture of repetition, of doing the gesture again. So that, for example, the object-concept "number" can be seen as rooted in the gesture of making a stroke, an undifferentiated mark, and then repeating it; likewise the object-concepts "equation", and "relation", are different conceptualizations of the gesture of pairing, and a "binary operation" the gesture of combining two things to make a third; the object-concept "function" is constructed out of two collections of things by repeating the gesture of pairing them; a "line" or "curve in space" realizes the gestural displacement of a body, of a thing; and so on.

These primary objects form the conceptual matrix from which mathematics develops, augmented by concepts invented to solve external problems of science and transformed through an internal dynamic of self-appropriation and recursive colonization that mathematics performs on its own products. In this, object-concepts can pair with each other and fuse to create new genres of objects and relations between them, as in a topological group or an entity in algebraic geometry; relations between objects can become objects in their own right, as in the action of counting becoming an infinite sequence, or the action of ordering things becoming objectified as a token of an order-type, giving rise to an ordering of types. In fact, any mathematical procedure or activity operating on objects can, in principle, become an object able to be operated upon; either by another object or by the same object recursively, for example, as in function of a function, or a category of categories. This self-enfolding dynamic of mathematics is without any apparent limit, guaranteeing a potentially infinite expansion of mathematical problems and thought from within itself.

The primary objects, and their transformations, mask any evidence of their embodied origins: once it has emerged, its movement folded into a formal mathematical object, a gesture disappears from view; its movement has become the mathematical attributes and agency the object possesses, its potential for reacting with and operating on other objects. To say more about the process behind the enfolding, whereby a gesture might give rise to and inhabit an abstract mathematical object, one has to consider the gesture before it becomes implicit in the object. This entails examining the response to problems outside mathematics; finding sites in the development of the subject at which a gesture is introduced into the language of mathematics as a solution to a problem; and examining its passage from the body to its entry and disappearance inside a new mathematical object-concept. This is precisely the project behind the mathematician-philosopher Gilles Châtelet's historical account of mathematical thought.

His essay *Figuring Space* playfully introduces itself as drawing the "moral of a vaudeville play with three protagonists whose disputes have animated various debates about ideas for two centuries" (2000: 1). It tracks the creation of mathematico-geometric space through the triple optic of mathematics, physics, and philosophy in the form of Gottfried Leibniz's concept of a flexible and elastic space and Gilles Deleuze's ontology of the virtual. Châtelet opens up the operative meaning, embodied origin, and intellectual potential – the virtuality – of such fundamental concepts as "point", "space", "dimension", "curve", and "line" through a series of historical case studies drawn from the work of Einstein, Maxwell, Grassmann, Ampère, and

Hamilton. The project is to rethink the geometric/scientific construction of space through gesture and diagrams, particularly the "patrimony of those [gestures] that set it alight and multiply it" (14). His aim for mathematics, understood as the creation of solutions to problems from sources external to it, is to reveal this patrimony.

Mathematicians think about disembodied, abstract entities. Châtelet articulates two prostheses they employ to overcome the difficulty of engaging with and thinking about invisible, immaterial "things". One – semiotic, public and formal – is the symbolic apparatus, "the official crutch of the literal text, which accounts for the carrying out of the operations and which guarantees the transmission of knowledge". The other – sensual, virtual and ruminative – is "a more subtle crutch, reserved for initiates, who are able to sense a whole network of allusions interlaced with the literal text and continually overflowing it" (8). The latter prosthesis is the one that interests him, namely the nature of this overflow; its virtual resources and their role in the creation/discovery of new symbolized operations and functions.

Refusing the Aristotelian division between movable matter and immovable mathematics, Châtelet insists throughout that mathematics cannot be divorced from "sensible matter", from the movement and material agency of unconscious as well as conscious bodies. Mathematics is an "embodied rumination", inseparable from the contemplative, a-logical and intuitive operations of thought.

Emphasizing the creative importance of what is before thought, what is sensed, he cites mathematician André Weil's observation that "Nothing is more fertile, all mathematicians know, than these obscure analogies, these murky reflections of one theory in another, these furtive caresses, these inexplicable tiffs; also nothing gives as much pleasure to the researcher [...] as when the illusion vanishes and presentiment turns into certainty" (7). Certainly, many a-logical, barely grasped mathematical "sensings" – allusions, hunches, intuitions, suspicions, feelings, convictions, premonitions and pre-knowledge – dance in and out of everyday mathematical thought.

Châtelet's project tracks the movement of thought, the transition from "figures tracing contemplation" to "formulae actualizing operations"; a process whereby "metaphors shed their skin", metaphysics catalyses thought, and the "passage from premonition to certainty is cemented". Pulsing through this movement, folded and implicit in it but invisible within the symbolized operations that enshrine it, is gesture. Citing Jean Cavaillès' insistence that one should rediscover the "central intuition [...] *graspable in action* of a theory", he takes from this a notion of gesture that is "crucial in our approach to the amplifying abstraction of mathematics". A notion, Châtelet insists, that "eludes rationalizing paraphrase, [...] metaphors and their confused fascinations and, above all, formal systems that like to buckle shut a grammar of gesture". Moreover, not only is gesture not formalizable, but it does not operate as a sign: it is neither referential – "it doesn't throw out bridges between us and things" – nor intentional – the grasp that it exerts is not pre-determined. "[N]o algorithm controls its strategy." It would be better, he says, to speak of a "propulsion, which gathers itself up again as an impulse, of a single gesture that strips a structure bare and awakens in us other gestures". He emphasizes that gesture refers to a "disciplined distribution of mobility before any transfer takes place: one is infused with the gesture before knowing it". Gestures are not, in other words, intentional acts; their effects cannot be understood or explained in terms prior to their action; their mathematical consequences are created "before one knows it" (9–10).

The propensity for a gesture to "awaken other gestures" enables it to "store up all an allusion's provocative virtualities without", as he puts it, "debasing it into abbreviations" (10). Virtuality concerns what is actualizable in an event, its potential, all the futures it could/might give rise to. Unlike the possible, which refers to determinations that are fixed but lack the conditions to realize them, the virtual is inseparable from tensions, problems, and open questions. And since for Châtelet mathematical creation is primarily the business of finding determinate solutions to problems, virtuality is the source for the novel abstractions solutions demand: "Virtuality invents and decides on a mode of elasticity: it prepares, cuts out and propels new plastic units. Virtuality awakens gestures: it *solicits* determination, it does not snatch it" (20).

To elaborate, Châtelet introduces a notion crucial to his account of the formation of mathematical abstraction, that of a diagram. But, in contrast to its usual meaning as a visual representation, he uses it in a radically different sense. He takes this from Gilles Deleuze, for whom it is related to the capacity of matter to become an active agent in the creation of form.[1] As such, it is intimately linked to the potential of "sensible matter" and the body's gestures to engage mathematical thought and produce novel mathematical objects. Specifically, he understands diagrams as arrested gestures, gestures caught mid-flight in the creation of a formal concept. "A diagram can transfix a gesture, bring it to rest, long before it curls up into a sign." Diagrams distil action and experience, they function as a "technique of allusions" that relays gestures, and he speaks of dynasties of them, "families of diagrams of increasingly precise and ambitious allusions", playfully adding: "For those capable of attention, they are the moments where being is glimpsed with a smile" (10). Diagrams are distinct from metaphors, though in practice, in the moment of innovation, they occur together as the common fruit of a gesture. Discussing "the screw as bold metaphor" in Maxwell's work on electromagnetism, he describes how it "endows the length with perforation power through a flick of the wrist [as] part of a whole set of diagrams and metaphors" (177). True, like metaphors, diagrams can "leap out in order to create spaces and reduce gaps", but unlike metaphors, they can "prolong themselves into an operation which keeps them from being worn out". Diagrams, then, are always unfinished business vis-à-vis the presence and action of gestures: "when a diagram immobilizes a gesture in order to set down an operation, it does so by sketching a gesture that then cuts out another" (10).[2]

Evidently, Châtelet is indifferent to diagrams in the everyday sense of visual representations of concepts or procedures; they form no part of his scheme of mathematical ontogenesis. Likewise the entire (visual) apparatus of mathematical symbols, which he diminishes as no more than a "crutch" necessary for the carrying out of operations and transmitting knowledge. This means that his understanding of mathematics has no purchase on its modes of internal growth, since the mechanisms of self-appropriation fuelling it are inextricable from definitions, notation systems, and the formal, symbolic and linguistic dimensions of mathematical thought.[3] In relation to the question of mathematical movement, then, I shall change tack and approach diagrams from a direction quite outside Châtelet's approach. Specifically, one that understands them as semiotic/linguistic devices, representations, or rather, presentations of mathematical object-concepts built into them.

To this end I shall shift attention to an algebraic language that conceptualizes mathematical entities as diagrams of arrows between objects. The language is that of *categories*. Introduced some 60 years ago, category theory has since rewritten the presentation of algebraic objects and colonized large areas of topology, mathematical logic and theoretical computer science.[4] As we shall see, the diagrams, far from freezing motion mid-flight on its way to curling up into a sign, are an essential element of the signifying apparatus itself.

Simply put, a category is a bunch of objects with arrows between them. The formal name for arrow is *morphism* (as in iso- and homomorphism). The objects can be any kind of mathematical entity – numbers, functions, knots, groups, relations, topological spaces, graphs, and (importantly) arrows and categories themselves – while the arrows must satisfy the three simple axioms illustrated in Figures 3.1, 3.2 and 3.3, each of which can be expressed syntactically or equivalently as a *commuting diagram*, meaning any two paths in the diagram with the same source and target objects are equal.

Our aim is to externalize the movement inside a category diagram. There are several features of such diagrams that commend them for such a project. In addition to presenting concepts, such diagrams are integral to expositing and deploying them: they allow one to think diagrammatically: one can construct a concept or prove a theorem by tracing paths of connections through a diagram and seeing that it commutes. More importantly here, they encapsulate movement. What is immediately striking about category language, in contrast to that of sets, is the suggestion of motion evoked by an arrow. Arrows are of course used universally as signifiers outside mathematics, as a semiotic for a movement, a transformation, a change of state from source to target, an itinerary, a causal chain of consequences, a flow of information,

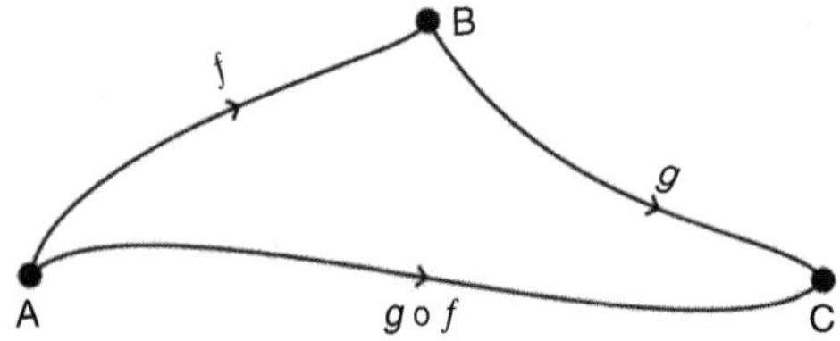

Figure 3.1 Composition axiom: the diagram commutes. *Composition*. If f: A -> B and g: B -> C are arrows from A to B and B to C respectively, then there exists an arrow $g \circ f$: A -> C from A to C which is their composition. In other words, a category is closed with respect to concatenation of arrows

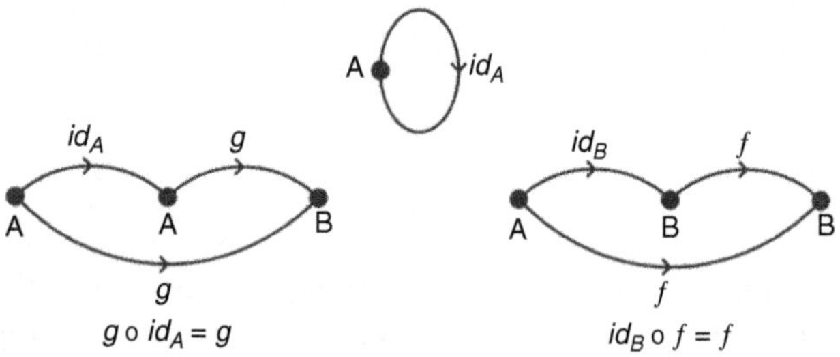

Figure 3.2 Identity axiom: the diagrams commute. *Identity*. Every object A has an identity arrow id_A to itself which is inert in the sense that for any arrows f and g going in and out of it, the equations $id_A \circ f = f$ and $g \circ id_A = g$ hold. Observe, id_A is the analogue for the operation of composition what zero is to the operation of addition and one is to multiplication

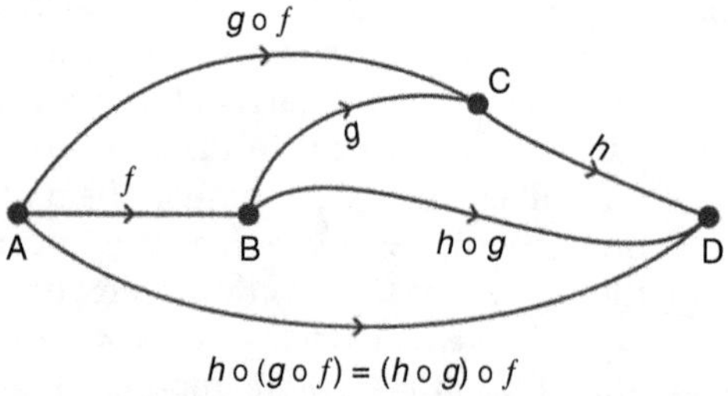

Figure 3.3 Associativity axiom: the diagram commutes. *Associativity*. The operation of composition is associative. If f: A -> B, g: B -> C and h: C -> D then $f \circ (g \circ h) = (f \circ g) \circ h$. In other words, the two possible paths from A to D are the same. (We follow standard practice here and omit the identity arrows from diagrams)

a path to be followed, and so on. In C. S. Peirce's vocabulary, arrows are at once icons, symbols, and indices of a directed dynamic.

Given a diagram of category theory, made up of objects, arrows, and equality such as, for example, one of the axioms presented here defining the formalism, how might we actualize the movement inhabiting it? We need to interpret the objects as fixed entities, the arrows as movements in some sort of space, and the equality relation between arrows as a relation actualized in this space. There are several possibilities. The space can be that of physical locomotion, the objects as fixed locations within it, the arrows actualized as possible paths, to be instantiated by particular routes from source to target locations. Or the space can be acoustic-aural, with objects as fixed rhythms or riffs, and an arrow indicating a sonic movement that effects a transition from source to a target rhythm. Or the space can be purely gestural, with objects as fixed gestural configurations, arrows understood as performers morphing from source to target gesture. Or the space could be chromatic, objects as fixed colours, arrows as modulations between them. For all the possibilities, the interpretation of the equality relation can be synchrony: namely, the equality between two arrows is interpreted as the simultaneous arrival of their actualizations at their target.

I'll consider here only the first of these possibilities, which interprets space as familiar physical space and arrows as locomotion within it. Further, we shall concretize the scheme by having locomotion be an embodied performance, that is, a movement scheme enacted by dancers. Thus, objects will be designated locations in a performance space, arrows interpreted as paths traversed by the dancers, and composition of arrows realized as one path followed by another. Equality between arrows will be realized as synchrony: if two routes from object A to object B are equal then the dancers materializing them arrive at B simultaneously.

Observe that the scheme – an arrow from A to B is realized by a dancer moving from location A to B – allows each of the axiom-diagrams to be performed. For the identity diagram the arrow id_A is interpreted as a dancer moving from A to A in a loop back to herself. For the composition diagram, dancer 1 moves from A to B to C, arriving at C simultaneously with dancer 2 moving from A to C. The pattern is repeated for the associativity diagram with four dancers: dancer 1 moving from A to B, dancer 4 moving from A to D, with dancer 2 traversing path A to C to D and dancer 3 the path from B to D, all arriving simultaneously at D.

Diagrams are themselves mathematical object-concepts and in fact the axiom-diagrams themselves are categories, that is, they satisfy themselves. As objects, they are, as one would expect, elemental: they are precisely how category theory defines/presents the ordinal numbers. The identity diagram is 1; the diagram of an isolated arrow is ordinal 2; the composition diagram is ordinal 3; and the associativity axiom is ordinal 4. Being able to perform the axioms suggests the possibility of performing other diagrams

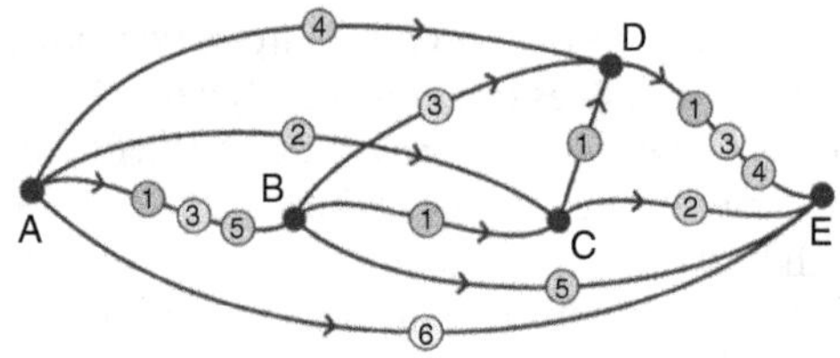

Figure 3.4 *Ordinal 5* with performers. Diagram by the author

of the same kind. The axioms count to 4. The simplest diagram to perform beyond them will be the next number: the diagram of ordinal number 5, which will again be a category.[5] Together a choreographer and I produced a dance piece, *Ordinal 5*, which performed the diagram of the ordinal number 5 as part of an event on topology organized by the Tate Modern in London.[6] The diagram annotated with the paths traversed by performers 1 to 6 is given in Figure 3.4; all of them are assumed to issue from A and converge, at the same moment, at E.

Observe, there's nothing special about either the choice of the dance medium to realize the movement inhabiting the diagram or the interpretation of equality of arrows as simultaneity. As indicated, the movement could just as well be realized in a sonic or gestural or chromatic space. In the first, the overall effect would resemble the structure of certain piano pieces composed by Steve Reich, and in the second that of the video installation *Quintet of the Astonished* by Bill Viola.

Finally there is the advent of digital gesture. Digital motion, technologies of motion capture, enable gesture to be recorded, electronically coded and digitized. And not merely recorded, but transmitted, studied, altered, processed, simulated, invented, and programmed; to be deployed/implemented across digital environments from touch screen/haptic interfaces, to remote surgery and the control of military drones. Several millennia after writing enabled speech to be captured, motion capture has provided gesture with its electronic medium. The latter is larger in technical scope than its predecessor: recording gesture encompasses recording speech, since speech, after all, is a particular (coded, auditory) form of gesture. What writing did to/for spoken language, digital capture is beginning to do for gesture. With the medium come new modes of visual representation, new kinds of diagrams. And how a movement inhabits or is captured by a diagram is embedded in the larger question: how are the contemporary practices of digital visualization producing, capturing and inflecting thought? But this is outside the focus here (see Rotman 2008).

To conclude: one can understand mathematical motion with Châtelet, as the movement of/within mathematical thought from the body's disciplined mobilities by way of their frozen forms to formalized abstractions. Or one can, through the kind of project I've just outlined, arrive at mathematical movement from the differently conceived visual diagrams of category theory. The two conceptions link mathematical and corporeal motion from opposite directions: the first from the material body through the Deleuzian diagram to abstract, immaterial object, the second from diagrammatically presented abstract objects to bodies performing in space. Each reveals, folded deep inside every mathematical object, the gestures of bodies; a fact that, among other things, is central to the ability of these very bodies to deploy mathematics to successfully map out and predict the world they physically experience.

Notes

1. More specifically, it has to do with the refusal of a hylomorphic understanding of material morphogenesis, that is, the imposition on supposedly inert matter of an external idea or pattern, and, an embrace of the contrary idea of the internal production of form from within matter itself. See DeLanda 2000.
2. Concrete examples of this "cutting out" process have been shown to arise in the mathematics classroom. In fact, inspired by and extending his essay, Elizabeth de Freitas and Nathalie Sinclair (2014) have theorized "the body" and its gestures to include its material environment, allowing an amplification of Châtelet's gesture-diagram account of mathematics to emerge in practical work with students learning mathematics.
3. The issue here is the manner in which the creation of mathematical "content" is or is not extricable from its symbolic apparatus. I claim that it is not. That, while appearing to describe (refer to, organize, name) a universe of entities external and prior to itself, the symbolic language of mathematics, on the contrary, has a causal or constitutive relation to their character and existential status. Thus the ordinals are not, as commonly understood, names of "numbers" that exist prior to them, but are co-existent with them, objects imagined to be named by them. See Rotman 1987 for a semiotic elaboration of this capacity in relation to the zero symbol, and Rotman 2000 for a wider discussion. For comment on a much earlier exploration of the crucial role of writing in the operation of mathematical thought and imagination by Paul Valery, see Krauthausen 2010.
4. More specifically, category theory conceives a mathematical object not as an isolated set but as an element within a collectivity – a category – of objects that are kin to it, objects linked to it by arrows. For the definitional and epistemological difference between set-theoretical and category-theoretical formulations of mathematics, see Rotman 2012.
5. By comparison to the diagrammatic language of categories, in set-theoretical language the ordinal number n is perforce defined as a set: specifically the set of all previous ordinal numbers starting from 0. Writing e for the empty set, and [x, y, ...] for the set whose members are x, y, ..., the following set s spells out the number 5:

$$s = [0, [e], [e, [e]], [e, [e], [e, [e]]], [e, [e], [e, [e], [e, [e], [e, [e]]]]]$$

6. My collaborator, who was primarily responsible for the aesthetic (individual gestures, movement style, affect) of the dance, was choreographer Jeanine Thompson (Theatre Department, Ohio State University). The piece, *Ordinal 5*, was performed, with music by Dan Scott, by students from the departments of theatre and dance at the University, at the Tate Modern, London in November 2011. A film of the dance produced by Janet Parrott can be viewed at https://vimeo.com/80409215 (accessed 12 March 2015). An earlier, proof of concept piece, based on *Ordinal 4*, was constructed by Nicolás Salazar Sutil and performed at Goldsmiths College, University of London (see http://www.salazarsutil.net/#!gallery/cff9 [accessed 12 March 2015]).

4
Virtual Choreographic Objects

Stamatia Portanova

Together with biological, cultural and technological viruses of all sorts, globalization has facilitated the exchange of a particularly infectious bacillus: dance as the tendency to follow a pattern of imitable steps. This chapter will analyse phenomena of contagious choreographic design as events of multiple reanimation: the same choreography can be given infinite new lives, as a bodily movement spreading across people and cultures, but also as a dance of digitally designed figures. The myriad flash mobs, game platforms and YouTube videos reproducing the choreographies of Michael Jackson's *Thriller* or Psy's *Gangnam Style* across analogue and digital worlds are only two symptoms of this choreutic virality. The discussion will focus on the digitalization of these two highly popular dances produced by the contemporary music, film and entertainment industries, in order to understand the peculiar way in which digital applications contribute to the generation and replication of what will be defined as *virtual choreographic objects*, steps adopted and adapted by people and cultures, movement ideas obsessively circulating and continuously mutating on the markets of an increasingly neurotic and technological capitalism.

Among the most recent theoretical frames adopted for the study of contemporary cultural and social dynamics, the *topological* approach seems to be of particular interest, for its tendency to subtract space from its essential definitions, and for its capacity to consider all phenomena beyond their supposed depth (or shallowness) of meaning. *Cultural topology* (the application of mathematical topology to the study of culture) describes territories and societies as surfaces without any essential homogeneity, n-dimensional spaces of constant variation where no supplementary dimension or neutral observation and judgement point can be found. In the following pages, this theory will be combined with Gilles Deleuze and Alfred N. Whitehead's respective concepts of the "intensive" and the "extensive continuum", as two philosophical re-writings of topology and complexity. The result should be a capacity to follow (rather than observe and judge) the trajectories of infinitely repeated dances, bringing to light *what changes* and *what stays the same* in them, showing their open malleability but also their magnetic fascination, revealing their potential to simultaneously move and be moved by millions of on- and off-line dancers.

Choreography as bodily design

It is now more than forty years since an eleven-year-old Michael Jackson was seen for the first time singing and dancing, together with his four brothers, a piece entitled *I Want You Back*. In the young Michael's body, the movements seemed to simply develop from a natural ability to relate to the score faithfully and creatively; from a capacity to spontaneously play with the music; from a sort of innate tendency to move in before and after the beat. Since this early debut and for the rest of his life, these qualities always accompanied Michael's performances, together with a sharpness and a clarity progressively increasing at each step, and with an incredible stretching capacity donated to him by a hard training that made everything in the anatomical system operate precisely for the dance. The topicality of Michael's dancing body was therefore the result of spontaneity and technique, exercise and naturalness, proportionately combined in a kind of careful bodily design.

As Sanford Kwinter (2008) points out, and contrary to the common sense use of this term, to design something does not coincide with a simple predetermination of its end results: more subtly, the operation consists of the engineering of initial parameters or constraints that are left open to various possibilities of actualization. The generative aspect of design alludes to the generation of novelty, diversity and complexity from the unpredictable combination of a few basic units and rules: it is also the way in which viruses work.[1] Imitating physical and biological phenomena such as bacterial infections and swarm systems, evolution and contagion, digital generative design, for example, transfers the creativity of matter to processes of algorithmic self-organization, shifting its main focus from pre-programmed objects to the openness of the process, and allowing a redefinition of the designed artifact in more dynamic terms. This concept acquires a particular interest when applied to choreography: we can thus arrive at a topological concept of choreographic design as an open process, a process whose initial algorithms envelop the potential of every moving body to fall over, or to deviate and transform its route into something else, something unknown. It is exactly here, at the point of convergence between rules and unpredictability, between composition and complexity, between impeccable execution and creative improvisation, that Michael Jackson's virtuoso performances originate.

Generative movement algorithms can in fact be the basic production units not only for an individual dance, but also for the choreographic design of a whole society. Every social group, Gabriel Tarde (1903) explains, is a complex series of minute events springing from the engineering of a few elements according to a few formulas: a collection of human movements, a set of shapes and folds on a computer screen, or a wave of physical vibrations, can thus all be defined as societies generated starting from an initial behavioural seed. From this perspective, the Thriller dancing body also emerges as a kind of designed society, not simply in the episodes of social aggregation continuously coagulating around its basic rhythm, but first and foremost in its consistency as a chain of perceptible and imperceptible, more or less choreographed bodily events. All physical movements, from a simple step to a complex choreography, can in fact be thought, in Tarde's monadological terms, as rhythmic processes generated in the connections between a body's molecules and cells, limbs and senses, movements and perceptions, particulate dances composing whole bio-physical societies. In their turn, the choreographic formulas that give the Thriller movements their particular contours are themselves a mixture of heterogeneous stylistic components: hip-hop, sock hop, Soul Train, disco, jazz dance, even tap and Charleston (Acocella 2009). "Letting the dance create itself" through his trained body, Michael also used to collaborate with other choreographers (such as Michael Peters, Vincent Paterson and Jeffrey Daniel) in order to organize all the corporeal and cultural fragments into a well-defined whole. The final choreographic definition of the Thriller dance, then, functioned as a generative inflection point to be each time metamorphosed into a different performative wonder. In the end, a distributed choreographic agency autonomously organized itself at the body's inside-outside limit, assembling motions suggested by both corporeal habit and historical memory, and engineering Michael's body as a dancing molecular society with always new shapes and consistencies.

The molecules of a dancing body can spread their rhythmic contagion at the speed of light. In Tarde's conception (2012), a viral infection cannot be explained as a whole being or a homogeneous structure, but is a molecular society in itself, a multiplying series of little events. A dance is an epidemic, as an epidemic is a dance. Imitated by millions of bodies in all kinds of places and situations, the molecules of Thriller have become a true epidemic whose germ is situated in the short film released in 1983 as a music video for the Thriller song, directed by John Landis and featuring a dance performance choreographed by Michael Jackson and Michael Peters. What is it that has given this video its capacity to ignite such a diffused obsession, and made it so viral? Imitation, for Tarde, is the contagious basis of all social behaviour. The true cause of the imitative process (or the origin of the viral series) lies in the "distinct" and "discontinuous" nature of the idea that provokes it: it is exactly because of their distinctness and discontinuity, and therefore of their clarity and doability, that some ideas tend to propagate themselves "like a wave of light, or like a family of termites" (Tarde 1903: 3). Emerging as a novelty or a cut in the choreographic field, the distinct and discontinuous nature of the Thriller dance generates a viral choreutic behaviour characterized by the easily identifiable and performable aspect of its ideal movement segments: the Zombie march, the jumps, the typical back and forth oscillation of the pelvis. Take the Zombie march. Here, the steps are identified through the discontinuous positioning of a few bodily parts and spatial points (left, right, left leg, right leg, hands and knees), and through a few distinct trajectories or lines linking them (turn, raise, crouch down, extend). In order to dance the Thriller dance, performers comment, it is certainly vital to feel the rhythm and enjoy the movement as an event fluidly running across your body. But, as tutorials and instructions also advise, it is first of all indispensable to recognize and learn by heart (or by body) its easily executable geometry of points and lines.

Freezing the continuity of the motion into a discontinuous series of body parts and spatial trajectories, or of points and lines, appears as a paradox or a contradictory movement vision. This paradox is not only identifiable with the ingenuous attempt by a group of Thriller fans to somehow regulate the performance, but finds correspondence in the topological vision of movement and change, a vision characterized by an ontological dualism. On the topological plane, an object (such as a raising leg in the Thriller choreography) is not defined by its geometrical shape (raising as a movement from point A to point B) but by a set of properties (such as the leg's capacity for stretching) that persist, and by the homeomorphisms, the continuous functions indicating the connected variations of those properties (stretching as a set of infinitely diverging and approaching points). According to Brian Massumi (2002), this vision makes of topology an extremely useful conceptual aid for analysing the continuity of process and transformation, and for illustrating how a form (such as a choreographic step) must be understood in terms of its deformations. And yet, a topologist would remind us, the concrete fluidity perceivable by our senses is accompanied by discrete invariants that can be abstracted from the particular differences of events and stay the same: a raising leg, in the gradual amplification of all possible stretchings, remains a raising leg, its own continuity being composed by the topological points of articulation of the movement (knee, ankle, toe, and so on). This topological aporia between process and point perfectly coincides, in its turn, with the possibility of philosophically thinking movement as an "intensive" or an "extensive continuum".

The intensive continuum of thought, Deleuze and Guattari write (2002), is a plane where de-individuated subjects and objects connect through a continuity of sensations (a plane where philosophical thought profoundly distinguishes itself from science). Philosophically thinking movement as an intensive continuum means feeling and dancing it, as Erin Manning suggests, as a continuous folding of imperceptible microperceptions and micromovements with each other: dance becomes a field of resonance, nebulized into an emergent relational system that is tuned via a certain technique. Each time the movement is reproduced, the technique needs to be reinvented, in order to respond to the shifting conditions of movement's emergence (Manning and Massumi 2014). Through this kind of intensive vision, we can perceive the virtuality folded in every performative act: difference and repetition enveloping each other, as if on a Moebius strip. And we can understand that transmitted from one Thriller occasion to the other is the movement's own capacity to think of itself as different, to actualize itself in different velocities and shapes that are not entirely dependent on the subjective intentionality of the dancer but spring from a unique field of conditions: from the stumbling steps and enthusiastic little jumps of a three-year-old Korean girl, to the awkward pirouettes of a team of hockey players. In apparent opposition to the practice of extracting the continuous connectability and indiscernibility, or the intense mutability of all bodies (Deleuze and Guattari 2002), Whitehead's (1985) "extensive continuum" is like an analytical, simultaneously mathematical and ontological schema that allows different relations to be woven between extended, divisible and computable actualities of the physical world. In the extensive continuum, thinking changes its status: relationality gives way to divisibility as another potential of thought, a potential for spatial and temporal systematization. Whereas in the intensive (or, as Whitehead defines it, "genetic") view, concrescence (the process as an emerging microscopic field leading to concrete moving entities or movement occasions) is beyond time, physical time makes its appearance in the "coordinate" (or discrete) analysis of extension (1985: 283). Spatiotemporal order becomes thus possible.

We already have an example of dance perceived as an extensive continuum, when the very first line of the Thriller choreographic score tells us: "Take four steps forward starting with your right foot first, then take four steps back with your right foot first" (*How to Do the Thriller*, 2014). When movement is thought in its extension, counting the steps and accordingly following the beat become the simplest and most efficacious ways to reproduce a routine in a detailed and accurate manner, and also to precisely regulate it. This simple but rigid metric has been quite successful in attracting innumerable imitations, revealing armies of Thriller dancers as lovers of bodily discipline, and showing the choreography as a ritual of socio-kinetic regimentation and metric control, a choreographic law composed by a topological network of points and numbers, lines and forms. And, as Bruno Latour suggests:

> Just once, before praising freedom, improvisation, tolerance, leeway, straying or disorganization, measure the greatness of metrology, the fragility of the (...) networks that make a minute fraction of commensurability possible. Yes, the form and the content have been permanently inverted: inconsistency is the rule, the rule is the exception. (Latour and Hermant 2006: 83)

In the end, the performance improvisations and the choreographic constants, or the intensity feeling and the extension metrology, apparently split the dance into two different events, conceived along the two poles of the topological aporia: the meter and the fold, the frozen and the open, the repetitive and the creative. One makes you see the dance in the continuity of its metamorphoses, the other in the discreteness of its reproductions.

It will be useful, at this point, to recur to the lens of Whitehead's "mereotopology", a topology of "regions" (rather than points) in which the ontological relation between A and B as two perceivable and thinkable regions allows us to bypass the dualism of process and point, of sameness and mutation.[2] According to Whitehead, in every occasion of experience (for example, in every step), an object is "prehended" by, or becomes part of, another object: prehensions are acts of opening, or of unmediated connection. What can be phenomenally observed are not isolated steps from point to point but prehensive events between body occasions (such as a raising of the leg, a bending of the knee, a flexing of the ankle) whose duration is a thick spatiotemporal slice. Discrete abstractions such as points, lines and surface forms simply constitute the logic and ideal limits, the centres, perimeters or contours of the durationally consistent movement slices.[3] Every gesture, every act of leg raising contains in its spatiotemporal volume a series of smaller motions (and is in its turn contained by other motions), in a relation of "gestures within gestures within gestures", all connected to each other and degenerating to a precise (rather than infinitesimal) point. This is the topological point of variation, the mutating invariant, when and where the leg raises – a point that is of the mind, simultaneously and as much as it is of the body.

As soon as you distribute something in space, Deleuze and Guattari (2002: 389) write, numbers appear. In the same way, as soon as a body distributes itself in space-time, points and lines appear. The mereotopological abstraction of movement is a vision that follows these points, but it does not start from them. On the contrary, to extensively abstract the Thriller dance means to see it as an aggregate of kinetic regions, portions of space and time mereotopologically divisible and connectable to each other through points. In order to conceive the experiential status of the point and its apparition, we therefore need first of all to conceive a movement that, in the body (rather than from it), becomes distributed between its regions or parts, rather than being directed by a unique central locus (consciousness, thought, mind) guiding it according to anatomy and physical laws. In other words, the thought of movement is not separated from it and situated in a different place, or temporally delayed, but coincides with it "in the act". This simultaneity of body/mind appears in the distribution of thought/motion across the body's parts, so that thought and action coincide in their bodily de-localization. Ever-changing and non-conscious points and lines, forms and numbers emerge from this de-localization, composing the rhythmic assemblage and connectability of the body's parts. The abstraction of movement into points is therefore not a subjective intentional deduction retroactively capturing movement after its occurrence, but a distribution of virtual movement centres dictated by the body's own divisibilities and connectabilities.

The superposition of all the divisions and connections a body *could* activate gives us the virtual structure of its movements, while a precise typology of divisions and connections forms the particular mereotopology of the Thriller choreographic design (for example when the regions of neck turning and shoulder raising are connected, through the mediation of the nerves, into a "shneck"), a series of movement bubbles spreading as a physical and cultural virus. A virality therefore not simply belonging to the metric succession of the points (where and when the knee bends, the leg raises, the head turns, the shoulder raises, the pelvis oscillates . . .), but especially to the generative potential that is contained, as Fred Moten (2003) would say, "in the break" between bodily regions. A virality spreading from the intervals, from those spaces left open for the ongoing (re)production of that performance in all its guises, and for a critical awareness of how each of those guises is always already present in and disruptive of the supposed originality (Moten 2003: 4). All the guises of the Thriller performance, from a *Big Brother* house task, to the choreographed performance of 1,500 detainees of the Cebu Provincial Detention and Rehabilitation Centre (Philippines, 2007), are thus potentially contained in-between the steps.

Affecting different social groups in their movement practices, the choreographic design of Thriller reveals society as a spreading of potent little ideas. On one hand, the virus is responsible for an epidemic of extensive motor coordination, and for the proliferation of myriad bodies simultaneously raising their legs and oscillating their pelvises around the world. According to anthropologist William H. McNeill (1997), coordination (and the subsequent possibility to transmit information in a linear and clear modality) is one of the most significant gluing factors in the social individuation of the human species, a capacity that consolidates the bio-genetic separation line between man and animal, while highlighting the borders between compact and coordinated human groups characterized by different rhythms and choreutic traditions.[4] On the other hand, beneath the level of a coordinated (or, in Whitehead's words, a "coordinate") biological and cultural evolution of dancing groups, there is another intense and subterranean level where the same movement idea acts like a ghost continuously brought to new life by its different reincarnations. Every collective, uniform and standardized habit, for Tarde, is in fact the product of a multiplicity of individual innovations.[5] Following the trajectory of an imitation, in Latour's reading of Tarde (2005), becomes thus an observation of how society weaves or designs a sort of contagious "behavioural network" based on the return of "virtual centres of gravity", ideal patterns attracting a repetition of movements that look the same but are always different and unpredictable.

Designing the social imaginary

How to account for the difference between the steps? In Deleuze's interpretation (2001: 172), the variations of an idea are mathematically explained through a certain formulation of *calculus*: not as a continuous variability or an infinite approximation, but as a series of static and ideal limit-points, that which modern mathematics has defined as "cuts". Neither a fixed quantity (for example, a rational number or a particular value) nor a changing variable (x,y or another general value), what makes an idea vary is an undetermined value which very precisely determines the relation in which it finds itself (d, a universal differential value). As an example, we can think of a raising leg as a relation between the *up* and *down* variable positions, a relation that varies according to a differential (or a "quantitability") of the motion. The ideal differential of the movement is the degree of variation of the *up* and *down* relation itself; a border between the unchangeable and the changeable in the different acts of raising a leg, a variation that is neither rigidly fixed nor infinitesimally approximate, but very precise in its multiple occurrences. Whitehead (1985) defines these differential degrees of variation as "adventures of eternal objects" that, entering the generality of "any" experiential relation, transform it into a particular qualitative contrast (d in relation to x,y as any *up* and *down* positions, transforming it into a low/high contrast). The strange temporal nature of the differential as a returning novelty that makes of every ideal repetition a difference, can be explained through the notion of the same eternal object always entering different occasions of experience, not in the sense of its residing in a transcendental nowhere (although Whitehead repeatedly underlines the closeness of his eternal objects to Platonic forms), but in the sense of a degree or a pattern that has always already been objectified, and will always be actualized again. A not-yet-seen memory. A dead living in the future, such as Psy's idea to represent the Gangnam Style dance (the gesture of riding a horse) as the future development of Korean Buddhist tradition, by superposing the angle of his own raising arms and legs to those of Joo Rim Hoon, a hero of ancient Korean folklore.

What is significant, in this case, is not the truth of the Gangnam historical antecedent, but the vortical past/future embrace that Deleuze has defined as a reciprocal relation of repetition and difference, a relation that posits the temporal dimension of ideas as a complex and layered plane, a plane where all kinds of design process can find the right time to make novelty emerge in an automatism. And yet, Kwinter notes, the way in which the capitalist design culture is conceived gives all its products the phenomenological flatness of that image industry "of which television, advertising, the cinema, and the shop window are only the most self-evident examples", a flatness which results from "a set of processes whose scope has become almost dimensionless, yet whose invasive intimacy borders on a kind of micro-violence" (2008: 23). It is, in other words, a defining characteristic of capitalist socio-economic violence that the extremes of generative complexity and monotonous flatness are attuned to each other, in a simultaneously hyperstimulating and repetitive time. Ideas and their difference, according to Kwinter, are continuously sublimated on the empty surface of images, which becomes like the contaminated tissue spreading an affective invasion. After the invasion, a vulnerable audience is immediately transformed (as in a Romero film) into a transmitter of superficial and repetitive tasks.

The most obvious successive step, after Kwinter's critical argument, would be the adoption of a theoretical paradigm that could save the audience from the net of commodified capitalist culture. Instead, it seems much more conceptually and ethically significant to give voice, as Moten auspicated, to the commodity itself. One way to revive the cultural commodity is by recurring to Scott Lash and Celia Lury's (2007) notion of a "thingification" of culture, the process through which the cultural things or objects of contemporary capitalism (for example, virtual choreographic objects) acquire the role of mediators in the construction not of identity but of difference. But before progressing along this object-populated path, it is important to note that the redefinition of dance as a series of objects is first of all a way to de-subjectify movement and to take it beyond its organic or phenomenological embodiment. At the same time, this definition intends to challenge the static character usually attributed to objects: choreographic objects (such as the left leg and right leg positions of Thriller, or the quality of a muscle's stretchability) cannot be considered as inert but constitute the changing potential of every dance. In the same sense, Lash and Lury oppose the static concept of *commodity* to that of the *brand* as a capitalist object come alive. Differently from the fixed, single and discrete nature of industrial commodities, the Gangnam Style is a *living* brand because it is constituted by a range of various products (videos, sites, gadgets, the steps themselves, the celebrities dancing them). It is discontinuous and distinctive, and its continuously shifting use-value is determined by its relations with different users in different parts of the world. This aliveness, associated with the beliefs and desires of the brand consumers, makes of branding one of the main tools through which contemporary capitalism can model and control people's ideas and movements.

At the same time, the aliveness of a virtual choreographic object such as Gangnam Style allows it to generate little clinamens or irregular, deviating trajectories in contemporary capitalist societies. Social irregularities, Arjun Appadurai (1996) notes, are caused by the fact that both cultural consumers and objects are in continuous circulation, and therefore not connectable to recognizable circuits or audience portions at a local, national or regional level. Let us remember that Gangnam is a popular and trendy neighbourhood of Seoul, South Korea's capital and the singer's native city, a place from which Gangnam derived as a local neologism mainly indicating a fashionable lifestyle. And yet, between August and November 2012, the Gangnam Style has been able, as a globally circulating virtual choreographic object, to expand beyond its semantic locality, and to generate a series of large-scale flash mobs involving people in cities as different as Atlanta, Seoul, Milan, Paris, Sydney, and Stockholm, revealing the tendency of people to move in ways that are strongly influenced by a mass-media imaginary trespassing national spaces or local meanings. As shown by Elias Canetti (1981) in his physical analysis of social dynamics, institutions, rules and practices tend to mould the social mass (the "crowd", as he defines it) by giving it rigid confines, by assigning precise spaces to its actions and events, and by limiting its exchanges with the outside, so as to guarantee a longer duration of established power systems: in this way, a mass crystal is formed. We can think of places such as dance clubs or theatre stages, or even of the home television set and the public square, spaces where each performative practice and each kinetic culture finds its rules and limits. At the same time, the steps of popular dance rituals trace a territory that allows the dancing body to escape, or to de-territorialize. A kinetic imaginary is composed, whose culturally and corporeally coded performative territory is in relation with the tendency of the dancing group to explode for crowd dispersion, and with the capacity of its virtual choreographic objects to become something else. In Indonesia, for example, the initiation of a Gangnam flash mob by a group of teenagers becomes the pretext to generate a situation of street confusion, and an occasion to transgress the municipal prohibition of jumping into the big fountain of the main square.

The flash mob events can of course take on different forms or levels of organization, from the spontaneous and uncoordinated street action in Indonesia, to a structured Gangnam performance under the Eiffel Tower. And they can be moved by different aims, such as when the street mob replaces the factory's conveyor belts, and the mob is harnessed as part of a commercial advert (the Wonderful Pistachios ad in New Orleans). But why is global capitalism so interested in dance, and in flexible creative bodies showing virtuosity and skill, Jussi Parikka (2010) asks? The answer is found by the theorist in the words of Nigel Thrift, for whom "dance can sensitize us to the bodily sensorium of a culture, to touch, force, tension, weight, shape, tempo, phrasing, intervalation, even coalescence, to the serial mimesis of not quite a copy through which we are reconstituted moment by moment" (2007: 140). "The serial mimesis of not quite a copy": propagation here really emerges as the key word to understand the capitalist investment in dance as one of its most reproduced practices. Capitalism, it is now well known and accepted, desires and fosters transgression. Dance, we have seen, is pure repetition, and every repetition is, as Deleuze reminds us, not a moral habit but a transgression or an exception by nature. Repetition is never a tamed resemblance, or a habitual return of the same, but always an unpredictable return of difference in itself, as in a viral infection. According to Eugene Thacker (2009), the transgression of viral contagions elicits a rigorous defence of the body's boundaries, despite the fact that living beings are certainly defined by their continuous exchange of matter and energy with their surroundings. So what more paradoxical defence tactics than those of the capitalist body against all transgressive contagions, a tactic mainly consisting of the incorporation of the most dangerous elements? It is the anti-body tactic as coinciding with the schizophrenic character already attributed to capitalism by Deleuze and Guattari, a schizophrenic form of socio-political and economic control where giving direction, tempo, aim, and even creative momentum to the most rebellious and revolutionary flows (the flows of dance included), becomes for the system more vital than their blockage.

With its simultaneously standardized and malleable nature, digital technology responds perfectly to capitalist needs. It is not by chance that we see today the Thriller virtual choreographic objects proliferate in the digital world, mutating into characters of the Sims world, or into a squad of animated lizards dancing the Thriller dance for the SoBe Life Water commercial, a video that reached more than five million YouTube and online hits in two weeks, while:

> the rhythmic reptiles ... slither[ed] their way across Manhattan and the nation, spreading the SoBe Life Water Thrillicious movement from coast to coast through audacious billboard displays. (BevNet 2008)

Technology is clearly at work here as a tool of power and colonization: the Thriller world invasion, from the thrillicious lizards, to the Xbox Michael Jackson's avatar, to a Thriller-dancing Shrek, is mostly a digital invasion. The return of the same appears as a concept built in the very technicality of the digital tool, as the original dance is recorded to the computer using video cameras or markers, and the same performance is then applied to various animated characters. In fact, the process of digitalization implies not only a reproduction of individual avatars, but also a passage of the Thriller virtual choreographic object from the mass-logic of the "publicum" (the spectacle) to the individual level of data, and then back to the mediasphere. On this level, the individual unit of choreography, the step, is divided among a crowd of differentials, a data crowd whose complex playing-out generates macroscopic shifts in the movements' general orientation: the step as a "dividual", as Deleuze (1992) would say. In the passage from individual step to dividual datum, the possibility of a standardized reproduction of movement progressively gives way to a more abstract level, as the microscopic gaps opened by technology in the choreographic design expose its topological continuity to the intrusion of many parasitic, extraneous rules (or errors) acting as magnetic attractors of difference at each and every portion of a step.

Abstracting movement to its skeletal informational structure and to its extreme cutt-a-bit-lity point, the simple logic of the digital algorithm also allows for multiple connections to take place. It is important here to reiterate that an object (technological, biological, social) is never the result of a linear algorithmic process of growth from simple elements and rules, but always emerges from a multiplicity of relations between mutating parameters, variables that are always already complex and complexly related to each other: shifting sets of variables, and their relations, are what truly determine the transformations of forms. In the case of technology, an important objection to this statement comes to mind: in the ontological basis of the digital, every new occurrence is the result of a previous operation, incorporated in a causal linearity. This simple observation of the digital in its linear essence obscures the digital in its complex behaviour. Digital objects are in fact characterized by relational capacities, and it is indeed another scale of relations that is superposed, or ties in, between digital dance steps and human dancing bodies, as the digitalization of virtual choreographic objects allows them to circulate and distribute gesturalities at an increasingly minuscule level. All technologies can, from this point of view, be seen as micro-relational: the microscope putting scientists in touch with the smallest particles of matter, photography and cinema weaving a perceptual line between human observer and the imperceptible inflections of movement, the Thriller Experience video game allowing dancers to play with the uncontrollable slippages of the music-motion tempo intervals. And it is today thanks to the digital that the human can enter into dialogue with a crowd of non-anthropomorphic, non-conscious experiences inhabiting its own steps.

Caution: contains ideas. Handle with care

Bodies of all dimensions and scales meet and mix by chance: every encounter is a new corporeal mixture giving a different quality to the world. It is by chance that you meet and mix with a flash mob in the streets of your city. Bodily encounters also emanate ideas, which Deleuze (2004: 10) defines as the surface effects of the real. The Thriller idea starts to touch and affect you (or we could also say, Thrilling thoughts correspond to Thrilling feelings). As a matter of fact, all ideas have an appetite for proliferation and tend not to stay with you alone but to multiply, to the point that Tarde (2012: 12) could define society as a "multiplication of purely mental agents in the world". Deleuze and Guattari (2000) also described the capitalist lifestyle as a "schizophrenia", a horizontal relational attitude that induces one to be affected and inspired by a multiplicity of ideas rather than guided by a unique dominating principle, to become multiplied into a crowd rather than remain the same individual. But it is also the case that capitalist imitation often ends up cataleptically following one or a few ideal generators, and "Thanks to the deep and intense stupor caused by their first miracles, each of their assertions and commands open[s] out an immense vent to (...) vast, vague, and impotent aspirations..." (Tarde 1903: 81–4). The schizophrenia gives way to a capitalist neurosis, as you find yourself part of an army of Thriller dancers exchanging instructions and technological tools in the aspiration to perfectly reproduce the same dance. A closer look at the experience through the mereotopological lens nevertheless shows a multiplication of different ideas: you do not copy or exchange but "steal" or "donate" something from each other, something unique or singular that is repeated each time and, at the same time, can have no equal or equivalent. It is the power of difference that affirms itself while you dance, against the law of your limited physical constraints or against the choreographic rule, as in Jackson's own improvisations, or in the quick flickering of a digital lizard's tongue.

This chapter has tried to practise (instead of solving) a double, mirroring vision at all levels of movement analysis: at a physical level, the conception of movement as a corporeal folding continuity has been reflected by that of choreography as a metric succession of numerable points connected by geometric lines and composing the forms of a dance. Mereotopology has then been deployed as the stereoscopic vision combining the two concepts into a system of connected movement regions thinkable as a series of virtual choreographic objects or movement ideas. At a social level, the tendency of virtual choreographic objects to choreograph the movements of large groups of people along the dictates of contemporary capitalist marketing has been seen as accompanied by the potential of the crowds to appropriate, move and change the virtual choreographic objects in various ways. And finally, at a technological level, digitalization has emerged as a movement codifier, and therefore as an amplifier of reproduction and imitation, but also as a facilitator for choreographic contamination and networking not only between human bodies, but also between a human and a microscopic level of movement experience. Even capitalism has in this sense been doubled between a schizophrenic multiplication of ideas and a neurotic fixation on a few obsessions. All these doubled visions have nevertheless converged into a concept of the movement idea as separate and indifferent to the occurrences of physical, social and technological bodies. Every idea, as Deleuze (2004: 8) would have it, remains cold, neutral and indifferent like a ghost, a film without volume that envelops a body, or an empty mirror that reflects it.

After arriving at the indifference of ideas to the actual peripeties of their attached bodies, it is now time to conclude by briefly exposing the extreme care with which ideas need to be manipulated (or, which is the same, with which they need to be thought). Ideas, there is no denying, have a phantasmatic role – in the sense that they do not represent corporeal actions or passions, but are simply their results ("pure event", in Deleuze's words). As bodily mirrors, they have a very fragile consistency. If you are an idea handler, therefore (if you are, for example, a dancer, or a thinker of dance), you have two main tasks. First of all, that of reaching a clear distinction, understood not in the sense of a separation between real and imaginary facts, but as one between events and corporeal states, bodies and mirrors: the task is to achieve a correct identification of the idea beyond all perceptual judgements. Deleuze (2004) is also very precise in explaining the importance of establishing a tight relation, and also of differentiating, between events and corporeal states of affairs: phantasms are produced by reality, but they also transcend it. They fly over it. For this reason, ideas should be considered as effects not producing or coinciding with, but simply following, a body's evolutions. Second, as an idea manipulator you are required to linger over the ideational surface over which these effects are produced. You must concurrently transcend the critique of all inside-outside identifications as to the ideal event's origin (body, senses, psyche, the environment, another body), since the topological property of the idea is to bring these into contact, in order for them to become one single, folded surface. A double causality is the logical form to which the idea is effectively submitted: the external and internal causes that produce it in depth, but also the quasi-causes which, on the surface, make it communicate with other ideas. When handling, therefore, remember: the event of an idea is neither active nor passive, neither internal nor external, neither imaginary nor real – it is the most impassible and ideal of illusions. Just stay on the surface, follow its developments as they follow you, and carefully, delicately start to dance with the myriad phantasms appearing on it.

Notes

1. In fields such as graphic or architectural design, "developments in scripting have opened the way to algorithmic design processes that allow complex forms to be grown from simple iterative methods while preserving specified qualities" (Meredith 2008: 3). Adopting generative software in architectural design, it is for example possible to imagine a whole human body moving in space and then replace it with a single joint: this technique is used to understand how the unit relates differentially to the whole building across a series of frames. In parametric design, this differential relation between variables is computationally processed by reducing it to the binary code, but at the same time by exposing code itself to a whole variable environment. Generative and parametric tools are usually used in architecture to map and manipulate the variables of a geometric form into a tuned formal solution: calculating every parameter as the algorithmic evolution of a point in space-time, and then relating multiple parameters among themselves (the movement of a body in space, for example, but also the progressive changes of building materials, the weather, economic and geographic factors and so on).
2. Mereotopology is a mathematical and philosophical concept that combines the mereological analysis of reality as made of parts and wholes with the topological notion of connection, with the result that the infinite regressions and progressions conceived by topology find their limit in the regions-parts.

3. In the same way, negating the process of the infinitesimal, Tarde (2012: 9) writes: "where will this process of elimination finish if not at a geometrical point, that is, at pure nothingness? Unless, (...), this point is a centre."
4. The acquisition of a sensori-motor capacity for coordination was already considered by Henri Bergson as the main trait distinguishing animals from all other living organisms, the two "extreme" realizations of the two variables (the static and the mobile) being represented by the immobile jellyfish and the dancing human.
5. This methodological statement does not correspond to an affirmation, by Tarde, of the necessity to reduce sociology to the study of individualism. Following phenomena of imitation does not imply a limitation of social analysis to the level of the individual, but neither does it link it to the study of a homogeneous structure (Latour 2005).

5

catch/bounce: Stack Overflows and Digital Actions

James Charlton

Movement is a fundamental sculptural concern – not simply through the overlap of the shared spatio-temporal concerns that have aligned sculptural practice with performance art practices, but through the central role of the body in regard to materials found in object-based practices (see Serra 1968). In the slipstream of 1960s and 70s conceptual art, the authorial singularity of the artist can be seen as yielding to material influence. No longer subjected unquestionably to the artist's will, materials become agents in the work. Challenging the oppositional duality of artist and material, object and subject are brought together in an inter-subjective act that repositions the sculptural artifact as residue – as after-fact more than arti-fact, as movement-object rather than static image.

Digital stickers

It is, then, not surprising that a sculptural practice inclusive of digital media processes would attempt to understand the specific nature of *the digital* and how it might be engaged on the same footing as corporeal media – that the material sensibilities and tacit knowledge that inform many sculptural practices should be used to inform our understanding of digital materiality.

In demanding to understand the inherent properties of the digital, a new-media artist might well have begun by assimilating the wealth of digital media theory in this field. However, approaching the question from a sculptural perspective concerned with materiality, my stance has been to reject the representational function of digital media as providing unmediated access to digital materiality, instead striving to understand both the ontological nature of *the digital* and a means by which it might be engaged within sculptural practice.

The term *digital* is commonly used as a qualifier of another object – for example *digital*-media, *digital*-network, *digital*-camera, and so on. As Florian Cramer points out, the digital has in fact become a post-digital sticker that can be applied to any object or concept in order to assert its technological superiority (Cramer 2013). My use of the term extends David Lewis's widely accepted definition of *the digital* as being a discrete representation in opposition to the analogue, which is seen as a continuous representation (Lewis 1971). However, I reject Wheeler's (1990) 'It from bit ontology' of the digital in favour of a 'Level of Abstraction' argument (Floridi 2009) in which the perception of an epistemic agent (presumably the artist) is seen as differentiating between discrete and continuous modes.

Rather than validating representation as a necessary consequence of the epistemic agency as proposed by Floridi on the basis of Levi Bryant's 'Latour Principle', I contend that *the digital*'s materiality should not be bound by representation alone. Rather, *the digital*, as proposed by Barbara Bolt (2004) in her counter-representational reading of Heidegger, should also be located in a dynamic non-representational space directly between artist and material, thus eliminating the necessity of any representational mediation by digital media. Building on this I propose an alternate definition of *the digital* as an underlying *structural method* that has to date been sublimated beneath technologies to which it is not exclusive.

Materiality as structural method

Although the contemporary revelation regarding the technological confluence of the physical and the metaphysical[1] lends currency to the idea that materiality is not contingent on substance, this construct has in fact long been understood through sculptural practices. As Jacob Lillemose explains: when Lucy Lippard labelled the conceptual practices of the time as *dematerialized* she was in fact making an argument not so much for the disappearance of materiality but for a rethinking of materiality in conceptual terms (Lillemose 2008).[2] Under such a post-media revaluation of non-technological practices,[3] materiality should then be taken as an immaterial idea rather than a corporeal state.

It is within this framework that materiality is conceived of as a *structural method*. *Structural methods*, like Jane Bennett's vital-matter-assemblages, are emergent properties that are distinct from individual actants (Bennett 2010). In a state of becoming, assemblages and *structural methods* emphasize the dynamic means through which parts are related, and from which the underlying materiality of practice is derived. Digital materiality might thus be seen as a method of practice that promotes discrete non-corporeal structures regardless of the ontological affiliation of its constituted parts.

Agency and co-constitution

In arguing for a materiality of substance-independent *agency* I draw on the flat ontological status of the *object* proposed by Speculative Realism (SR), which places immaterial thoughts/actions on an equal footing with material agents within a Latourian-like network (Harman 2013). This equity between the physical and the metaphysical supports a treatment of materiality as a non-corporeal state and distinguishes it from material substance not just by a parallel etymology[4] but, as both Kant and Heidegger suggest, as a materiality that can be differentiated from substance.[5] While Kant's treatment of *materie* and Heidegger's treatment of the *thingness* of things supports a notion of substance-independent materiality, these positions also maintain a correlational position that conflicts with SR's flat ontology and denies objects *agency*. In his argument for ontologically independent objects, Harman reconciles Heidegger's *readiness-to-hand*[6] and posits objects as irreducible agents (Harman 2002). With *agency* positioned as the immaterial hybrid interaction of actants, materiality exists as an *object co-constituted* in each moment by an active *agency*.

It is the ability of an active agent to overreach itself while remaining embodied[7] that causes it to emerge as a method by which materiality is actualized. Materiality, then, is both an independent agent and an agent dependent on the *co-constituted structural method* of the actant network that realizes it.[8] *Co-constitution* also provides a model for an investigative strategy capable of clarifying ontological, epistemological and phenomenological relations as developed by Luciano Floridi in his Levels of Abstraction (LoA) argument (Floridi 2003)[9]. In his integration of ontological commitment and epistemic resilience, Floridi positions phenomenological awareness as central to agent networks. However, I contend that Floridi's schema privileges an anthropocentric epistemology and fails to resolve issues of causation and intentionality central to developing a digitally inclusive artistic practice.

Accessing the digital

Having briefly outlined ontological and epistemological frameworks for the consideration of digital materiality, I want to return to aspects of new-media theory that I initially bypassed, in order to propose a means of gaining practical access to the materiality of the digital. While both Laura Marks (2000) and Vivien Sobchack (1992) enable us to understand the role of embodied experience in film and, like Deleuze, allow us to break down the image into *movement-images*,[10] there is still no means for reconciling a shared agency between the analogue and digital without recourse to some form of representation or concealment within some LoA-based black-box.

In *Notes on Gesture*, Giorgio Agamben's analysis of Deleuze's *movement-image* provides a framework for establishing a connection between analogue and digital, by proposing gesture as a particular event that 'is neither production nor enactment, but undertaking and supporting' (Agamben, in Agamben and Casarino 2000. Gesture in this sense becomes a structural object – a network of relationships within a shared agency between subjects[11] capable of providing a practical resolution to the problematic lack of phenomenological access identified by Floridi.[12]

Further support for the role of gesture in mediating between subjects is presented in Katharine Young's[13] investigation of gestural inter-subjectivity (Young 2011). Young locates Merleau-Ponty's 'intentional arc' (Merleau-Ponty 1962) as a category of gesture that loops through the body; loops through the body and the world; and loops through the body and the virtual world (Young 2011). Although Young's use of the word *virtual* is undefined, she allows us to establish a relation between corporal and non-corporeal (analogue and the digital) subjects and establishes an inter-subjectivity that is consistent with Aden Evens' functioning of the digital fold – 'the mechanism by which the digital connects to the human and the actual' (Evens 2010). While for Evens the *fold* is seen as a technological function exemplified by CTRL-ALT-DEL and COMMAND-OPTION-ESC keyboard functions, he allows for other ways through which the digital might engage with the actual without resorting to representation. The *fold*, like a gesture, is really an inter-subjective overlap that allows not simply humans and computers to *touch* but for the digital to be actualized and accessed as a material.

Although gesture is still not shown to possess materiality, materiality is brought forth by inter-subjective agency that 'call(s) into question linguistics and informatics models of gesture and opens up new ways to understand gesture and agency as embodied a-linguistic experience(s)' (Sha 2003). Inter-subjective gestures 'suspended between manipulations of objects and manipulations of one's own body' (Young 2011: 67) inhabit all actants in a network simultaneously. Rather than being an emblematic or representational construct, inter-subjective gesture is the means through which we might engage the digital while maintaining our analogue existence.

32-*Bit Catch*

As the body extends beyond itself through the inter-subject gestures, it occupies a liminal state – or rather, as Nathaniel Stern explains with his Implicit Body framework, the body is in a constant state of becoming (Stern 2013). Building an argument around Massumi's analysis of movement and affect, Stern posits the body as constituted by its relations with subjects – subjects that it cannot quite reach.[14]

Reminding us of Zeno's arrow paradox[15] – a *reductio ad absurdum* of the mathematical impossibility of an arrow ever reaching its target – Stern critiques a conception of time as a point, rather than a line, that parallels the construct of the point vector in which movement is defined by start and end points, rather than by the line-duration itself.[16] Zeno's paradox then serves as a mathematical counterpoise for the simple action of throwing a ball that is found in 32-*Bit Catch* (2013). Although perhaps surprisingly it is not the ball that occupies the duration-less moments along the vector path;

rather it is the body that becomes the arrow that 'never quite reaches the target'.[17]

In *32-Bit Catch* a ball is thrown against a wall and caught again thirty-two times.[18] The arm is shown disembodied by the framing of the camera. It is *an* arm not a specific arm. The body it belongs to is never declared, although it is in fact the arm of the artist. The video starts with a blank wall indicated by the optically distorted corner towards the left of the screen. The footage is highly compressed and the almost monochrome image is reminiscent of pre-digital video works.[19] Only when an arm enters the frame on the right of the screen are we aware that the video is in slow motion. The hand holding a ball slowly primes itself to throw. In the moment preceding the release of the ball the video cuts – we hear a dull thud and the ball is suddenly hitting the wall on the left and bouncing back. Until now the video has been silent. No sooner has it bounced back than the video cuts again and suddenly the ball is back in the disembodied hand.[20] This cycle is repeated thirty-two times, punctuated by the sound of the ball hitting the wall.

There is never the narrative potential in *32-Bit Catch* that we find in its obvious reference points – Richard Serra's *Hand Catching Lead* (1968) and *Surprise Attack* (1973). Unlike *Hand Catching Lead,* we are not aware of content or action outside the frame – all potential within the work is completely self-contained. The image is discrete – even its repeated narrative is contained within a single phrase – catch/bounce. This contrasts with the leaden sheets that constantly pass through the frame in *Hand Catching Lead,* and the disembodied voiceover in *Surprise Attack,* and reminds us that this cinematic image, at least, is not encapsulated.

Initially *32-Bit Catch* may seem to operate in complete contradiction to Zeno's paradox as the vector has been edited out. All we are shown is the ends of the vector – the points, the moments of contact between the ball, the hand and the wall. This is certainly true if we focus on the trajectory of the ball. However, we never actually see the ball in flight along its trajectory – rather we are shown the points of contact between the ball and body and between the ball and wall. The moments of physical overlap in which the two subjects form an inter-subjective third subject, the *ball-body/ball-wall.*

In the thirty-two moments within each *ball-body* moment, the inter-subjective gesture is realized not just by the literal grasping of the ball, but in the agency that each subject exerts on the other beyond the point of contact. As the ball bounces off the wall and establishes a 16-frame trajectory back towards the body, the arm anticipates it. It reaches beyond itself – not in the literal sense of reaching out for the ball, but in the projection of itself forward in time into the trajectory of the ball. In this way the body is ahead of itself in time and open to 'a future with no present to speak of' (Massumi 2002).

But it is important to note that this is not an anthropocentric reading of the inter-subjective subject that dwells with bias on the body. While

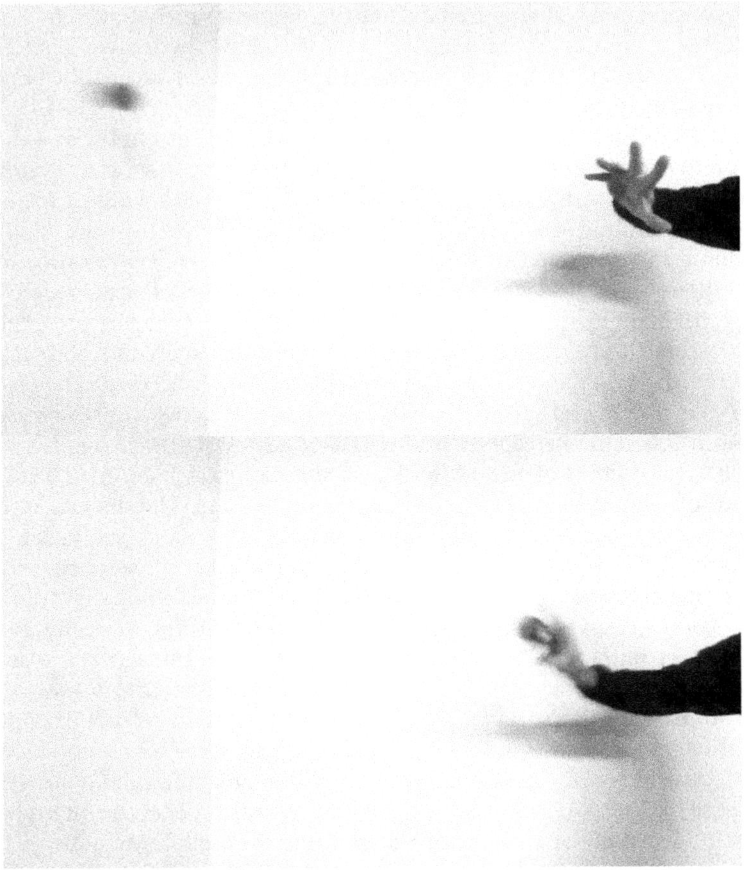

Figure 5.1 James Charlton, 32-*Bit Catch* (2013). Still images from video

the ball does not throw itself against the wall, neither does the arm reach out for a phantom image in space – the ball in its 16-frame trajectory subjects the body to its agency – the spring of the ball, the physics of the ball hitting the wall, etc. The *ball-body* materiality has an inseparable agency of infinite recursion without interrupt from a *stack overflow*[21] in which a material exceeds its subject's potential. As it generates its own potential before itself in time it can never exceed its buffer potential. Unlike Evens' *fold,* which proposes that the interrupt or failure is a means by which humans and computers come into contact, the co-constituted materiality of the *ball-body* is not a true-false binary construct entrenched in an anthropocentric materiality that situates humans and objects as finite oppositional resources. Rather, materiality is an infinite Boolean union without loss that forms a new discrete subject-of-the-moment instantiated by inter-subjective conjunction. As objects always withdraw themselves from each other (Harman 2013), inter-subjective actions can never exceed the bounds of their constituting subjects.

In 32-*Bit Catch,* inter-subjective action is shown to us almost as a single image that oscillates between catch and bounce to construct movement. Like Zeno's arrow, the *body-ball* is motionless in these moving images – to move, something must get from one point to another. In each of these motionless-moving-images we remain in a specific point – catch/bounce, catch/bounce... Each inter-subjective moment, like Deleuze's *movement-image,* is a moment that is past itself as soon as it comes into being. As Massumi acknowledges, it is Deleuze who opens up this connection between the body and its position, specifically when he breaks down the *movement-images* and specifies the *affection-image* as occupying 'the interval between incoming perception and outgoing action; it is, one might say, in the interval itself' (Deleuze 1996). The *movement-image,* like Zeno's arrow, is a kind of relay system between discrete mobile units. The discrete mobile units of catch and bounce are pieced together as we watch the video. The hand never throws the ball. The ball never bounces back. Instead the viewer enters into the structure of the work to complete the action, to draw the line that is never given forth by the two points. In constituting the vector in this way we become agents in the work – a work that, as I have tried to show, comprises discrete instantiations of mobile images. It is a work that draws us into its structural schema in order to complete itself. As part of its digital structure our analogue function is not surrendered – we do not become digital. Rather, we extend beyond ourselves while remaining in ourselves in an act of digital *co-constitution.* Being denied an anthropocentric interruptive agency that falsely offers hope of materializing the digital, we suddenly find that we are in a constant state of interruption and caught within the infinite recursion of the system in which we find ourselves. In fact, without our being interrupted there is nothing to interrupt. We, too, are *movement-images* beyond ourselves in time.

Leaking Into Ourselves

The manner in which the body reaches beyond itself to engage in inter-subjective acts is not just an externalization of self – a leakage of the body into the world – but a constant state of analogue being. To be in the world means to be beyond oneself but it also means that the world is in us. Inter-subjective actions reach back into the interior of the body as readily as they reach out beyond the body. While Harman reminds us that the body is in constant excess of itself in this way, he also warns us that it is dangerous to think of the body being privileged when this is in fact the state of all objects (Harman 2013). As we see in 32-*Bit Catch,* the inter-subjective agency of the ball extends itself beyond its own spherical confines. It, too, reaches

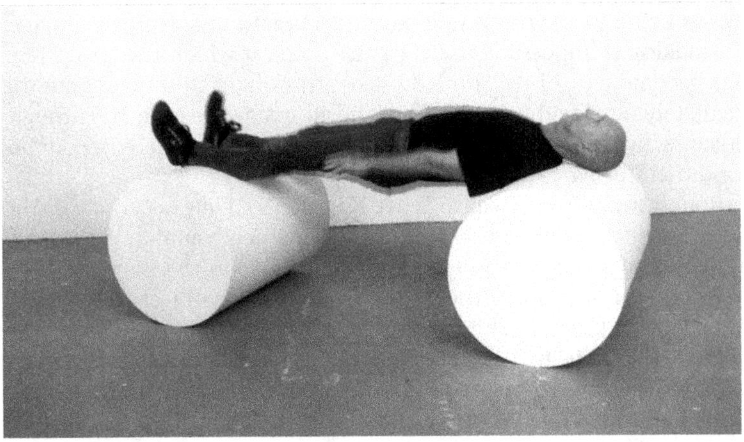

Figure 5.2 James Charlton, *Leaking Into Ourselves* (2014). Still image from video

out to intersect the body in a non-reductive respiration that is articulated in *Leaking Into Ourselves* (2014). The intersection of one object with another doesn't necessitate reduction or loss on the part of either object. An object is in excess of itself but never exceeds its potential.[22]

In *Leaking Into Ourselves*, with only my head and feet resting on them, my rigid body is suspended horizontally between two cylinders that are positioned parallel to each other across the screen. Unlike *32-Bit Catch* there is no disruptive interrupt. No punctuating sound or cinematic cut for the body to bridge. Rather the static body occupies the vector space between two points designated by the cylinders. Like the *motionless-moving-images* in *32-Bit Catch*, the video, playing in real time, has a sense of suspended time. Nothing seems to happen. If not for the twitch of a foot this might be a static image and we might even doubt the authenticity of the action as performative.

As the video plays, we become attuned to even smaller movements of the body until, straining to maintain its rigid form, the body appears to blur and then vibrate with each breath. The head and feet remain sharply focused, pinned to the points that define the vector space in which the torso is suspended. This ghosting of the body[23] appears to extend the torso outside itself in time while remaining grounded through its immobile points of contact. The body here, after Rotman, is taken as the indexical 'I', that is both a self-enunciating and self-subjectifying *ghost* of itself that is not simply specific to the 'digital forms of self-reference' but to the body itself (Rotman 2008). It is as if each breath diffuses the body into the world – extending it beyond itself. This is quite literally what happens in the act of breathing, which breaks down the normal binary ways we think about our body and quite literally extends us beyond ourselves. While we readily speak of things as being inside or outside our body, breath dissolves this binary formula, making it impossible to detect the point at which the air we breathe in has become part of us. Similarly it becomes hard to differentiate the air immediately outside the mouth from the air inside the mouth. Through the inter-subjective act of breathing we leak into the world, extending our bodies beyond themselves.

Although strikingly similar to Bruce Nauman's *Failing to Levitate in My Studio* (1966), a double exposure, still image in which Nauman is shown both lying between two chairs and slumped in failure on the floor, *Leaking Into Ourselves* does not propose the object as an anxious object as suggested by Parveen Adams (1998). While, for Adams, Nauman's *failure* is like an inter-subjective gesture in that the body is outside itself, it is also seen as lacking shared subjectivity. The inter-subjective act is generative and has no such anxiety. Embedded in an economy of excess, the body, like other SR objects, is inexhaustible in this way.

While one could hardly define the reflexive act of breathing as a gesture, this is precisely what the body does when intentional inter-subjective actions are required to engage with works. The use of exhalent devices in other recent works[24] such as *TradeAir* (2004) calls into consciousness the act of exhaling as a gesture that activates the work. In *TradeAir*, blowing into 3D printed receptor vessels that house small fans attached to micro-motors, the volume of air exhaled is calculated and transmitted across localized and cellular networks to inflate forms with the installations. The viewer quite literally gives shape to the work through a gesture that places the body outside itself without loss. Suddenly, in this framework, we have a new object that is both self-contained in the form of the work, yet constituted by and part of another object – the body of the participant. Both the work and the participant are objects that exist autonomously and discretely, yet are made up by a continuous and sustained engagement with other objects. Through the inter-subjective gesture the body has, without surrendering its analogue self, become digital in the terms defined earlier.

The digital is no longer mediated or represented at all. It has become its own subject through a body-object action whose materiality is located in the dynamic inter-subjective space that, like Bolt's materiality and Deleuze's *affection-image,* is a space of interval rather than position – movement rather than stasis. It is through movement, then, that the digital emerges as a material.

Error messages

This apparent tension in the proposition of a discrete digital that exists as a result of a continuous movement is not as contradictory as it first seems. In fact, even as a technologically monopolized term, the digital is portrayed as both rigidly concrete and fluidly mobile. A supposedly stable media capable of delivering unlimited and flawless instantiations across spatially disparate networks, the techno-digital is seen as being both as permanent as stone and as transient as breath. This is achieved not through any miracle of technology but because of the *structural method* by which objects extend themselves beyond their analogue self to *co-constitute* materiality.

This state of becoming is a materiality of the 'instant itself' (Deleuze 1986), rather than substance. Made epistemically available through inter-subjective *structural methods*, digital materiality, rather than being the property of an object, is manifest in each moment of overlap between agents. As materiality is not the exclusive property of an object, nor the purely perceptual construct of humans, it both becomes itself as an object without recourse outside of itself, and is as inexhaustible as any other object. In this way, digital materiality shares something in common with a technological prescribed state. Capable of being both discrete as an object and continuous in its constitution, digital materiality breaks down oppositional anthropocentric frameworks. Instead, objects are seen as having both analogue function in and of themselves, and yet digital function in the co-constituted *movement-image* instantiated by any given inter-subjective action. The Boolean union of analogue objects is infinitely recursive, as every object is in excess of its potential relationship with other objects. While programmers continue to struggle with overflow errors, this state of material failure is not a function of *the digital* but of the technological. Objects themselves are never overloaded by the inter-subjective networks in which they participate, as they possess an agency of infinite recursion without fear of interrupt from a stack overflow.

catch/bounce
catch/bounce
catch/bounce...

Notes

1. Ongoing academic debate builds from Kevin Ashton's coining of the term *the Internet of Things* in 1999 (Ashton 2009).
2. This proposition has been elucidated through analysis of seminal works – specifically Allan Kaprow and Robert Morris. See Charlton, *Acts of Materiality* (2013).
3. In *The Post-Media Condition*, Peter Weibel asserts that the impact of new media necessitates a radical re-reading of media to the extent that 'all art is post-media art' (2012).
4. As explained by JeeHee Hong, material and materiality are ambivalent terms that refer both to physical and non-physical matter (Hong 2003).
5. The philosophical concept of substance is an *a priori* condition for our experience.
6. *Zuhandenheit*. This argument is first laid out in *Tool-being* (Harman: 2002) and later developed by Levi Bryant into object-oriented ontology in 2009.

7. To reinterpret Young's reading of Merleau-Ponty: co-constituted objects such as materiality loop through objects, loop through objects and the world, and loop through the objects and the virtual world (Young 2011).
8. I have argued elsewhere that gesture in the sense defined by Agamben (1992) becomes a structural object – a network of relationships in a shared *Exhalent Gestures* agency between subjects that Young (2011) defines as gestural inter-subjectivity (Charlton, 2013).
9. For Floridi such a phenomenological position results in an epistemologically resilient ontological commitment that can be resolved by a co-constituted materiality derived from objects of a flat ontological order (Floridi 2003).
10. 'Gilles Deleuze has shown that cinema wipes out the fallacious psychological distinction between image as psychic reality and movement as physical reality' (Agamben and Casarino, 2000).
11. Theoretically, gestures can occur in a network of non-human subjects but as I am interested in the ability of embodied practice to engage with digital materiality it can be assumed that all gestures have at least one human agent.
12. See Rotman's analysis of gesture (in this volume), focusing on the connection between a mathematical (not computational) sense of the virtual as gesture. This provides a useful basis upon which to theorize gesture in a technologized mathematical sense, i.e. in terms of the numerical and discretized language of the digital.
13. Young cites the work of David McNeill as the basis for his analysis of gestures. McNeill's taxonomy of gesture concentrates on expression of speech that accordingly takes a language-based approach to gestures that is not the focus of this inquiry (McNeill 1992).
14. A position that parallels Harman's *sensual objects* which he contextualizes in regard to aspects of Husserl's intentional objects (Harman 2013).
15. So called after Zeno of Elea (ca. 490–30), motion is shown to be impossible by breaking time down to a series of points that are in themselves motionless and thus can never mathematically reach the target (Salmon 2001).
16. 'This is how we have unfortunately come to view the subject, and the body, through contemporary theories of construction' (Stern 2013).
17. Étienne-Jules Marey's later chronophotographic investigations of falling objects (Marey and Pritchard 1895), not unlike Zeno's arrow, present visual mathematical models of movement in which balls are presented as motionless moments along a trajectory – a treatment that is extended in the inter-subjective duration-less moment found in *32-Bit Catch*.
18. 32-bit is a reference to the basic unit of information in computing.
19. In particular, Vito Acconci's *Centers* (1971) is brought to mind.
20. Technically what has happened is that the video has been edited based on the sound. On the impact with the wall and later with the hand, 16 frames either side of the point of contact are shown – thirty-two in total. The rest is edited out.
21. Stack overflows occur in computers when the allocated memory of software is exceeded – typically resulting in a crash or error message.
22. These construct of the object aligns closely with Harman's analysis of *real* and *sensual* objects (Harman 2013).
23. Achieved by offsetting opaque frames slightly.
24. This devices have been used in an ongoing series of works including *AirTravel*, *Blow* and *Be My Mirror*.

Part III

Movement Inscription/Movement Performance

6
Movement and 21st Century Literacy

Maaike Bleeker

This chapter addresses *corporeal literacy* as a perspective on the relationship between movement and the digital. Digital technology marks a new step in what Mark Hansen (2000) and others describe as *technogenesis*: the co-evolution of humans and technology. This co-evolution does not imply that technology estranges us from our embodied engagement with what we encounter but that it intimately intertwines the perceptual cognitive capacities of our bodies with technology. This situation requires a critical awareness of how this intertwining defines what emerges from our interaction with media, and how our own doing is implicated within the affordances of these media.

Characteristic of the most recent phase in this co-evolution are types of interaction that happen when media operate without necessary or direct connection to human sense perception and conscious awareness. "What gets stored by today's media are no longer human experiences themselves but bits of data that register molecular increments of behavior and that do not in themselves amount to a full picture of integral human 'lived experience' " (Hansen 2015: 40). Hansen's understanding of media is unlike 19th and 20th century media discourses derived from photography, audiotape, video and film, insofar as there is no longer a direct correlation between what media capture, or record, and human experience. A characteristic of what Hansen calls "21st century media" (which would not apply to the examples of 19th and 20th century media mentioned above) is that they do not record human experience but data. In this context, he observes, mediation can no longer be understood to name the technical inscription of human experience but rather describes "the task of composing relations between technical circuits *and* human experience" (43). As a result, 21st century media "affect the materiality of experience at a level more elemental than that of perception: more precisely, they impact experience by shaping the ongoing worldly production of sensibility that constitutes the sensory confound out of which perception proper can arise" (46).

This situation requires a new approach with regard to what has been termed media literacy or visual literacy. These terms have been proposed both to describe the skills and competencies to engage with media, and to refer to (the importance of) a heightened awareness of the workings of media. Both aspects are also part of the original meaning of literacy as it emerged in relation to alphabetic writing and print, and where literacy denotes both the skills to read and write, and the condition or capacity of engaging with written and printed texts in a well-informed and critical manner. The shift observed by Hansen – from 19th and 20th to 21st century media – indicates that with regard to the workings of 21st century media such critical awareness cannot stop at what these media show and tell us, but will have to include our own embodied engagement with them. What these media show or tell emerges from our engagement with them: it emerges from how we enact what they afford us to do. Furthermore, as Hansen observes, at least part of this happens at a level more elemental than that of perception. In this situation, being literate involves a heightened awareness of perceptions and actions, of how these emerge in such a way that they are implicated in the medium, and how media perform their task of "composing relations between technical circuits and human experience" (Hansen 2015).

Deleuze's (1986) account of the 20th century medium of cinema, and how cinema makes movement part of perception and thinking in new ways, already points in this direction. In watching film, Deleuze argues, our sensorimotor schemas make it possible to grasp the continuity of movement and actions shown on screen. We are capable of grasping the movement of, for example, a character on screen as a continuity unfolding over time. This relates to how we bring our sensorimotor schema to bear on what is depicted in the image – as these schemas are also constitutive of how we enact perception in everyday life. Our perception of movement emerges from our perceptual encounter with the affordances of the medium. This potential of film would become all the more important with the discovery of the moving camera and montage. For example, when a camera moves through a room we see an image of an immobile room while we experience movement. We experience movement even though we do not see a body moving in the image. Our familiarity with the effects of movement and changing points of view makes it possible to follow the changing point of view of the moving camera or to grasp the logic of the cuts and jumps of montage and thus extract an abstract motion out of the succession of images.

Deleuze's account of how montage and the moving camera generate experiences thus already points towards media that can engage with experience at a level that is more elemental than perception. The moving camera and montage can make us experience movements we do not perceive but abstract from the images we encounter. This capacity for abstraction is fundamental to how media shape spectatorship and how viewers are able to make sense of what is presented to them. At this point, Deleuze's explanation of cinema foreshadows what more recently is confirmed by enactive approaches to perception like, for example, Alva Noë's (2004). The possibility to abstract objects of perception out of a continuum of impressions is given in the structure of our embodiment and did not come about with the invention of cinema. Movement, more than static objects, may alert us to how perceiving is a matter of abstraction, in the sense that abstract movement is also constitutive of perception that is not mediated. Movement never *is* at any one moment. It exists in continuous transformation and must be lived-through in a succession of sensory impressions. Yet to become an object of perception, movement has to be abstracted out of the succession of impressions in which it is experienced because movement is always more than the individual impressions. Therefore movement is the example par excellence of what Brian Massumi (2011) terms "lived abstraction". Massumi explains abstraction as "a technique of extracting the relational-qualititive arc of one occasion of experience – its subjective form – and systematically depositing it in the world for the next occasion to find, and to potentially take up into its own formation" (15). From this perspective, what we call objects are actually lived relations. He quotes Deleuze, observing that: "The abstract is lived experience. I would almost say that once you have reached lived experience, you reach the most fully living core of the abstract... You can live nothing but the abstract and nobody has lived anything else but the abstract" (Massumi 2011: 15).

21st century media build on our bodies' capacity for abstraction in ways that allow for new kinds of experiences to emerge, including experiences of things we do not perceive. Unlike cinema, where movement is abstracted out of images that show us the world as seen from a moving point of view, our own corporeal movements have now become increasingly part of how we engage with media. We ourselves produce the movements with which we scroll through information, navigate the web or interact with games and other entertainment. At the same time, of course, the scope of movements and what they effect is determined by technology. Technology affords certain interactions and what is effected by them. Increasingly so, technology is also capable of incorporating our anticipated moves and choices in what it allows us to do, thus taking us along in feed-forward processes that give us the impression of giving direction to procedures that are actually anticipating our behaviour.

In this situation, we need to become more corporeally literate in the sense of developing an informed and critical awareness of our own movements and perceptions as both our own *and* as implicated in the media we participate in. By extension, we need critical awareness of how our sense of what we encounter and our sense of self as movers and perceivers emerge as a result of interacting with media technology. This is not to say that we as subjects are simply produced by media. But it does mean that we need to take into account how our sense of self as subjects of behaviour and thought emerge from this interaction. Here, the condition of literacy as it emerged in relation to the technologies of writing and print provides a useful perspective on how we might conceive of our co-evolution with technology and the possibilities of critical engagement with the implications and possibilities of this condition.

Writing and print operate on the bodies of their users through the practices they afford, the routines, patterns of movement and gestures involved in using them, and through the perceptual activities that are mobilized by the medium or are part of the background conditions for the possibility of the medium. These effects are not only a matter of representation, but also of *enactment*. They result from how media facilitate behaviour and engage their users in patterns of action in perception. The same might be argued about more recent developments. The control mechanism of an iPad or a touch-screen computer, for example, is designed to meet our potential to perform certain movements. We are capable of performing these movements and therefore we are able to use the technology. However, in using these movements to interact with the iPad or touch-screen computer, the sensorimotor skills at work in engaging with these devices become part of something they were not before. They become part of modes of navigating through information, of finding the right piece of music, of scrolling through lists of data, of organizing and making connections between diverse materials, etc. As a result, the skills involved in performing these movements become part of how our bodies are able to make connections, relate to what they encounter, and make sense of it. These new technologies mediate in new modes of handling information and by extension they hold the potential of changing our modes of understanding and thinking, including our understanding of what knowledge is and what it means to know.

In the following, I will elaborate on how becoming more corporeally literate will involve developing a new kind of awareness of movement analogous to what alphabetic writing allows with regard to scripted language – an awareness of something we participate in and from where a sense of perceptual self emerges. I will take as my starting point a proposal made by Brian Rotman, to consider the possibilities of motion capture as a new form of writing analogous to but also different from alphabetic writing.

An enactive approach to literacy

The technology of writing and print has profoundly changed our relationship to the language we use in ways that, over time, have alerted us to language as a medium that we participate in, and to utterances in language as being both our own and afforded by the medium. Writing, Walter Ong (1989) observes, instigated a certain distancing towards language that would deeply influence the understanding of language and more generally give rise to new analytical modes of thinking. The structure of the alphabet mediates in an understanding of language as being composed of (and analysable in terms of) a limited set of smallest units (letters) that can be combined to form bigger units (words) that in their turn can be combined into sentences and texts. Writing and the collection of texts in libraries also made it possible to study texts time and again, in different circumstances and in relation to other texts in ways that were not possible before. This allowed for new types of overview to emerge as well as new types of more abstract knowledge.

The technology of writing, Ong argues, even more than providing us with a means to capture, store and transmit spoken language, quite literally changed our minds. Importantly, Ong's account of the impact of the invention of the technology of writing and print does not focus on what language per se does to us. What concerns Ong is not what words do to our minds but how the transformations of modes of managing knowledge, thinking, and of "self" – of human identity – can be brought about by extensive interaction with the medium of writing and print. These effects of interacting with the medium of writing and print, Rotman (2008) observes, are not only a matter of what written texts represent but also of how they afford enactment. They result from how these media facilitate behaviour and engage their users in patterns of action in perception and how this opens what Rotman terms "dimensions of the virtual". Although the virtual is often associated with recent developments in electronic media and computer technology, Rotman argues that in actual fact the phenomenon is ancient and emerged from language (spoken and written) and its mediation in a succession of dissociations that, like Ong, Rotman understands in terms of a restructuring of consciousness.

'Spoken language is a remediation of the pre-linguistic system of human gestural communication, and the passage from a dumb, haptic, self-pointing "me" to spoken "I" is a shift from actual to virtual, from a proprioceptive self to its symbolic form. If X is me, then virtual X is spoken "I", and a belief in a ghost spirit of the self (...) arises from experiencing the two self-designations against each other' (Rotman 2008: 118). Language mediates in the emergence of a distinction between the body speaking over here and an "I" represented in language. In speaking, this "I" is still directly related to the speaking body in the act of speaking. This connection is broken in writing:

> Like any medium, writing projects an arbitrary or universal user of itself. This mediological parallel to the indefinite pronoun "one" is a virtuality, an abstract would-be figure distinct from any potential or actual reader or writer located in space and time. In the writing of this "I" this figure, hypostatized as an agency circumscribing all possible referents of the pronoun, can, under appropriate conditions, become the putative, self-referring originator of the text containing it – an invisible, absent writing agency, detached from the voice, unmoored from any time or place of origination, and necessarily invisible and without physical presence. (Rotman 2008: 118)

Different media present us with different kinds of "I"s or subject positions to relate to in our use of the medium. The incorporation of these virtual users in our modes of enacting perception, Rotman (2008) observes, gives rise to new forms of action, communication and perception, as well as transformations in our sense of self. One way in which these transformations manifest themselves is through bodies and bodily experience in reading and writing alphabetic language that develop capacities of renegotiating relationships between the "me" of their gestural and proprioceptive self and the medium-specific "I"s or subject positions discussed above. One key feature of the mediation of language by writing is the utterance in written language, which is disconnected from a speaker. In writing it is no longer self-evident who is uttering the utterance. Who the "I" speaking is has to be deduced from the writing. This sets the stage for a new awareness of language as something we participate in and that we can use to express ourselves, but that does not originate from us.

Motion capture as gesturo-haptic writing

With motion capture it is possible to capture the movements of the body, abstract these movements from the moving body, store them as data sets and make them available at other moments. Motion capture allows movements to be disconnected from the moving body, and also for movement to be cut into smaller entities and recombined, all of which is much like what alphabetic writing does to spoken language. Unlike alphabetic writing, however, motion capture is not a symbolic notation but proceeds through a direct capturing of movements of a body expressing itself.

Motion capture, Rotman argues, holds the promise of a new and more sophisticated mode of writing because it captures something that is absent from written language and this is the movements of the body expressing itself. Rotman refers to evolutionary neurologist Terrence Deacon who observes that:

> Surprisingly auditory processing of speech sounds does not appear to be based on extracting basic acoustic parameters of the signal, as a scientist might design a computer to do, before mapping them onto speech sounds. Speech analysis appears designed instead to predict which oral-vocal movements produced them and ignore the rest. (Rotman 2008: 23)

This leads Rotman to conclude that we listen not to speech sounds as isolable sonic entities but for the movements of the body causing them. Alphabetic writing involves a kind of double leap as a result of which we are twice removed from what we are actually listening out for in speech, and this is movement. Writing has the means to notate (part of) the sounds of spoken language by means of symbols instead of a direct capture of the sound of speech. What is written down is only part of what is communicated in speech. Completely absent are the movements of the body speaking. Motion capture is a gesturo-haptic form of writing that promises to avoid this double leap. For this reason, Rotman argues, motion capture offers a more sophisticated, gesturo-haptic mode of writing that:

> extends to gestures the same kinds of conceptual and pragmatic mobility, spatio-temporal dislocation, freedom for the context of their production, and analytical transparency as the notational system of alphabetic writing afforded human speech. For as we know, the alphabet by allowing (insisting) that words became self-standing objects, discrete items of awareness that could be isolated, studied, compared, replicated, and systematized, gave rise to grammar, written discourse and literature, and a science of linguistics. Likewise gestures with respect to their digitally captured forms: they too are now being identified, individualized, examined, replicated, and synthesized as discrete autonomous objects of conscious attention. The opportunity is thus opened for such newly digitized and objectified gestures to emerge from the shadow of speech, to be "grammaticalized" and give rise to a gesturology, the theoretical implications of which would extend to a rare re-thinking of the status of human corporeality. More practically, such a gesturology might serve as an enabling semiotic frame allowing the gesturo-haptic to function as a medium for Sign to possess what it has so far lacked – a stock of iterable gestural artefacts constituting a "literature". (Rotman 2008: 50)

The way motion capture handles movement extracted from the body is analogous to the way alphabetic writing captures spoken language (as accounted for by Ong and others). Therefore, Rotman argues, motion capture might similarly give rise to new ways of engaging with and understanding movement and even give rise to a new kind of literature.

To understand how such grammaticalization of movement could also give rise to new kinds of literacy and in order to become more corporeally literate it is important to consider not only the similarities observed by Rotman between what alphabetic writing did to spoken language and what motion capture can do to movement, but also the differences. Alphabetic writing as a way of notating the sounds of speech shares with 19th and 20th century media a capacity to document the perception of spoken language. Unlike 19th and 20th century media writing does not offer us a record of perception but a symbolic notation of language as perceived in speech. Motion capture, on the other hand, is an example of what Hansen calls 21st century media. In this case, what is captured is not movement as perceived but an abstraction in the shape of data sets. In order to make data perceptible to humans, additional mediation is required to compose relations between technical circuits and human experience. This is also the case with motion capture.

Motion capture does not result in a representation of movement in another medium but in a data set contained within a data processing computer. In order for human beings to perceive the movement, what has been captured needs to be made perceptible again and in a way the user can perceive, for example by means of a visualization of the movement, or a re-incorporation of movement data within the animation design of a digital doll or avatar. However, this means that what we are facing when looking at the visualization of movement or the doll moving is precisely that: a *visualization* of movement, or a moving *doll* or *avatar.* In this situation it is easy to make a mistake similar to the one Rotman observes in the haunting of language after the incorporation of writing, namely a conflation of the "I" seen moving (the doll or avatar) with the "I" of the movement executed. Such conflation, it may be argued, is in the case of avatars and dolls precisely what is intended. However, in order for movements to become objects of a gesturology, as imagined by Rotman, they will have to be understood in relation to an "I" moving that is not the doll or the avatar but an "I" that (if we continue the analogy with what writing does to spoken language) can only be abstracted from the enunciation (an "I" that in writing is marked in the text by the written sign "I"). This would require a cut similar to the kind of distancing evoked by speech and writing with regard to the subject of expression (i.e. the "me" of gestural expression learning to perceive itself as an "I" of speech and then becoming the "I" of writing). Now then, this cut is significant with regard to the subject as it emerges from motor perception. What seems to be required for the kind of gesturology Rotman proposes is that we learn to distinguish between an "I" implicated in the movement and the "me" experiencing the sensation of moving. This involves developing a new kind of awareness with regard to the movements we perform.

Rotman gives the example of Stelarc who sent readings of his arm movement via the Internet to the wired arm of a remote host, making this host perform his arm movements (2008: 46). Technology thus helps mediate a situation in which the host literally performs a movement that is not hers. The explicitness with which she is made to perform the movement of someone else evokes a technically induced differentiation between the sensation of moving, which is hers, and the movement, which is not. This example helps us understand that becoming corporeally literate is not a matter of learning a new movement language, like for example learning the language of ballet. What is required is the development of a different awareness with regard to the movements one performs: an awareness of one's own movements as implicated within and resulting from the interaction with what one encounters. This involves a shift in our understanding of movement from how it can be understood in relation to the intention of the mover as the "author" of the movement to a sense of "I" emerging from the ways in which our bodies' sensorimotor capacities engage with what we encounter. How this involves a different relationship with one's own movements is illustrated by an experience described by Brian Massumi in his *Parables for the Virtual*:

> It is with some chagrin that I confess to having sat confidently in my temporary office at the Canadian Centre for Architecture for no less than two months looking at the wrong street out the window. I was looking east onto Rue St. Marc. But I was seeing north onto Rue Baille. I am sad to report that there is no resemblance between the two scenes. Something that was seriously disorienting me was happening in the time it took me to get from the side entry of the building to the door of my office. But that's just the half of it. The something seriously disorienting that was happening as I snaked my way through the corridors overpowered the evidence of my eyes. It was completely overriding the clear-as-day visual clues available to me out the window of my office. The sudden realization that my north was everyone else's east was jarring. True, I hadn't paid much attention to the scene. But I wasn't just not paying attention. When it hit me, I had the strangest sensation of my misplaced image of the buildings morphing, not entirely smoothly, into the correct scene. My disorientation wasn't a simple lack of attention. I had been positively (if a bit vaguely and absent-mindedly) seeing a scene that wasn't there. It took a moment's effort to replace what positively hadn't been there with what plainly was. (Massumi 2002: 178)

Massumi moves upward along various floors in the building; he arrives in his room, looks out of the window, and sees the street he knows to be at one side of the building. He does not realize that his route through the building actually brings him to the opposite side of the building and that therefore what he is seeing is another street. It takes some time until he realizes which street he is actually looking at. Until that moment there is an expectation as to which street he can see outside his window. This mistake is not the result of him getting lost in the building. As he describes, he is perfectly able to orient himself within the building. It is also not the result of him having been blindfolded while going through the building, or of an inability to see the street outside. He knows where he is and yet he doesn't. He realizes that there are two different systems at work in how he orients himself in space, in how space becomes available to him. He writes:

> Thinking about it, I realized that I could make my way to and from my office to the building's exit without error, but if I had been asked to sketch scenes from the corridors or map the route, I couldn't have done it with any accuracy. I had precious little memory of the way. I must have been navigating on autopilot, using some form of basically nonvisual memory. If I put myself mentally through the paces, I felt twists and turns coming

> one after the other with variable speed. I was going on a bodily memory of my movements, one of contortion and rhythm rather than visible form. (Massumi 2002: 178–9)

Massumi is perfectly able to find his way through the building abstracting his clues for navigation from how the building allows him to move. His experience illustrates how this is not a matter of objectifying the world he encounters but of his body responding directly to the affordances of the building without his conscious awareness. This is what Mark Hansen calls, after Maurice Merleau-Ponty, the "workings of the body schema". Thus, "whereas the body image characterizes and is generated from a primarily visual apprehension of the body as an external object, the body schema emerges from what, with autopoetic theory we have called the operational perspective of the embodied organism". And he adds: "originary, preobjective process of world constitution that, by giving priority to the internal perspective of the organism, paradoxically includes what is outside its body proper, what lies in the interactional domain specified by embodied enaction" (Hansen 2006: 38–9).

Body image and body schema refer to two different systems that nevertheless in behaviour are related and interactive. The body image is "a system of perceptions, attitudes and beliefs pertaining to one's own body". The body schema on the other hand is "a system of sensory motor capacities that function without awareness of the necessity of perceptual monitoring" (Gallagher 2005: 24). The body schema emerges from the operational perspective of the embodied organism. That is, it emerges from the potential of the organism to engage with what it encounters, or from the organism's potential to couple with the world, as Merleau-Ponty describes it. The fact that the body schema functions without our being aware of its functioning awareness does not mean that the world available to us is given in the structure of our embodiment only. The body schema develops out of the body's interaction with the environment. It evolves from the experience of moving, doing, touching, and from there it shapes our perceptual field. How this happens and therefore also how the world is informed or even generated by this encounter depends on the capacities, habits, and abilities of the organism.

Massumi's experience illustrates how the coming into being of the world as he encounters it through bodily interaction happens largely outside his awareness as a result of sensory motor capacities that function without awareness of the necessity of perceptual monitoring. He only becomes aware of its functioning after realizing the difference between where he finds himself and where he imagines himself to be. This capacity of the body to engage with what it encounters and to couple with the world becomes increasingly important to understanding the functioning of 21st century media and how media communication no longer transmits records of perception and experience but captured data in ways that exceed human perception and operate by "reconfiguring the very sensory field within which our experience occurs" (Hansen 2015: 45). One can become aware of these mediations, if only indirectly by increased understanding of how our actions and perceptions are implicated within the affordances of the medium and thus by an increased awareness of our own behaviour as lived abstractions afforded by the medium.

In Stelarc's experiment this mediation and distancing is induced by means of technology that literally makes someone perform movements that are not her own as a result of which this person is made to abstract an understanding of her own movement outside the motor issuance of the movement. Another example of how we might develop such awareness with regard to our own movements could be William Forsythe's *Improvisation Technologies*. Forsythe's improvisation technologies are both a set of tools to train dancers and a means of developing an understanding of the work from an observer's perspective. This way of thinking movement is explained and visualized in a way that demonstrates how choreographic movement involves a logic of spatial relationships, which the dancers are asked to relate to in their movements. The dancer's movements are, at least in the case of Forsythe's technique, the result of a constant process of decision-making according to sets of rules. Improvisation thus involves developing a kind of double relationship to the executed movements in which one is at the same time doing the movement and conceiving this movement from a position of analytical reflection, and therefore from a place abstracted from the logic of spatial relationships. The technique, one might argue, constitutes an actualization of what Massumi describes as lived abstraction and in this way Forsythe's analytical tools can evoke a reversal in attentional awareness with regard to one's own movements, thus raising conscious awareness of how objects, images, and more generally the world as we perceive it are defined by lived relations.

7
Movement Scripts: the Materialisation of Movement through Digital Media

Lise Amy Hansen

Movement is made digital through creative and critical decisions that affect the way our physical actions are codified. The main focus of this chapter is the process whereby physical movement is given written representation in order to be processed by a digital medium. This process of technologically specific inscription – of laying down in written form a human and machine-readable signal or issuance of human movement – is dealt with especially in terms of what I call "movement scripts". I will address these in terms of the creative possibilities that digital inscriptional practices can afford despite the condition of most writing technologies, which, in broad terms, is iterative and normative. I will attend to the way movement scripts can be tailored to the moving body as a sensate origin of action, by describing the development of digital tools for movement data visualisation. These scripts position the materialisation of movement as a critical practice navigating both the immaterial nature of code and a physically performed movement material.

The either-or logic of digital movement scripts recognises some movements and exclude others, according to how digital systems are designed and put to work. In this sense, movement scripts comprise boundaries that decide *which* movements are to matter as well as *how* they are to matter. My aim is to explore and build on such binary communication and its performative potential. I suggest that providing a variety of movement representations presents an opportunity to trouble and challenge our current use of digital media and to take into account how we may move *through* and not just *for* the media. As Nicolás Salazar Sutil has argued, the overlapping of anatomical body and digital technology entails more than the opportunity to improve performance (e.g. speed or efficiency): it is also a risk-taking opportunity to create new ontological mergers that might be rendered "poetically, aesthetically, choreographically" (2011). In addition, it is important to raise questions through such artistic exploration, and to use this "kinetopoiesis" – the creative making of digital things through movement (Salazar Sutil and Popat, this volume) – as a reflexive mechanism that can provide critical awareness of technology's influence in the way we move and the way certain movements matter through normative processes of coded inscription. But if digital media can make us move in new ways, how is the potential to be explored? My enquiry situates movement as an expressive material, thus connecting to the artistic processes of choreography, and the creative skill of generating new movement expressions.

Digital material

Materials play a central role in theory-building. Timothy Lenoir (1998) has argued that communication technologies are fundamental as they mediate and provide stable representations with which to theorise. In this sense, media constitutes more than a supplement to theoretical concepts. In fact, material representations "actually constitute the signifying scene in technoscience" (1998: 12). Furthermore, material representations not only mediate, but can also be understood as an embodied enactment – a materiality constituted through an act of performance, of being materialised in some contingent way. Brian Rotman (2000) proposes that a simple function of mathematics – counting – is an ordering procedure that makes sense as it is enacted, taking into account the style and manner of a number's role and function as we count with reference to our fingers. In this sense he challenges the view that numbers are objective and abstract entities, and he urges us to reconsider whose fingers are doing the counting. In other words, Rotman is considering an abstract entity – the number – as a culturally performed convention. These two approaches make a case for technoscience as a practice whereby material representations are fundamental to the constitution of those objects that communication technologies perform.

This concern for materiality can be problematised, however, and from the opposite side – from the side of that which is not material. This is the perspective taken by Nigel Thrift, for instance, whose argument focuses on the immateriality of representation. He argues that software has come to play a part in all aspects of everyday life, changing our embodied experience. And yet: "software very rapidly takes on the status of background and therefore is rarely considered anew" (2005: 241). Human experience where the body is extended through screens and software is further troubled by the way the body moves according to its surroundings. Digital media challenges known repertoires of moving, by introducing processing of information different to what our minds are capable of calculating. Mark Hansen (2006) describes this disparity between our body schema (how we experience our "inner" motile body) and our body image (how we see our visual self outwardly) as "body-in-code", a dynamics that can be extended, shifted or shaped by digital media. However, Hansen also points out that such data is stored beyond our direct perception: "below the threshold of representation itself" (2000: 4). Therefore data requires tools for it to be perceived by our senses. Data needs to be materialised: "However immaterial [digital information] might appear, information cannot exist outside of given instantiations in material forms" (2011: 1042). The way this happens is of particular interest when it comes to understanding movement data, as data has no necessary form (Galloway 2011).

Here then is the premise of my investigation: to navigate between the material and immaterial views of movement. I argue that light can be shed on the role played by digital movement by describing digital tools for materialising movement – tools that include the moving, sensate body in the shaping of code. Here, I refer to material not only by physical attributes. In fact, I will use a broader definition of "material" – meaning the various elements that contribute to creative practice: "Form is the way material builds things; to build a thing, we form materials" (Hallnäs et al. 2002: 157). In this sense we can see that when designing movement-based digital interactive settings, a programmer's material is similar to a choreographer's. The material is movement – which then constitutes a shapeable, temporal and visual element in a designed event. In developing an understanding of movement as a digital material, then, we need to explore qualitative differences. For more conventional materials such as wood or fabric this sense of quality is more available and more immediately known by touch, for instance. As far as movement data is concerned, understanding the quality of digitally mediated movement remains more elusive. If we are to understand the possible parameters within which change occurs in the material that is digital movement, then we need to attend to how movement data is presented; how movement data is *visualised*.

Digital movement

Physical movement is a resource that brings forward the potential for communication of bodily qualities such as age, ability, health or history, and movement qualities such as intent, intensity or frequency, as well as cultural and social relations. These are in part physical and in part expressive in that both age and intent, for instance, can be expressed in temporal, visual qualities performed in a movement. As Daniel Stern has observed:

> We naturally experience people in terms of their vitality. We intuitively evaluate their emotions, states of mind, what they are thinking and what they really mean, their authenticity, what they are likely to do next, as well as their health and illness on the basis of the vitality expressed in their almost constant movements. (2010: 3)

Movement data is particular in that it may include these corporeal qualities, yet is abstracted from the body (as well as from a specific time and space).

In making sense of such data, we rely on tools of representation. How the body is accounted for by the various technologies of representation available today determines how we can work with the body as a "mark-maker". However, whilst we are highly skilled in deciphering movement with our eyes, to do so computationally is still in its infancy (Bevilacqua 2007). Carrie Noland argues that when we move we are informed not just by society and the learned and coded ways of making meaning, but that our movement is also informed by the *performance* of the movement – the issuing forth of the movement, what earlier I referred to in terms of Rotman's "enactment". In this sense the moving, sensate body plays a role in producing knowledge, habits and sensations through the agency of movement itself. Noland further argues that it is here that we find the opportunity to recursively iterate, correct and indeed push the boundaries of what constitutes a mark:

> We can leave our marks in the wrong place, invent private or countercultural mark systems, or use mark making as an exploratory project, investigating how our bodies might move differently and thereby achieve materialization and cultural legibility in unexpected ways. (Noland 2009a: 215)

In this sense there is potential for digital media to draw upon our skills, not as a singular attribute of a body, but as a system of relations involving our presence in a configured socio-material environment (see Ingold 2000).

Materialising acts

When exploring digital media via body-based practices and when building knowledge of how digital tools may influence movements, at least two sets of skills must be simultaneously applied. As Mark Coniglio puts it: "Those creating content-driven work must forge ahead in parallel with those who are creating materials-driven work [...] and simultaneously develop theory, technique, and content" (2006: 83). This means that there is a shift from the use of current digital media to the practice of shaping this media, and from digital material as a given, to something to be shaped through individualised and personalised practices.

I suggest we can start to examine the role digital movement plays in everyday life by challenging the way new media accounts for our movements. This leads to the question of how the public in general adjust to new media, and how the comprehensiveness and complexity of movement scripts come to be acted out. Richard Coyne (2010) argues that this is an incremental development involving elements such as repetition and familiarity, aggregation and rhythm, which together constitute our social and spatial relations. Space can be construed by fine-tuning our connections through pervasive digital technologies. However, such tunings are performed acts: they do not happen to a body, but are acted out by a body.

Exploring movement data through the *Sync* tool

In what remains of this essay, I will describe processes and tools that engage with an appreciation of both the material and immaterial aspects of digital movement, drawing on my own practice. It is worth noting that research projects in the performing arts now increasingly develop their own tools to prise open new ways of representing movement.[1] Within this context there has been a shift from usefulness, function or optimisation, to attending to the expressive and artistic potential of inscription via motion. One such material exploration can be seen in the well-known *Synchronous Objects* project.[2] William Forsythe's performance *One Flat Thing reproduced* was translated and transformed through a collaborative effort involving experts from several fields into twenty visual objects, each examining an aspect of this specific performance. Another exploration of scriptable movement material is the *Choreographic Language Agent*,[3] a software environment that generates variations of choreographic instruction by mutation, selection, and parameterisation. The software functions as an intelligent agent creating unique solutions to choreographic problems, leveraging computational capacities to augment creative decision-making processes. One other project worth mentioning here is the *Reactor for Awareness in Motion (RAM) Dance Toolkit*,[4] which focuses on real-time movement data visualisations. The tool allows for the creation of a set of different surroundings called "scenes" for dancers, where movement data from body-worn sensors is visualised and made available in real-time by a digital interpretant known as the "actor". Overall, these research projects focus on movement for the stage through an exploration of the unique blend of physical and mental processes of which choreographed movement is comprised. Drawing on these works as exemplars, my own practice offers ways of materialising the representation of movement dynamics by displaying different calculations of movement, and by providing a means of comparing and relating the data back to actual movement.

Sync[5] is a collaborative research and development project developed by the author at the Oslo School of Architecture and Design, in collaboration with Hellicar and Lewis. Through a wide variety of parameters, *Sync* allows for a range of visualisations to be generated (real-time or recorded), informing digital interactivity and performance. The tool is capable of identifying fifteen capture areas from a Microsoft *Kinect* sensor corresponding to feet, knees, hips, shoulders, elbows, hands, torso, and head (see Figure 7.1). As a visualisation tool, *Sync* was designed to scale up visuals in order to allow users to work with life-size movement data at a 1:1 ratio (see Figure 7.2). We did this to leverage the sophistication with which we perceive movement. Relevant parameters such as acceleration, deceleration, frequencies and repetitions are thus clearly visible as the corresponding mark or traceform is drawn. We also decided to give the option to link some sets of tracked points: hands, shoulders and hips to give the opportunity to gauge a change in posture. *Sync*

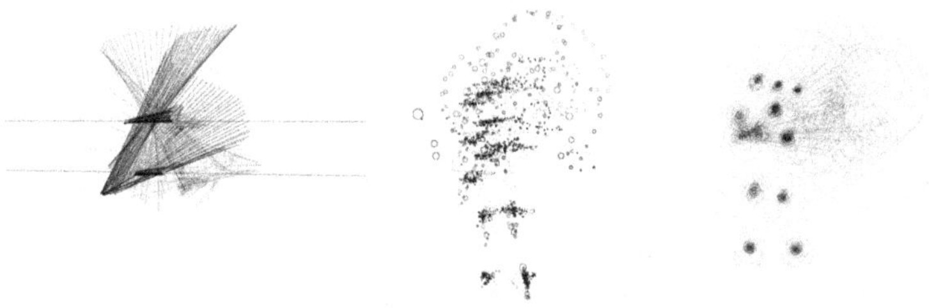

Figure 7.1 The *Sync* tool allows for movement data to be visualised live or repeatedly through recorded data. In this collage, a hand wave is visualised in three different ways. Image © LA Hansen

became a tool for teasing out how to account for speed, direction or less tangible notions such as the hesitation of a person approaching.

Sync focuses specifically on the *movement dynamics* of a human figure. Dynamics is calculated as the rate of change in the position of each point relative to the *Kinect* sensor. The mark is visualised by a line, circle or banner where each visual mark expands for acceleration and contracts as the tracked point decelerates. The various movement marks are selected in a menu, where one can also set the parameters for how the change in each point appears. There are options for the sensitivity of the mark to change relative to the dynamics as well as the size and density of the mark and the length of tracked movement. The tool's menu also allows for the movement marks to foreground different movement qualities, such as direction, non-movement or location. Crucially, whilst *Sync* generates dynamic visuals in real-time it also records the data (as well as a video clip of the movement sample). *Sync*'s recording function allows for playback and a detailed analysis of a movement phrase whereby the data may be recalculated and presented differently for each new call on the same data. In this way *Sync* provides dynamic visuals that can contribute to the process of interpretation and meaning production.

In addition, the dynamic visuals can also contain movement calculations and thus these visuals present movement relations not necessarily available to the naked eye. Because the visuals can be calculated in near-instant ways they can provide opportunity for comparison to physical movement. This comparison is key to understanding the data as a performed or enacted object (as opposed to a perception of performed movement). Comparison also allows us to see how the calculated visuals make sense, relate and influence the actual, original movement. This is particularly relevant as creative processes involving digital media (i.e. contemporary choreography) are conditioned by how live movement can be modelled and what it can represent. *Sync* can thus challenge the expertise of dancers whose trained and controlled expression has been honed over decades in mirror-covered studios.

It can provide visual feedback on how movement can be traced and in turn it can help generate new movements and new movement expressions through an embodied digital interface. *Sync* also gives the opportunity to recognise what is particular to a certain body or movement phrase and thus tailor digital media to a specific body (as opposed to everybody). This gives the opportunity to work on personalised movement. For instance, *Sync* might be suitable in scenarios where technology needs to be tailored to particular people's movement such as in stroke rehabilitation. Other individual interventions could be considered in work carried out with children on the autistic spectrum, such as in the Reactickles[6] work (Keay-Bright and Gethin-Lewis 2011).

In sum, *Sync* makes a case for inscription via motion as a direct process – a process unmediated by notation or kinetic symbolism, as well as a dynamic and personalised process, which is a determining factor in the design of new media. In this way, *Sync* alerts the user to the myriad choices available when exploring the dynamics between the immaterial movement data and the instantiated movement. Exploring the data visuals compared to live movement is central because one can make creative, pivotal, and conditioning decisions as to how the data will be visualised, and how it will appear as an object of representation. Is height or age accounted for in the data? Do signs of hesitation or ability appear in the movement data presentations? Can one register intent or particular expressions? In order to handle relevant movement qualities in the data, the particularities of immaterial digital information coupled with physical performed movement need to be considered. The sheer variety of ways a single movement phrase may be represented shows that we need to "see" the data in order to know how to apply it in meaningful ways. Therefore, *Sync* allows for the visual exploration of movement data to raise critical questions and perhaps even evolve a reflexive practice around the processes of inscription via motion. Ultimately, *Sync* is an effort to show that the process of bringing the moving body to rest in representation is an open-ended one, and representation is not the same as standardisation, as is the case with movement notation systems. New tools are needed that can help us question the way bodily qualities or movement qualities can come to be represented, and what choices are available when engaging creatively with the process of inscription via motion.

Concluding remarks

Our environments are not passive backdrops in which we live our lives; rather they are increasingly active and digitised configurations in which we choose how to act and where our everyday routines are played out. As these environments incorporate scripts written prior to the acts that activate them, it is important to critique their performative potential. It is necessary to consider this because the understanding of movement data shapes what is written for our physical selves, and these designs in turn shape our movements and by extension our lives. Movement has become a field for human-machine communication, but since both human movement and its machine-readable representations are often invisible from our everyday experience, it is necessary to visualise movement as data to gain access to this knowledge. In this essay I have suggested that to understand the role of movement data and its potential for meaning-making, there is a need to examine the strategies for categorisation and interpretation in the visualisation and application of movement qualities through digital media. However, to see movement as data is not just a step toward analysis for the generation of data intelligence. *Sync* and the other tools mentioned here enable a visual understanding of how to creatively engage with an altered digital corporeality through an understanding of the limitations, prescriptions and potentials of our coded bodies. In this sense, movement visualisation can also be a step toward critical intelligence, which prevents us from taking the transformation of human movement through technology for granted, whilst prompting us to investigate, intervene and debate the ongoing objectification and materialisation of human movement.

Notes

1. See for instance Motionbank's *Piecemaker* (http://motionbank.org/en/event/pm2go-easy-use-video-annotation-tool [accessed 16 March 2015]); Openended Group's *Field* (http://openendedgroup.com/field/ [accessed 16 March 2015]); BADco's Whatever Dance Toolbox (http://badco.hr/works/whatever-toolbox/ [accessed 16 March 2015]); and a Transmedia Knowledge Base for contemporary dance, TBK's CreationTool (http://www.clunl.edu.pt/PT/projecto.asp?id=1555&mid= [accessed 16 March 2015]).
2. See http://synchronousobjects.osu.edu/ (accessed 16 March 2015).
3. Developed by Openended Group in collaboration with Wayne McGregor. For more see: http://www.openendedgroup.com/artworks/cla.html (accessed 16 March 2015).
4. Created by Yoko Ando and YCAM InterLab, including Yoshito Onishi, Satoru Higa, Motoi Shimizu and Kyle McDonald. For more details, please see http://interlab.ycam.jp/en/projects/ram/ram_dance_toolkit (accessed 16 March 2015) and the RAM Dance Toolkit wiki: https://github.com/YCAMInterlab/RAMDanceToolkit/wiki/Overview (accessed 16 March 2015).
5. *Sync* application is available for download (http://kinetically.wordpress.com/sync-download/ [accessed 16 March 2015]) and the source code is available (https://github.com/HellicarAndLewis/Sync [accessed 16 March 2015]).
6. See http://reactickles.org/ (accessed 16 March 2015).

8

Performativity of Movement: Coding, Segmentation and the Valorization of User Behaviour in the Development of Smart TV Technology

Wan-Gi Lee

In the preceding chapters of this book, the question of how human movement is coded and formalized as data for reconfiguration of motion possibilities in the movement arts was also addressed in terms of a codification of everyday movement through technological mediation. In this chapter, I would like to pick up on this challenge in order to suggest "segmentation" as the core method for data processing and analysis of up-to-date technological firms, through which the sensory and physical movement of the user is translated into meaningful data for technical (re)configuration. With this behavioural coding, the movements, experiences and sensations of the user are turned into a technologically engineered signifying practice, which is then narrativized into value discourse and strategies for product development and marketing activity. Standing in proportion to the collective information, each movement of the user bears a performative function, in the sense that user movement can be coded, yielding input-output relations that will be analysed by technology development firms. Thus, the technological performativity is embodied in the communication of the movement, enabling continuous technical (re)arrangements. In digital environments, a technology firm should be approached as a strong interpretive apparatus through which user behaviour is incessantly registered, evaluated and reinterpreted. The collective behaviour accumulated becomes information for segmentation of users that provides product platforms and market forecasts.

Segmentation has been widely adopted in business and marketing ever since the 1950s. It has been put into practice as a critical operation enabling articulation of the signifying practices of behavioural coding into the identificatory categories according to which the identification of user groups and differentiation of product ranges are guided and structured. By examining actual segmentation processes through behavioural coding and motion technology in developing the latest smart TVs, this chapter observes the current stage of technological development to be a symbiotic stage, combining corporeal gestures with technical performance, leading to market advantage through differentiated digital technologies. For technology firms, the value transcoding function of segmentation does not only condition technologically mediated action and cognition in value production. It also transforms social relation with the insinuation of qualitatively different criteria for individual performances, necessitating further value realization. However, as this chapter underscores, such differentiation presupposes a de-differentiation of users by containing different social positions, which is a critical target of the technological operation of de-politicizing the social subject. Technological differentiation in this respect replaces political issues with identity politics without changes in the crucial relationship of ownership, as it also raises the problem of inequality and the unequal ownership of data between users and technologies.

The location of value in digital technology

We are in the era of radical changes within the milieu of digital technology, as the technology itself is now living *in* us, propelled by the thorough connectedness of the digital devices we are using. Every movement of the user is datarized and quantified. Any small gesticulation with the devices, whether it falls under haptic, optic, acoustic, or sensory movement, is being ceaselessly registered and translated as a meaningful signal for technology, and stored as so-called "big data". The technology lays open unlimited possibilities of redefinition and representation of data for further value realization. It enables, through stored data, descriptive and predictive approaches towards the objectives set up by technology, vastly enhancing new analytics that perform intelligent inference and predictions regarding users' behaviour and action. On an unprecedented scale, human physical corporeality as well as delicate cognitive responses now seamlessly interact with, and are thoroughly embedded into, the interfaces created by technology. Devices, software, hardware and interaction with users are all then converged into the so-called digital "ecosystem"[1] as the new mode of value (re)production.

A big data environment presupposes connectedness, which inevitably opens boundless leeway for data processes of collection and interpretation, particularly given the interconnection of digital devices such as mobile phones, tablets, and TVs through the Internet and through customized platforms devised by technological firms. With this connectedness, "smart" devices go beyond their own functions. They take on the function of *registers of data* – registers of cognition and action. Every movement with the devices, be it active Internet search, use of apps, GPS navigation, watching a TV programme, or even just a simple phone call, is recorded in a system that stores, monitors, interprets and evaluates the movement as data. Every movement in the connected digital environment then becomes a trace of existence, a datum of a user as a digital being. Even the passive use of a device, like static keep of a mobile phone or turning off Internet TV is meaningful as information for registering the idle period between uses. Each user's behavioural pattern of usage, duration, location and preference are, consciously or unconsciously, thoroughly exposed to the system on an unprecedented scale. In this sense, a corporeal movement with a connected digital device gives the system a *ubiquitous structure* through which users' behaviours and actions are thoroughly datarized, providing a source that facilitates unrestrained data mining. Sophisticated data process analytics and algorithms are applied to this structure, and these can then redefine users' movements in accordance with whole new dimensions of value production. A property connoting descriptive and predictive value as a guiding force for cognition and action is thus through this data turned into a source for further value realization and technical revamp.

In exemplifying the technical coding of segmentation, we will raise two fundamental questions pertaining to the existential condition of the subject in this connected digital environment. We will then address the functions of the field that the subject is ushered into in the name of the seemingly neutral term "ecosystem". First, we will scrutinize the location of value by which the social positions of users are re-evaluated in digital technology, on the understanding that technology aims to redefine and represent users above all as (un)bearers of value from the perspective of economic value production. Secondly, and in operation of value definition and production, we will approach segmentation as the objectifying field of value and the legitimate relation of value production. This theoretical concern also addresses the problem of inequality and the unequal ownership of data between users and technologies. These are the critical questions that cultural theorists need to pose when analysing technical coding and interpreting systems for the value production of the information and technology industries. This chapter will, within the frame of analysis, examine the ways in which sensory and physical movement of the user is transferred and signified into data through the latest smart TV functions. According to Nissenbaum, whose theoretical interests focus on information coding systems and value, the driving force of informational and technological development is not the rationale of technology as such, but *value*. It is from this that the inner demand of technology is guided and implemented. She writes:

> Common to both research categories is the direction of causation: Information technology changes the world, and some of these changes challenge previous commitments to values and principles. Yet the idea of values embodied in computer and information systems suggests motion

> in the opposite direction, from value to technology. Values affect the shape of technology. Briefly, the values that systems and devices embody are not simply a function of their objective shapes. We must also study the complex interplay between the system or device, those who built it, what they had in mind, its conditions of use, and the natural, cultural, social, and political context in which it is embedded – for all these factors may feature in an account of the values embodied in it. (Nissenbaum 2001: 117)

The very act of accepting the system can, in this sense, have "moral or political properties" (Nissenbaum 2001: 118). There is a clear necessity of revealing value as the factor that channels technological platforms. The term "value" is, however, in technology and markets, often mystified and thus ambiguous, mainly due to the fact that the term itself is hard to define and can be easily appropriated by the economic and political fields. In the technology consulting industry, the seemingly neutral term "value" is interestingly yet symptomatically defined as profit, which is "synonymous with economic surplus, not net present value".[2] What is at stake in this definition of value is value as a broad entity that can presuppose and subsume the latency of value in potential surplus. Value is "not net present value" but profit that can be expected and anticipated, and subsequently guided towards further surplus. Value as profit, therefore, "must be imagined before it can be extracted" (Tsing 2004: 84). Value can, in this sense, acquire a meaning as the "guiding force of cognition and action" beyond the economic (Lee 2008: 180), since it is anticipated and advanced through the systematic intervention, coordination, and organization of the economic. This can, within the scope of our analysis, be suggested as the objectifying field of value production.

According to Simmel, value is "a [critical] category through which our perceptions become images of the world" (Simmel 2004: 59). Value is determined in relation to the utility and scarcity of an object (Simmel 2004: 205). However, the determination presumes a process of valuation in the objectifying field. It is here important to note that it is "a comparison of demands, i.e. the exchangeability of its objects" (92) that creates the economic value of an object, rather than an individual demand. Therefore:

> A value can be formed in proportion to utility and scarcity, but the economic value of an object can only be acquired through the exchangeability of an object, according to which the value can be measured, evaluated, and ranked. Only through the formation of an objectifying field of comparison, where each value is measured and evaluated, can objective valuation arise. (Lee 2008: 185)

Concerning value and the actual process of value-formation, we observe that what is at stake is the existence of the "objectifying field of comparison" through which the exchangeability of an object as economic value is eventually measured and evaluated.

We can assume, following Simmel's analysis, that developers, project managers, and designers in the current connected digital climate are the ones that create the platform for the value coordination and comparison through which all data is measured and evaluated, thus acquiring meaning for value production and realization. The market for various data analytic tools such as Database Management Systems (DMS), Demand Side Platforms (DSP) and other big data analytics programs by big software companies like IBM, Oracle and Software AG is expanding massively. It was reported that in 2012 the big data analytics software market was worth $100 billion with an expected annual growth of more than ten per cent.[3] With systematic use of these tools, digital product developers and marketers can produce "added value" by taking a predictive approach based on users' current buying behaviour and preferences. By incorporating DSP and Data Management Platform (DMP), a big data analysis and marketing company insists that "its added value has the ability to tap deep audience analysis that can, for example, look at shopping-cart data, see a pink sweater there, then offer matching pants". The company also "offers tools for additional analysis and reporting and lets its clients buy other kinds of data".[4] Here we get a glimpse of how the functions of their tools – the objectifying field of value production and realization – can realize value by translating current data provided by user movement into possible predictive measures. The behaviour of the user as a meaningful movement, as data, is here clearly assumed as a latent possibility for further value realization. The data is, in this regard, a meaningful movement that can subsume even the latency of future value for further potential surplus value through a predictive approach based on interpretation of current user behaviour. Value, as a guiding force of action and cognition for further movement, is consequently anticipated and advanced through the big data analytics with systematic coordination and organizational tools. These, then, function as the objectifying field of value production. Therefore, by articulating disparate and diverse user movements into the coordinates of analytics, the developers "want to follow all the breadcrumbs and put them in one place" in order to "analyze what is happening properly".[5]

Segmentation as signification of value

It was the American economist Wendell Smith who, in the 1950s, first proposed the notion of market segmentation. He underscored, in his article "Product Differentiation and Market Segmentation as Alternative Marketing Strategies", the necessity of capturing market diversity as consumer demands became more varied. He believed that by setting up the relevant marketing representation tools to be used for user identification, companies could also diversify consumer groups through an effective application of segmentation in identifying objective and distinctive value groups, leading to effective target marketing.

Segmentation is based on the developments on the demand side of the market and represents a rational and more precise adjustment of product and marketing effort to consumer or user requirements. In the language of the economist, segmentation is disaggregative in its effects and tends to bring about recognition of several demand schedules where only one was recognized before (Smith 1956: 8).

Smit's "recognition" of diversified consumer groups and marketing strategies is based on the idea that segmentation meets the needs of market diversification and selective marketing, which lays the ground for a descriptive approach towards the identification of diverse groups of users. In 1968 Russell Haley criticized Smith's segmentation model for being descriptive, thereby insisting that the segmentation-tool should have a predictive function, locating causal relationships that would anticipate future purchase as well. Ever since, the study of market segmentation has been focused on an analytics that combines descriptive and predictive methods, and from which current analysis of customer behaviour is closely coded, interpreted and defined for executable predictive measures to funnel future buying. With segmentation at the heart of marketing technique, users' profiles – of demographic, geographic, behavioural and cognitive character – are categorized, identified and redefined. This more traditional approach to customer grouping has gone through a qualitative change with the introduction of more refined methods of analytics arising out of the advent of the Internet. In the 1990s, the then new marketing model of Customer Relationship Management (CRM) brought about a more systematic way of customer data processing in which data-collection and analysis are technologically integrated under one management tool. Effective customer segmentation and targeting led to customer retention and value increase for companies. The analytics is regarded as a "breakthrough" in marketing both in this sense and in terms of the large volume of data that it has capacity to deal with. With the mediation of advanced technology and a connected environment, companies could practise more delicate and diversified approaches for direct yet customized promotions and advertisements, reducing both cost and response-time through various selective approaches which had not been feasible before then.

There is a set of basic components whereby each component is closely planned and streamlined, with the aim of locating target customer groups. The proces, "Create a Database → Analysis → Customer Selection → Customer Targeting → Relationship Marketing → Privacy Issues → Metrics" (Winer 2001: 91), is initiated by creating a customer database in which records of customer profiles, purchase histories, and web page visits are stored. This information then becomes object-data for analysis of customer selection and targeting, in order to identify groups of loyal customers for maximizing customer retention and profit rates. The goal of the analysis is "to increase 'conversion' rates, the percentage of browsing customers into actual buyers" (Winer 2001: 95). The various analytics are made to trace and measure types of customer visits, buying patterns, trends in search queries, etc. With the application of these analytics, companies are able to perform customer selections, where various groups are segmented to differentiate between desirable segments and undesirable ones. Subsequently "promotions or other purchase-inducing tactical decisions" are heavily applied to the former (Winer 2001: 95). Having identified the desirable target groups through customer selection, the opportunity arises for companies to build relationships with more direct marketing methods, such as personalized emails and promotions, through which the "new mantra of 1-to-1 marketing" is put into practice (96). Relationship marketing includes, for example, tailored customer service and loyalty programmes. These measures are regularly reset and updated with diversified metrics in accordance with the needs of the firms.

As our theoretical concerns are focused on the descriptive model of segmentation on the basis of which the predictive approach is made to operate as extended intelligence of technology, we will here briefly exemplify with a case study, so as to give a more specific picture of the way in which up-to-date smart TV companies segment and then signify target customer groups in – importantly – embodying customer-identities for specific product lines. In accordance with user-interaction with the interface, such as access to preferred apps, duration of using time, frequency of access to and use of contents, etc., a certain group of users is created and signified for streamlined product development and marketing narratives.

Partnered with Nielsen, the digital video and advertising firm, Yume conducted a survey – allegedly the first report on smart TV ad effectiveness– with more than 600 respondents, all in the US. The aim was to understand the effectiveness of and responsiveness to smart TV advertisement. What is interesting in this report is the way in which smart TV users have been segmented. In the respondent profile, smart TV users are identified as "young, higher income, and professionals" with 39 per cent of users having a household income over $90K. Equipped with buying power, smart TV users are described as "tech savvy, will pay a premium for latest technology, and are influential among friends in the consumer tech product category". This descriptive identification of users is further concretized with the identification of five lifestyle factors: "Techie", "Frontier", "Premium", "Social networking", and "Family oriented".[6]

These descriptive approaches are not just for the sake of categorizing users; they also show an active intervention by technology in guiding predictive narratives for further cognition and action. The descriptors are ontological building bricks in the representation of users with a technically engineered signifying practice. This descriptive approach is beyond technical coding of user identity, considering that the identificatory segments themselves function as predictive communicative codes with users being guided by value. The descriptions become the guiding force of cognition and action towards new values, new products.

Segmentation is subsequently carried out, in combination with the five factors, as technically motivated evaluations of lifestyle attributes. As a result of the imputting process, four categories of segmentation are drawn as the distinctiveness of users: "affluent techie", "social youngster", "traditionalist", and "family oriented midlife". While "affluent techie" and "social youngster" – occupying 22 and 25 per cent respectively – are enthusiastic about and responsive to new technology and have a high and medium-high household income, "traditionalists" are identified as older and belonging to a lower income group with a lesser interest in technology and smart TV ads. This is therefore the group that is the least targeted of the segmentation categories. The last segment, "family oriented midlife", forms the biggest portion of users with 35 per cent. Like the "traditionalists", the group is old and not familiar with technology but they are nevertheless an important target group since, being the group with the highest income, they have the strongest buying power. We can see from the segmentation practice that the choice of description for each group closely associates the rationale of technology with further value realization, implying a predictive approach to each group. More importantly, what is at stake is that segmentation here conditions not just technologically mediated action and cognition in value production, but also aims at the transformation of social relations with the insinuation of qualitatively different criteria for individual performances. The separation of a performing group from a non-performing one is rife with political implications, encoding undesirable target groups as conservative and less progressive.

According to the philosopher of language J. L. Austin, "the issuing of the utterance is the performing of an action" (Austin 1976: 6). The "uttering [...] is, or is a part of, the doing of an action, which again would not normally be described as, or as 'just,' saying something" (Austin 1976: 5). In this sense, we can suggest that the technologically mediated linguistic building up of a user group, or the interpellation of a user group, is a strong performative utterance. This is so because it insinuates future action beyond description in the embodiment of the identity of the groups. The term "performative", Austin continues, is used in a "variety of cognitive ways and constructions, much as the term 'imperative' is" (Austin 1976: 6). As the word "perform" stems from the word "action", this signifies that "the issuing of the utterance is the performing of an action" (6). Technology makes continuous utterances and the utterance becomes "a code both legislative and communicative" (Bourdieu 1992: 44) in the transformation of social relationships through the implication of qualitatively different criteria of economic performances, determined through behavioural coding of segmentation, in relation to the ubiquitous structure of technology.

In postindustrial societies, Lyotard argues, the normativity of laws is replaced by the performativity of procedure legitimizing "the complexity of argumentation". This process of performativity replacing "the normativity of laws" in order to optimize the system is symptomatic of a technological coding system, functioning as "the route of data storage" in guidance of "the operativity of information".

The performativity of an utterance, be it denotative or prescriptive, increases proportionally to the amount of information about its referent one has at one's disposal. Thus the growth of power and its self-legitimization are now taking the route of data storage and accessibility, and the operativity of information. The relationship between science and technology is reversed. The complexity of the argumentation becomes relevant, especially because it necessitates greater sophistication in the means of obtaining proof, and that in turn benefits performativity (Lyotard 1984: 47).

Each utterance acts as a condition of performativity in proportion to collective information. And the performativity embodies its function with the communication of an utterance because "every utterance should be thought of as a 'move' in a game" (Lyotard: 10). Here an individual can access the system only if he or she agrees to enter it as a player or a performer. In the system, any fluctuation is registered, monitored, and evaluated as a "move" in an endless game; and the moves accumulated become data, which is stored and provides a probability forecast. Thus, it contributes to the stability of the system and allows it to produce effects in reality.

Segmentation through behavioural coding is in operation as the objectifying field of value onto which colliding values are coordinated and represented in the redefinition of the performative standards of the new main values against old and inferior ones, here exemplified by "affluent techie" versus "traditionalist". With regard to this, the technologically engineered segmentation of each segment group carefully guides the meaningful signifiers that can subsume and extract the latency of value of each group for the sake of further potential surplus value. This is so because it functions as a predictive approach as well. As a result of this systematic coordination of segmentation, we witness value as a guiding force of action and cognition where further movements are anticipated and advanced. The performativity of the economic in technology is, as Donald MacKenzie grasps, here initiated through an "interweaving of 'words' and 'action' – of representations and interventions – that the concept of 'performativity' is designed to capture" (MacKenzie 2008: 5). The performativity of technological interpretive segmentation can, in this regard, be seen as the power of intervention and representation that leads to practical action, if such coding fails to describe the reality (Lee 2011).

With communication code enabling continuous technical (re)arrangements, technology firms become strong interpretive apparatuses by which user behaviour is incessantly registered, evaluated and reinterpreted; and the accumulated collective behaviour becomes information for the segmentation of users, providing product platforms and market forecasts, connoting user action and cognition towards new frames of value production. From the perspective of cultural analysis, the practice of segmentation has two important implications. First, human behaviour becomes a meaningful code of the economic, and it thereby acquires an intrinsic and latent value that is to be translated and transmitted into a socially meaningful narrative. Second, and more importantly, the fundamental operation of segmentation by descriptive and predictive representation of users' behaviour and cognition becomes the powerful performativity of the economic. This is because such representation amounts to active intervention of economic hypotheses, channelled into practical action as a structure of meaning-formation, with "conditions of perception" (Hall 2004: 121) as a process of technological coding operations.

Performativity of motion control in smart TV technology

As we can see, each movement of the user of digital technology acts as a condition of performativity, standing in proportion to collective information. The performativity embodied in the communication of movement can be compared to a move in a game. This way of activation – through the performativity of user movements that differentiate and optimize the system – is a salient feature of digital technology. Technology interpolates every individual as data, or as code in the system as a whole. Movement is registered, monitored, and evaluated like one move in an endless game. The accumulated moves become data, which provide a probability forecast and contribute to the stability of the system, enabling its effectiveness.

Technology differentiates performers according to levels of engagement, and for the optimization of the system to take place. It is necessary for the user to be subsumed under the ceaseless chain of behavioural codes set for value production and technological rearrangement. In digital environments, the activation of technological rearrangement and development is different in that digital systems reconfigure both the player and the way that the game is played. This is an important point about the interaction between digital media and users in the digital "ecosystem".

Current levels of user engagement with smart devices have been enhanced immensely by the facilitation of contemporary technology directed at, above all, motion recognition. And there is enhanced translatability of smart TV equipped with the computerized interpretive function of corporeal gestures including motion, voice, and even recognition of face and facial expression. At this level, the segmentation criteria for identification of the user are not confined to traditional distinctions of age, occupation, or social class, as aforementioned. In the beginning we proposed that the current stage of technological development was a symbiotic stage, with a combination of corporeal gestures and technical performance that leads to market advantage through differentiated digital technologies. By exemplifying the up-to-date functions of smart TV in motion and sensory detection and recognition, we will observe that smart TV technology is now capable of engaging with the user at a kinetic and sensory level of intelligence, both for individuation of service and for market diversification. Smart technology expands its functions beyond conventional use. More importantly, it connects the bodily engagement of the user to the experience of the sensible in guiding and implementing new technology.

Samsung, for instance, put forward the catchphrase "smart interaction" when launching a series of its own flagship smart TV line-ups in 2012. The Korean firm is the dominant manufacturer in the global smart TV market, with more than a quarter of world market share as of 2013. The clear target of "smart interaction" is to incorporate the user's motion and voice, since the essential function of smart TVs is to take up and respond to users' sensory and physical movement in order to interpret it in using the product. With simple voice commands, the user can turn the TV on and off, change channels, and adjust the volume. Motion control is also integrated as a crucial function. With simple hand gestures, such as waving and clenching, the user can also change channel, volume, and browse the Internet. The TV can also recognize the face of the user, thus replacing user registration processes such as ID and passwords. By recognizing the user's face, voice, and motion with its in-built camera, the smart TV can provide tailored customer service to match individual tastes and preferences. With the incorporation of the sensory and physical movement of the user, smart TVs replace the traditional remote-control with motion control. Kinetic movement therefore works as an inseparable part of this new technology. This tendency is reconfirmed through recent product launches. Showcasing new flagship models in 2014, Samsung announced it would expand voice control to twenty-three countries, in order to take the initiative in the smart TV market. It also stated that more varied finger gestures would be included as a significant feature of the motion control in the new product. "Thanks to finger gesture," the company insists, "users can change the TV channel, adjust the volume, find and select what they want to watch just by using their fingers. They can also go back to the previous screen and stop the video by monitoring their finger counterclockwise – a more intuitive way to control the TV." The intention of the development is to make "the service more convenient by focusing on the most frequently-used functions and taking into consideration user's usage patterns".[7]

LG, Samsung's Korean arch-rival, also focused heavily on smart TV products in 2014, featuring "Voice Mate", which can allegedly process "natural speech recognition" combined with a motion-control function that is similar to that of Samsung, yet compatible with a remote-control. LG also specifically emphasizes the "intuitiveness" of their smart TV as well. The smart TV operation system and home screen, according to the firm, become more intuitive with the integration of voice and motion control. Kinetic engagement with the device opens up a new level of experience, which allows direct interaction with motion control rather than the traditional remote-control. Technology is, in this respect, symbiotic with the sensory and bodily experience of the user in the activation of the new system. Traditionally, TV has been recognized as a representative lean-back device, which is passive, and which exemplifies one-way communication. But with the incorporation of sensory and corporeal movement in the operation of TVs, the experience becomes more responsive-interactive, and thus the convergence of varied services such as TV programmes, video on demand, streaming, and Internet browsing is made possible.

Panasonic also exhibits the same tendency, implementing the "Viera Home Screen" as a major feature of their new smart-TV platform in which the personalization of the home screen is accentuated. The home screen functions as "a personalized gateway", because "the customizable [screen] makes it fast and easy to access the content". Again it is emphasized that this is "a more intuitive home screen", whereby Internet-based services are integrated with voice command and motion control in watching TV.[8] Combining user tastes and preferences with face recognition functions, smart TV offers customized services as a distinctive point in its marketing. As we have seen, the aim of incorporating sensory-physical movement into the platform is critical for helping users to participate with more ease, thus expanding this service area. The critical target of motion technology is to increase the level of participation of the user. Gesticulation, simple voice, or even facial expression can provide data that can be interpreted as a meaningful signification in technological expansion. As such, user engagement and the customized service are closely related and are critical factors.

Customized tailored service would not be feasible without an increase in user participation, which also raises the problem of privacy. The integration of motion control in smart TV is intended to maximize the translatability of the sensory and bodily movement of the user as meaningful data, connecting technology to the bodily experience of the user – and this is a general trend in household appliances, not just in smart TV. This is the way in which sensory and bodily movement turns into data and functions in an even more sophisticated way – as the intelligence of technology.

In concluding this chapter, we need to think briefly of the position of the subject within this new environment. Interaction with digital media can provide users with the possibility of defining themselves as the subject of technology. At the beginning of this chapter, I pointed out two things concerning the social positions of users, and the possibility of a democratic facilitation of the technological interface. Regarding the position of the subject in digital technology, we analysed the subject as having been (re)defined as the (un)bearer of value from the perspective of economic value production. Second, in our exemplification of the segmentation process, we saw that if we understood the digital ecosystem as a structure of dominance, the redefinition process could then combine unequal ownership of data with technical development and market expansion, thus justifying the operating system as the legitimate objectifying field of value. It is worth reconsidering Butler's discussion of the process of subject construction:

> Subjection is, literally, the *making* of a subject, the principle of regulation according to which a subject is formulated or produced. Such subjection is a kind of power that not only unilaterally acts on a given individual as a form of domination, but also *activates* or forms the subject. Hence, subjection is neither simply the domination of a subject nor its production, but designates a certain kind of restriction *in* production, a restriction without which the production of the subject cannot take place, a restriction through which that production takes place. (Butler 1997: 84)

Butler indicates that since subjection stimulates the subject, it is neither merely domination nor its production. *Motivation* activates voluntary performances on the part of the subject, generating a subject-effect, whose performativity is, however, *restricted*. The technological interface, like the smart TV home screen, affords the user direct access to the system, but the route to that access is controlled. We should also be aware that the seemingly free and direct access to the system is often made possible in exchange for, for instance, compromising data privacy and civil rights, which are critical in the formation of subjectivity in the digital era. As a White House white paper suggests, although "[b]ig data technologies will be transformative in every sphere of life", they also raise "considerable questions about how our framework for privacy protection applies in a big data ecosystem. Big data also raises other concerns. A significant finding [...] is that big data analytics have the potential to eclipse longstanding civil rights protections in how personal information is used in housing, credit, employment, health, education, and the marketplace".[9]

While maintaining the hope of the possibility of democratic participation in the digital ecosystem, we should carefully monitor the technical interaction between users and the interface, as well as data processes within digital products and analytics. Technologically engineered signifying tools and practices can then be more commonly agreed upon and developed without an abuse of personal data and privacy. If technology contains and restricts users within "the dominant-hegemonic position" (Hall 2004: 124) of performative sets of technological platforms, it will result in a de-politicization of the social subject.

Notes

1. In regard to the changed digital climate, Nokia CEO Stephen Elop recently remarked that we shall see a more collective interface as the ecosystem becomes more at stake, since the momentum of development is to develop user interface, rather than the devices themselves, mainly in that "[t]he battle of devices has now become a war of ecosystems, where ecosystems include not only the hardware and software of the device, but developers, applications, ecommerce, advertising, search, social applications, location-based services, unified communications and many other things". Nokia CEO Stephen Elop's "Burning Platform" Memo, *Wall Street Journal*, 9 February 2011. (Available at: http://blogs.wsj.com/tech-europe/2011/02/09/full-text-nokia-ceo-stephen-elops-burning-platform-memo/ [accessed 17 March 2015]).
2. "The Social Economy: Unlocking Value and Productivity through Social Technologies", McKinsey Global Institute, 2012.
3. "Data, Data Everywhere", *Economist*, 25 February 2010.
4. "A Clear Picture: Company Defines How They Add Value to the Ad", *MIT Technology Review*, 12 September 2013. (Available at: http://www.technologyreview.com/view/518556/a-clearer-picture-companies-define-how-they-add-value-to-the-ad-tech-space/ [accessed 17 March 2015]).
5. Ibid.
6. "Understanding the Smart TV Viewer", Yume Report 2013. (Available at: http://www.yume.com/smart-tv-viewer [accessed 17 March 2015]).
7. See: http://global.samsungtomorrow.com/?p=31581 (accessed 17 March 2015).
8. See: http://www.panasonic.com/promos/learn/smart-viera/ (accessed 17 March 2015).
9. "Big Data: Seizing Opportunities, Preserving Value", White House White Paper, May 2014.

Part IV

Technology and Materiality

9

Moving, Withdrawing and the Uncanny

Sita Popat

> *Ubiquitous computing will require a new approach to fitting technology to our lives, an approach we call "calm technology."*
>
> Weiser and Seely Brown in *The Coming Age of Calm Technology* (1996: 1)

Mark Weiser and John Seely Brown's claim at the end of the last century for a future in which ubiquitous computing becomes a "calm technology" suggests that computing will be so deeply embedded in contemporary developed societies that we will not even notice its presence as distinct from other aspects of the world. Almost two decades after they made this claim, our smart phones and our computers are part of our daily lives, as are our washing machines, the computer-controlled circuits in our cars and many other computational objects ubiquitous within our day-to-day existence. These are more tools to be added to the portfolio that has developed since first a stone was used as a hammer – are they not? Yet seldom in human history have there been such dramatic advances in a specific type of technology over such a short space of time as in the digital era. Even before these advances achieved their full speed, in 1954 philosopher Martin Heidegger was calling for questioning of the essence of technology and its impact upon human relationships with the world. He was deeply concerned that technology was distancing humans from the world and even from their understandings of their own essence as humans. At the end of his essay, he proposed that art might be the saving power that would enable the essence of technology to be revealed.

In this chapter I explain Heidegger's position and point out the flaw in the way he approaches art as the saving power. I propose, more specifically, that aesthetic movement is the most effective way to enable humans to develop an understanding of technology's essence, since it offers a unique critical standpoint through embodied engagement. By "aesthetic movement", I mean movement within an art context (dance, performance, installation art, etc.) that focuses on the act of moving and thus draws attention to the body and its relationship to the world. Movement of this nature highlights embodied experience, requiring negotiation with the environment and/or the dancing partner that is based on the conscious act of moving. I argue that aesthetic movement can problematize the essence of technology, calling into question notions of Cartesian subjectivity that separate mind and body and remain residually present in Heidegger's thesis, and thus revealing technology's essence via the mechanism of the uncanny.

Heidegger's questioning

The first move is to return to Heidegger's *The Question Concerning Technology* and review his argument for the purposes of this chapter. Heidegger's question is focused on the essence of technology, about which he writes, "the essence of technology is by no means anything technological" (1977 [1954]: 4). For him, our inability to engage with the essence of technology, as opposed to the technology itself, leaves us vulnerable to its impact, as we do not understand fully what effects it has upon our relationship with the world. He elaborates further by explaining: "we shall never experience our relationship to the essence of technology as long as we merely conceive and push forward the technological, put up with it, or evade it" (4). In her book on *Technology and Touch* (2013), Anna Cranny-Francis offers a helpful expansion:

> Heidegger's rather cryptic formulation argues that technology is not simply the devices or machines produced as a result of techno-scientific research; instead the essence of technology is the thinking, the domain of knowledge and its formative discourses, that generates the devices and which also, therefore, is where the actual nature of that technology (its underlying assumptions, beliefs and values) can be found. (37)

This "domain of knowledge and its formative discourses" is made up of the combination of two forms identified by Heidegger (from Plato) as *technê* and *epistêmê*. Technê Heidegger describes as "the name not only for the activities and skills of the craftsman, but also for the arts of the mind and the fine arts" (1977 [1954]: 13). Epistêmê is not defined specifically in Heidegger's essay but he refers the reader to Plato, for whom it is a form of conceptual knowing or understanding that is coupled with technê's applied knowledge of activities and skills. Heidegger explains that together technê and epistêmê mean "to be entirely at home in something, to understand and be expert in it" (ibid.).

The technê of the craftsperson or artist is the gathering of materials together and then the revealing of the object or artwork from those materials. Thus it is a mode of "bringing forth" (ibid.). The word "technology",

Heidegger argues, has its roots in technê, and technology does indeed reveal. However, this revealing is not in harmony with its environment and so it is not a matter of "bringing forth". Heidegger explains that the earth is "challenged" to give up its fossil fuels (a point driven home as we face a future in which the reserves of oil and coal are becoming seriously depleted), the fields are "challenged" to produce ever more grain with the natural nutrients in the soil enhanced and replaced technologically. So the natural resources "yield" their resources to humans rather than being part of the balanced relationship of "bringing forth" of pre-technologized mining and farming. Resources are stockpiled or held in standing reserves for human consumption. The world becomes defined by what humans can do with it and how humans choose to order it, until the point where all objects cease to exist as phenomenologically encountered and the "real" is concealed behind its representation as standing-reserve, ready for use. Heidegger (27) says that even "man himself", whilst believing himself to be "lord of the earth", becomes defined as standing-reserve also (as exemplified by the ubiquitous term "human resources" in business and management). Man (*sic*) does not even perceive, Heidegger warns, that his own essence has been thus concealed from himself. All of this, Heidegger claims is a result of our withdrawal from acknowledgement or awareness of the ways in which technology withdraws us from the world, and thus from our own essence as humans. We withdraw from our own withdrawal.

Heidegger ends his essay by proposing a possible answer to his questioning about how we can come to understand the essence of technology:

> Because the essence of technology is nothing technological, essential reflection upon technology and decisive confrontation with it must happen in a realm that is, on the one hand, akin to the essence of technology and, on the other, fundamentally different from it. Such a realm is art. But certainly only if reflection on art, for its part, does not shut its eyes to the constellation of truth after which we are questioning. (35)

Art and technology are linked by their shared roots in technê and so art can reflect upon those aspects of the essence of technology. However, art maintains its relationship with humans and the world of "bringing forth" whereas technology has taken up its mantle of "challenging forth". This provides art with a way into questioning of the essence of technology and gives it what Heidegger calls a "saving power" (ibid.).

The saving power of art

This subtle differentiation between technology itself and its essence is fundamental to Heidegger's philosophy and it leaves his relationship to technology largely ambiguous, "neither celebrating nor outright condemning" (Salter 2009: 28). Chris Salter, new media philosopher and design artist, aligns Heidegger's claim that the arts might save us from the process of withdrawal from withdrawal with "what Gilbert Simondon, the French philosopher of technology described in 1958 as 'a system of defense designed to safeguard man from technics' " (ibid.). However, Heidegger was not so much attempting to safeguard as to lead us to question, as the title of his paper indicates.

Almost in direct contradiction to Heidegger, Salter asks why performance discourse continues to debate the impact of technological mediation when the physical, computing and life sciences "juggle daily with the increasing panoply of effects caused by the merging of mechanical, computational, biological and ecological forces" (ibid.: 29). Heidegger's position suggests that the sciences have simply accepted the withdrawal from withdrawal and gleefully elided technology with the state of "being", ignoring the essences of both technology and man. Theirs is a technê-based understanding that avoids the epistêmê in the very way that Heidegger seeks to overcome. In contrast to the sciences, performance discourse sits within the realm of the arts – and art, according to Heidegger, can reveal truths about the essence of things:

> The [art] work, therefore, is not the reproduction of some particular entity that happens to be present at any given time; it is, on the contrary, the reproduction of the thing's general essence. (Heidegger 1977 [1954]: 26, my insertion)

This quotation is taken from Heidegger's essay on the *Origins of the Work of Art*, written prior to *The Question Concerning Technology*, and it is clear that Heidegger's claim in the latter essay for art as the saving power lies in his argument that art can reveal essences and thus combine technê with epistêmê. Heidegger argues that the artist's decisions when creating the artwork form the "world" of the work. Against this "world" are revealed the unknowns or irreducible elements of the work and its context, which Heidegger calls the "earth". The irreducible weight, roughness and colour of the stone are revealed in the sculptor's shaping of it; the aspects of the daily life of the peasant woman are revealed through Van Gogh's choice of techniques and colour in the painting of her worn, earth-stained shoes. This process enables the viewer to see the truths of what is known and what is not known; what is crafted by the artist and what is irreducible. Thus, according to Heidegger, the contemplation of artworks reveals truths about the essences of things. Art philosopher Paul Crowther (1993: 113) describes the artwork as "a self-contained portion of the world which invites us to contemplate its sensuous particularity", and it is in this process that Heidegger finds the essences revealed. The artwork allows us to glimpse aspects of things hidden in other forms of encounter. Yet Heidegger offers a heavily romanticized interpretation of the life of the peasant woman drawn from Van Gogh's painting of her shoes:

> On the leather lies the dampness and fullness of the soil. Under the soles slides the loneliness of the field path as evening falls. In the shoe-tool vibrates the silent call of the earth, its quiet bestowal of ripening corn and its unexplained self-denial in the desolate fallow of the winter field. (Heidegger 1977 [1954]: 34)

This poetic description demonstrates a fundamental flaw in Heidegger's own argument. His revelation of the essence of the peasant woman's shoes is deeply embedded in the early 20th century's withdrawal from the natural world through pastoral romanticism in the face of an increasingly technologized environment, but Heidegger does not notice his own perfect example of withdrawal from withdrawal. How, then, can art be the saving power if all it does is reflect back to us the essence of our own withdrawal?

The uncanny

Heidegger's claim for the saving power of art is that it sets up the "world" that is cognized and sets forth the "earth" that is uncognized within the work, revealing that which is "undisclosable" in the tension between the two. Elements of this claim have resonance with definitions of the uncanny, since the "undisclosable" is the gap between known and unknown where the uncanny resides. Psychoanalyst Sigmund Freud explores various definitions of the term "uncanny" and comes to rest on the multifaceted German word "heimlich":

> [...] this word *heimlich* is not unambiguous, but belongs to two sets of ideas, which are not mutually contradictory, but very different from each other – the one relating to what is familiar and comfortable, the other what is concealed and kept hidden. *Unheimlich* is the antonym of *heimlich* only in the latter's first sense, not in its second. [...] (*unheimlich*) applies to everything that was intended to remain secret, hidden away, and has come into the open. (Freud 2003 [1919]: 132)

Freud's two definitions of "heimlich", when taken together, seem to chime with Heidegger's "world" and "earth". Their conjunction raises those questions for the perceiver about "what is" that Heidegger declares vital to the revelation of the essence of things.

In his discussion of uncanniness in horror movies, Curtis Bowman draws upon Heidegger's writings on the uncanny in *Being and Time*:

> The main element of the uncanny is not-being-at-home in the world. We lose, so to speak, our ontological balance and become unsure of

> ourselves and of our understanding of the world around us. The main form that this can take is the realization that more possibilities exist for action and understanding than we ever thought. Perhaps we revise our ideas in light of experiencing the uncanny; perhaps our ideas remain the same. The point is that the experience of uncanniness forces us to appropriate or reject accustomed ways of thinking and acting. In this way, we make some small step towards authenticity. (Bowman 2003: 72)

"Authenticity" here is the quality (or qualities) that makes the entity what it is – the truth that reveals the essence, in Heidegger's terms. By forcing us to address what we think we know in combination with other possibilities that might now be apparent, the uncanny leads us to question the essence of the entity. Herein also lies the saving power of art, not in that it presents us with truths about the essences of things but in that it lifts the veil just enough to provide glimpses of other possibilities and thus to set us off balance in relation to the world. Heidegger suggests that Van Gogh's painting of the peasant woman's shoes reveals essences, but his own interpretation of that painting demonstrates the flaws in his argument for the saving power of art. Adorno (1997 [1970]: 269) claims that art offers a way to experience empirical reality through "the epiphany of its shrouded essence". Art enables essence to be intimated; it alludes and obliquely glances rather than shining a light upon the essence of things (the latter being more akin to how Heidegger describes his interpretation of the painting of the peasant woman's shoes). Thus art enables possibilities to be glimpsed and our ontological balance to be disturbed enough to require us to question, but it does not impose any given truth and nor does it identify those essences specifically. It simply allows us to see a little more than we saw before. In this, it holds some of the uncanny qualities that Bowman explores in relation to horror movies. Indeed Adorno suggests that art gives empirical reality its due in "the merited shudder in the face of it as in the face of a monstrosity" (269). However, this remains a mode of engagement with the artwork that is grounded in disinterested contemplation, based on cognitive processing of the experience that distances the viewer from the work. Epistêmê remains divorced from technê as we develop a cognitive understanding without a sense of the activities and skills. And so we remain potentially subject to the withdrawal from withdrawal that Heidegger demonstrates in his response to Van Gogh's painting.

In 1970, Masahiro Mori used the term "the uncanny valley" to model human responses to robots (2003 [1970]). He noted that the degree of familiarity (which we might call "heimlich"-ness) that humans experience towards robots increases as the robot looks more like a human. So a toy robot with arms and legs produces greater feelings of familiarity than an industrial robot that has no limbs. Mori compared this level of familiarity to a child's teddy bear. Yet he argued that there was a point at which the robot could look too much like a human whilst still being clearly not human. At this point the robot becomes suddenly uncanny (in Freud's terms, "unheimlich") – familiar and yet unfamiliar – and thus frightening. This, Mori suggested, corresponds to the uncanny nature of corpses and mannequins. Mori called this steep fall from familiarity to unfamiliarity the "uncanny valley" and he argued that movement increases the degrees of both familiarity and unfamiliarity as it suggests life and agency. The moving and human-like but still obviously non-human robot (e.g. Honda's humanoid robot, Asimo[1]) has a higher level of familiarity than the non-moving toy robot. However, the moving robot that looks much more like a human (e.g. the robot Geminoids developed by Hiroshi Ishiguro Laboratories[2]) is far more uncanny than the still one, just as a moving corpse is more frightening in horror films. The reason for this extension or stretching of the uncanny valley in relation to movement lies in the empathic response of the viewer. The term "empathy" was coined in the 1880s by German aestheticians "seeking to describe and analyse the depth of the act of viewing painting and sculpture" (Foster 2011: 10). The concept was deeply connected to kinaesthesia and the aestheticians sought to capture the physical connection between the viewer and the artwork "in which the viewer's own body would move into and inhabit the various features of the artwork" (ibid.). In the twentieth century, this concept has developed to the point of projecting "one's three-dimensional structure into the energy and action of the other" (ibid.: 11). This refusal of the Cartesian division between object and subject, observer and observed, is an important factor in the effect of the moving robot upon the observer in Mori's uncanny valley. In Heidegger's terms, these robots are technological devices and when they become too human-like, they challenge the divisions between human and technology, between subject and object, between known and unknown. In an empathic response that creates a humanist nightmare, we recognize ourselves and yet not ourselves in the robot. We are pulled unceremoniously out of withdrawal from our withdrawal from the world, as the very essence of ourselves as human beings is called into question by technology.

Heidegger's claim that we are in a state of withdrawal from our withdrawal from the world is grounded in his belief that we do not yet understand the essence of technology. His suggestion that art might be the saving power through which we can discover the truth is only partly effective, since the art forms that he describes are designed to be viewed from a position of disinterested contemplation. He looks for Cartesian-style objective truths in art, rather than a lifting of the veil to reveal other uncanny possibilities that might throw us off balance and cause us to question (which seems odd, given his desire to question technology). Epistêmê and technê remain disconnected and thus the development of understanding is partial and fragmented. However, Mori's analysis of the uncanny valley demonstrates the potential of movement to break down the Cartesian subject/environment divisions that protect us in our state of withdrawal from withdrawal, as it requires engagement and empathy with, rather than contemplation of, the technology. Movement within an aesthetic frame, as dance or as participation in an art installation, encourages attention to the subjective experience of that engagement. The process of negotiating bodily movement within a technological environment requires the reconnection of epistêmê and technê since it draws upon embodied understanding in which cognitive and practical knowledge are deeply embedded. I propose that the saving power for which Heidegger was seeking can be found far more readily in artworks that involve movement, as the next section of this chapter will explore.

The saving power of movement

When speaking of the uncanny in relation to performance and technology, one might be tempted to turn first of all to Australian-based artist Stelarc. Stelarc's cyborgian artworks question the boundaries and interfaces between body and technology. He has merged his body with robotic technologies to add a *Third Arm*, moving independently of his other two arms and being controlled by muscle contractions in his abdomen and legs. His *Ear on Arm* project has included multiple surgeries to implant cartilage beneath the skin on his forearm in the shape of a human ear, designed both to hear and to transmit sound. In these and many of Stelarc's other works, relationships between body and technology are brought sharply to the fore through the "unheimlich" appearance of his augmented, cyborgian form. At Brunel University in 2007, my own experience of touching and moving the flesh of the ear on Stelarc's arm and feeling the firm flexibility of the cartilage beneath his skin caused a shiver to run down my spine, as the misplaced human body part was brought home to me in a tactile experience. Cranny-Francis highlights the importance of touch in understanding technological devices and "the way that their close touching of (connection with) the human body confronts us with our expectations and assumptions about what it is to be human, by undoing the conventional divide between humans and technology" (2013: 4). Yet, like Mori's robots and corpses, it is the movement in Stelarc's work that drives furthest the "undoing" of this divide. Stelarc's enmeshment with robotic materials results in movement that is sometimes under his intentional control and sometimes defined by tiny motions in parts of his body over which he has little or no conscious command, e.g. heartbeat. The weight and limitations of the robotic constructs affect his balance and the range, quality and manner of his movement, making it difficult or impossible to distinguish body and technology. Stelarc's naked flesh, woven about with metal, rubber and plastic, struggles to navigate a space (*Muscle Machine*) or complete a task (*Third Arm*), or simply moves due to involuntary muscle contractions caused by electronic stimuli applied by remote Internet participants (*Ping Body*).

In each case, the combined movement of flesh and technology is where the differentiation between human and machine becomes impossible and the distinctions between physical substances are rendered irrelevant. The process of negotiation involved in the combined body/technology finding ways to move calls into question the essences of both body and technology. Mark Poster proposes that Stelarc's performances suggest "a new bringing forth, in Heidegger's sense, of the human and of being that renders obsolete not so much the body but the culture of instrumental technology, the Cartesian position of the human body as the ruling force, deploying nature and machines at will" (Poster 2002: 29). It is in the movement that truths about human and technological essences are revealed.

Stelarc's work draws heavily upon the uncanny, knocking us off "ontological balance" (Bowman 2003) by challenging conceptual and embodied engagement with his augmented body and offering new possibilities for human/technology interfaces. However, my initial question in this chapter was about the calmness of technology in relation to ubiquity. This kind of calmness relies upon a relationship with the technology that is similar to Heidegger's well-known scenario of the hammer. The hammer lies ready in the carpenter's workshop and can be picked up when required for the task. The carpenter's relationship with the hammer is as an extension of him/herself, since experience has bred familiarity that requires no specific conscious engagement with the hammer, provided that it is ready to hand when needed. Yet this resides within a deep understanding of the other options available to the carpenter – the other tools in the workshop. Without this understanding, Gunkel and Taylor explain, one ends up with "the homespun saying 'when you have a hammer, everything looks like a nail' ". They argue that the problem is magnified "when considering the use of technologies that rely upon an integrated system of mutually referential technologies such as the digital matrices that surround us today" (Gunkel and Taylor 2014: 2). Stelarc's work questions and offers possibilities for body and technology on a grand scale where shock is a significant factor in the "unheimlich" unveiling of hidden truths. The process of watching him move as a body/technology cyborg offers glimpses into both an epistemological understanding of the essences of body and technology and an insight into the technê of body/technology relationships in practice. Other artworks use movement in different ways that are no less revealing.

Revealing and unveiling

In 2007/8, a team of researchers from performance and computing at the University of Leeds and members of Shadow Robot Ltd created a robotic dancing partner called SpiderCrab. The robot consisted of a single arm, just over four metres in length, suspended at one end, approximately three metres above the floor. The other end moved freely in the space below the suspension point. The suspended end was reasonably substantial in width, tapering along the full length of the arm to a thin point at the moving end. It was constructed using air muscles, controlled by pumping compressed air into and out of soft sheaths of stretchy material, which meant that it could move its own weight but be pushed aside by a human quite easily.

SpiderCrab was designed in response to a challenge to create a robot that would be socially and aesthetically acceptable to people, that might lead in the future to robots becoming more prevalent in the home.[3] The aim was to overcome the uncanniness of a moving robot that can cause mistrust in close proximity. As the dance researcher on the team, I worked with the computer scientists to explore options for how this robot might move. To some extent we needed to counter the large size of the robot, which towered over the human dancer and engendered feelings of alarm and fear. We wanted to find ways to make SpiderCrab feel like a partner in an improvised dance experience, so it needed to be responsive and sensitive to the human dancer whilst also maintaining its independence and offering creative input to the partnership. Initial ideas based on aesthetic stances or positions were quickly overturned in favour of an approach to movement qualities. We decided to track the quality of the human dancer's movement and either reflect or contrast that quality within the robot's movement in the simulation of an empathic response. In a system extrapolated from the effort element of Rudolf Laban's Movement Analysis, we identified four spectra: sudden/sustained; light/strong; bound/free; indirect/direct. For the purposes of the computer programme, these spectra were interpreted as linear scales against which movement quality could be mapped at a basic level. The human dancer put a green band around wrist, upper arm, thigh, ankle, hips or head. Two cameras at the back of the space tracked the green band and the computer programme analysed the movement in relation to the four spectra to generate four movement parameters (one for each spectrum). Spidercrab's movement was generated using either the same four parameters as the human dancer (which we called "following") or contrasting parameters, i.e. at opposite points on each spectrum ("opposing"). The positioning of the robotic limb in space was random and there was no attempt to copy or respond to the spatial positioning, shape or direction of the human dancer's movements.[4] The result of this process was that the human dancer experienced the robot responding to her/his movement like a partner, sharing the mood and quality of the movement either by picking it up directly (following) or by offsetting it through contrast (opposing).

SpiderCrab was presented in a number of different venues and interacted with trained dancers, academics from arts and science disciplines, and the general public. Its presentation at the Designing User Experience (DUX) 2007 conference was enhanced by the discovery that the green beer bottles served in the bar were picked up by the tracking system, which meant that SpiderCrab would sometimes become distracted by a passing beer bottle. The interest that this independent behaviour added to the movement experience led the research team to programme an option in which SpiderCrab would choose to ignore the parameters defined by the dancer's movement and pick its own randomized parameters for a short while ("innovating"). This meant that it "led" the dance at times, rather than always following.

The "unheimlich" within the "heimlich"

The most interesting movement experiences arose when dancers were able to spend longer with the robot. Coming to trust the robot was a key element of the dancer's experience, and this could only be achieved through extended periods of moving with it and coming to know it. Initially SpiderCrab's appearance was perceived as alarming by most dancers (including me) and it was necessary for them to experience the physical ease with which they could push the robotic arm aside before they felt comfortable approaching it alone. Just as learning to improvise with a new human dancing partner for the first time is a slow process, so coming to know SpiderCrab took time. Dancer Isabel Jones spent significant periods moving with SpiderCrab over the course of the project (see Figures 9.1 and 9.2). She has an extensive history of working with people on the autistic spectrum and with other physical and learning disabilities and she referred repeatedly to this when describing her experiences of dancing with the robot:

> I didn't always want to face the robot [. . .]. I wanted to sense its presence through other parts of my body like my spine and back, trying to allow confidence to develop. Of course when I use a word like "confidence" it appears as if I am giving the robot human attributes, but strangely this is what struck me; I likened my approach to the interaction with SpiderCrab to the work I do with people with autism. Approaching and moving with the SpiderCrab demands sensitivity, it draws on the same skills as when I work with people who are different or may move differently to me. Not knowing what programme it is set to e.g. following, etc., adds to this, because I have to learn without preconceptions how the robot behaves and this process teaches me how to explore and be open to new possibilities. (Jones cited in SpiderCrab project evaluation, unpublished, 2008)

Jones's comments indicate the search for new possibilities and the process of movement exploration and negotiation necessary to find them. She was conscious of her anthropomorphism in this quotation but she made no apology for it because it is part of the process of learning how to move with the robot as a dancing partner. In fact this process of anthropomorphization could be interpreted as a Cartesian response to the "unheimlich" or unknown of the moving robot, by making it "heimlich" or known through

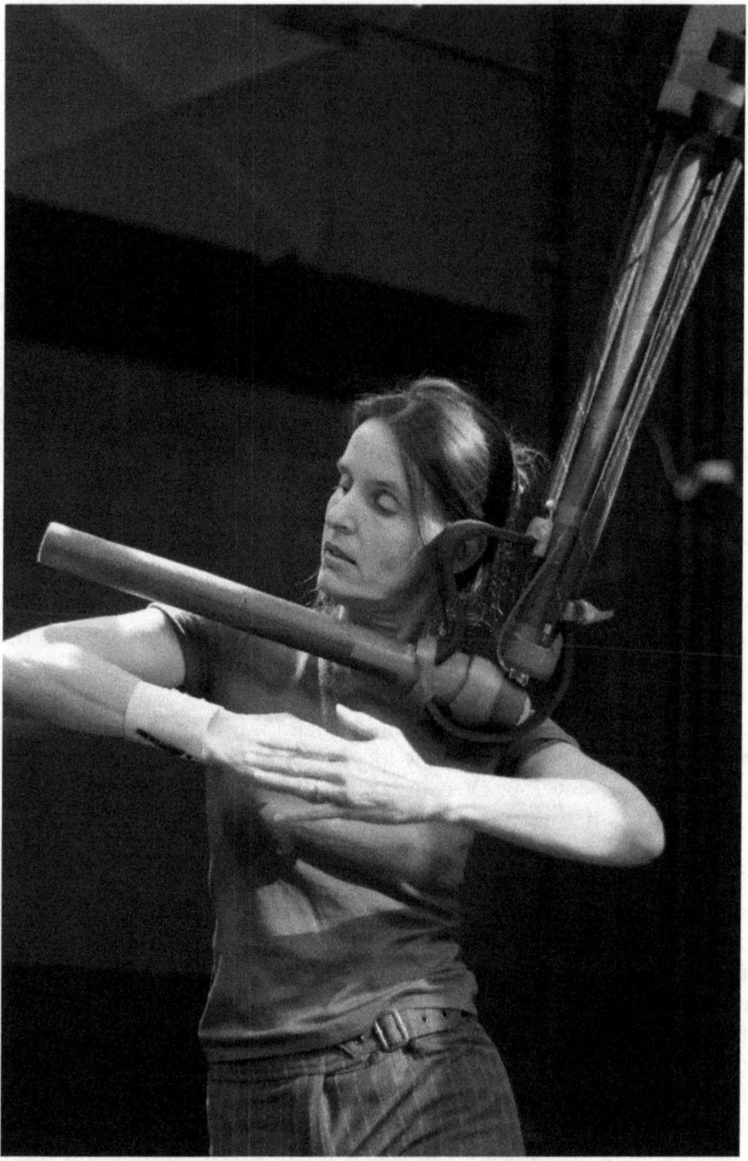

Figure 9.1 Isabel Jones dancing with the SpiderCrab robot. Courtesy of © Pixelwitch Pictures/University of Leeds

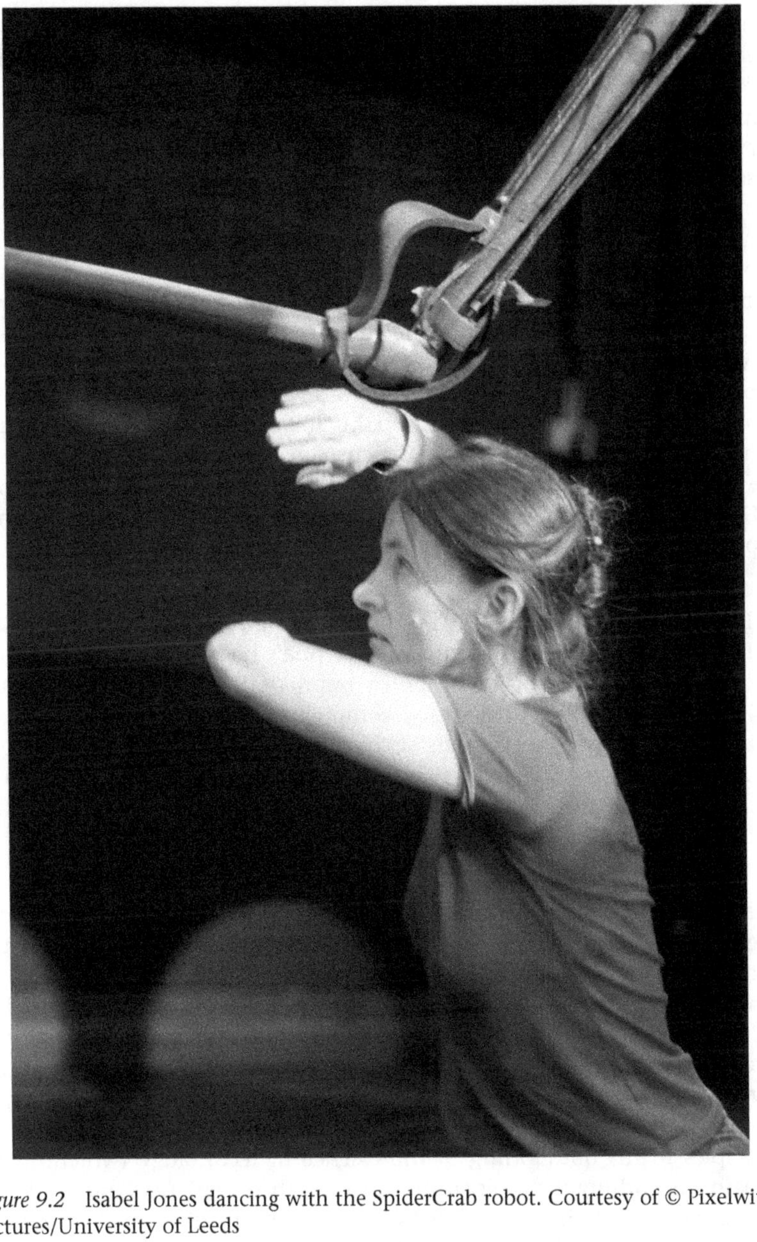

Figure 9.2 Isabel Jones dancing with the SpiderCrab robot. Courtesy of © Pixelwitch Pictures/University of Leeds

comparison to human behaviour. The synthetic agency in the robot's movement responses is more readily processed by the human dancer if it is perceived in relation to human behaviour. Jones knew that there was no human agency in the robot. However, she recognized something in its movement to which she could respond as if to another dancer. In Mori's description, the closer that the robot comes to looking and moving like a human, the more frightening it becomes. SpiderCrab did not look like a human at all but the essence of its movement drew upon aspects of Laban Movement Analysis, which can be applied equally to human and non-human movement. The process of negotiation through aesthetic movement enabled Jones to explore the robotic interface through movement, to recognize the qualities within the robot that were human-like but to appreciate its difference. The movement-orientated computer programme provided a connection or meeting point for dancer and robot where the dancer could explore the essence of technological movement by addressing the "unheimlich" within the "heimlich", the unknown within the known.

Dancer and new media philosopher Susan Kozel writes about her own experiences of dancing with interactive interfaces:

> Paradoxically, it is familiarity, or being at home, that provides scope for the unexpected. The idiosyncrasies of one's body and the unexpected qualities to emerge from sustained improvisation are matched by the idiosyncrasies of every responsive computer system, for which the process of calibration is as open to ghosts in the machine as is a human's mood and physical state subject to swings and flux. A deep physical understanding of how a system is likely to respond arrives through "dwelling" within the system, in the sense of being "absorbed" in it. (Kozel 2007: 76–7)

Kozel's approach to exploring the human/technology interface is very different to that of Stelarc. Stelarc's shock tactics thrust the viewer into the uncanny valley, unveiling "unheimlich" new possibilities to reveal questions about the essences of body and technology on a grand scale. Kozel searches the boundaries of that which is "heimlich", familiar or homely, in order to reveal what remains hidden or concealed. This process of unveiling is the discovery of small unknowns within the coming-to-know. This could easily be taken as a return to the detail of individual technologies, against Heidegger's principles in his questioning of the essence of technology (which, as he explains, is nothing technological). However, the experiences of both Kozel and Jones suggest that "dwelling" or being "absorbed" with the technology (and thus inevitably with a specific technology or technologies) enables an embodied coming-to-know that combines epistêmê and technê to reveal truths about the essence of technology in a way that is beyond what Heidegger's disinterested contemplation of artworks can achieve.

Negotiating technologies

I have used the word "negotiation" repeatedly in this discussion and there is no accident in this choice. The process of negotiation requires understanding that is multilayered. It requires both a practical understanding of the options available and a conceptual knowledge of the implications of those options. (The choice of options without a sense of the larger framework becomes a process of random selection rather than negotiation.) Both Kozel and Jones describe the need to start slowly when one is improvising movement with either another human being or a technological interface, since the options need to be discovered and explored. Heidegger uses the term "dwelling", which Kozel defines as "being ever-open and present, existing in any given moment" (2007: 77). Tim Ingold draws heavily upon Heidegger's definition of this term when he writes about the "dwelling perspective". He describes dwelling in a house as the process of living and being within the particular environment. He explains:

> Human children grow up in environments furnished by the work of previous generations, and as they do they come literally to carry the forms of their dwelling in their bodies – in specific skills, sensibilities and dispositions. (Ingold 2000: 186)

Heidegger's contemplation of artworks as a disinterested perceiver places him outside of the thing that he is viewing. He can observe whatever he wishes within the artwork and consequently it can simply reflect back to him his own values if there is insufficient uncanniness to offer new possibilities. The viewer negotiates the artwork in relation to his/her life experiences, but cognitive processes remain primary. Movement requires something different, and it is significant that both Kozel and Jones talk about "dwelling" with the technology as part of the experience of dancing for long periods with interactive systems. The process of negotiation in engaging in movement with another cannot be purely objective or cognitive, as it requires the practice of moving and thus physical interaction with the "other". Cartesian positions of subject and object become obsolete in the process of being there together and negotiating that togetherness through movement. Finding out what is and is not possible and exploring the limits of those possibilities through extended practical engagement develops knowledge of the technê of that tool whilst at the same time contributing to an epistemological understanding of embodied engagement with similar or other systems or tools.

Much discomfort in relation to robots and avatars is perceived when the robot or avatar appears too much like a human and calls into question the differences between the essences of human and technology, dropping us into Mori's uncanny valley. Humans are increasingly embedded with technologies including pacemakers, false limbs, even contact lenses. Stelarc tries to show this to us, but the Cartesian perspective that separates body and technology is deeply challenged by the indivisible cyborgian construct. Positive emotional responses to technology tend to lead to anthropomorphization, as when Jones was talking about SpiderCrab, although negative responses tend to refer to the technology as "it". Yet the statement from Mark Weiser with which this chapter began suggests that the world is already cyborg if technology is ubiquitous and we experience it in developed countries as an indistinguishable part of everyday life. I noted earlier that Kozel and Jones were both engaged in the process of seeking the "unheimlich" within the "heimlich" in the process of dwelling with the technological interface. Freud noted that "heimlich" could also mean that which is "concealed or kept hidden", to which "unheimlich" means the revealing of that which is meant to remain hidden. Heidegger's position is that we are complicit in our withdrawal from withdrawal from the world because we disguise that very process from ourselves. The "unheimlich" of the robot that resembles a human too closely is perhaps as much about the fundamental revealing of essences that we do not want to acknowledge as it is about the confusion between technological and human essences. Aesthetic movement enables us to explore the concealed essences through negotiation between human and technology without having to confront those essences directly in a way that the Cartesian self still struggles to do.

Heidegger points us towards the particular route of art as the saving power but he is unable to pursue it fully himself, since he is trapped within a representational theory of art that was prevalent in his time. Also he could not have conceived of the kinds of technological systems and robots that are common today. A technological device like SpiderCrab, or indeed many of the digital performance and art installations today, would have been unimaginable to him. Aesthetic movement of the type now often used in dance, live art, installation art and environmental performance shatters the Cartesian prison of Heidegger's perspective by engaging the viewer empathically and often physically in the experience. Consequently it is only now that the full potential of aesthetic movement to reveal the essence of technology becomes itself revealed. It seems likely that the cyborgian nature of the world will enable the technology to become calm because of our dwelling within it, as Ingold describes the children in relation to his "dwelling perspective". Our lived experience of the technology develops both technê and epistêmê through embodied engagement, but we are best able to reflect on that process and understand its essences through aesthetic movement.[5]

Notes

1. See: http://world.honda.com/ASIMO/ (accessed 10 June 2014).
2. See: http://www.geminoid.jp/en/index.html (accessed 10 June 2014).

3. SpiderCrab was part of the *Emergent Objects: Designing for the 21st Century* project funded jointly by the Arts and Humanities Research Council (AHRC) and the Engineering and Physical Sciences Research Council (EPSRC).
4. Later in the process, positional tracking was used to allow the tip of the arm to "pay attention" to the dancer by pointing towards her/his location in space. However, this was only partially developed due to time constraints.
5. I would like to thank the members of the Booth Group for their invaluable advice in the preparation of this paper, particularly Paul Taylor and Calvin Taylor.

10

Digitizing Dance Costumes: a Case Study of Movement and Materiality in iWeave

Rachel Fensham and John Collomosse

Complex questions about embodied interactions with digital technologies, as well as how subject-object relations are transformed in the digital environment, require the testing of new scenarios and the collaborative production of analytical and creative outcomes. Under an Engineering and Physical Science Research Council (EPSRC) scheme designed to engage engineers with "bridging the gaps" in conducting and communicating their research, the iWeave project aimed to involve computer vision scientists, archivists, dancers and dance theorists in re-animating the "traces" of an historical dance costume archive.[1] Previous collaboration with the National Research Centre for Dance (NRCD) at the University of Surrey had focused on a project designed to index videos of dance performance and make visual searching of them available online.[2] Building upon that success in support of cultural heritage, there was a motivation to explore the application of 4D performance capture to make other artifacts in the archive available digitally – in particular, costume.

NRCD is a small library-based collection of dance materials, whose visitors are not typically able to interact with their delicate and historic costumes for reasons of preservation. One of the aims of iWeave was therefore to develop a demonstrator for an Interactive Wearable Archive enabling users to "virtually try on" costumes from the NRCD physical costume archive. Such a project, however, was to require multiple processes of translation between the material and virtual domains of these dance objects, including visual research, rehearsals and improvisation as well as the decoding and recoding of data, writing algorithms, and training computer systems in pattern recognition. We wanted to see how the dancer's wearing of a material costume might become aligned with user interactions in a digital costume. But the project also provided an opportunity to test in practical terms the utilization of methodologies of scientific and arts research based on the premise that interdisciplinary academic research is a kinetic way of theorizing and making knowledge. Therefore, as a pilot project, iWeave was of necessity interdisciplinary and exploratory in its intentions.

A few words here about some theoretical assumptions. Within a framework of movement analysis, the project's motivations were based in a conception of materiality that is not static but mobile, and which traces history as it becomes virtualized. Our inspiration for such a framework came from Paul Carter's 2004 book *Material Thinking*. Inasmuch as the complexity of material histories represented by the costume were the "warp and weft of invention" intended to shape this project, and given that the original costumes had been created in "a craft environment", the project involved a weaving together of multiple strands of creative work, technological and social practices which, as Carter writes, means that: "its direction, like that of the shuttle,... [is] a product of the forming situation that impelled its motion" (Carter 2004: 15). Another contention is that the haecceity of stuff, or the materiality of things, often tests our assumptions about what we know, and how we think, through the effects of embodiment. Research methods, in these terms, become a means of following cultural objects that move dynamically between their conceptual force and a continuum determined by their social uses. This study of digital objects necessitates what sociologist Celia Lury calls "inventive methods", not least because so much of our social world consists of sociological and technological permutations that render data, material contents and history into patterns which "invite recognition of how methods participate" in the making of knowledge (2014: 4).

Working along with Janis Jefferies and Paul Stenner on topics that have a residual system of self-organization, respectively in relation to textiles and mathematics, Lury finds that it becomes relevant to consider repetition and multiplicity as pattern methodologies (Chapters 9 and 10 in Lury 2014: 125–35, 136–46). The study of pattern acknowledges that materials construct immanent, time-bound sets of relations in groupings that may have degrees of regularity, while subject also to variation; they exist as sets, and have properties that are both sensory and historically or socially constituted. And they can move from the material to the immaterial and back again at any point in a given trajectory. Because this project was concerned with two originating forms of aesthetic production – textiles and choreography (as a form of embodied aesthetic movement) – dependent upon both design and repetition, pattern has been a useful way to conceive of their complexity. As architectural theorist Ana Araujo explains, pattern is "intrinsically connective, rather than contained, bridging between architecture and fashion, fashion and mathematics, mathematics and textile design, textile design and biology, biology and architecture and so on" (2007: 16). Given the interdisciplinarity and critically divergent processes of thinking embodied in the virtual rendition of a moving dress, this particular investigation has been carried out by a dance historian and a computing engineer. This chapter thus aims to weave together, connecting rather than containing, these different perspectives on the patterns of costume as a digital object.

Starting points: costume objects and moving images

Fragile costumes in historic collections are valuable reference points for social and historical understanding. Think, for instance, of any historical museum in which uniforms, dresses or children's clothes are put on display, and then think of the memories or unique qualities that they evoke. In the NRCD, a collection of silk dresses and coloured scarves has been stored and preserved in the archive of a British modern dance genre called Natural Movement, created by the Manchester dance educator Madge Atkinson between 1910 and 1950 (Carter, in Carter and Fensham 2011: 22–40). These coloured and decorative costumes were worn repeatedly by her students, then stored until the following year before being deposited many years later at the NRCD. Now fixed, held, catalogued, preserved, photographed, put away – these objects produced a density of signification in trace and repetition that produce counter-narratives about the values related in modern dance history. While photographic records may document their use as costumes in various dances, the costumes also provide important evidence of relationships to designers, textile manufacturers and a wider social reform movement in dress and bodily conventions for women (Fensham, in Carter and Fensham 2011: 82–97). They are therefore simultaneously aesthetic objects and signs of technological change in society; thus they bear witness to forms of evidence that exceed the dress.

Material changes to a costume may include its recording, or documentation, and it makes sense that this recording represents a first order virtualization. For more than a century, visual technologies such as video, film and photography have provided researchers and audiences with a range of ways to experience the qualities of movement that can be captured in an image. In the prehistory of cinema studies, for instance, Eadweard Muybridge's "equidistant snapshots" of animals in motion attempt to transfer points in a sequence of movement, and thus to represent the dynamics of continuous flow. As important as the transitional states of linear movement, Muybridge also experimented with capturing the rotational possibilities of a tridimensional body, and some of his more interesting montages show the twisting of a cat after its tail, and the turning of a draped woman with her scarf. The haunting image of a woman dancing retains a trace of the novelty associated with modern dance as a new mode of embodiment. Not surprisingly, as Erin Brannigan notes in her study of dance on film, for philosophers such as Bergson and Deleuze, the fields of dance, ballet and mime epitomize "fields of innovation" during the early twentieth century because they released "values which were not posed, not measured, which related movements to any-instant-whatever" (2011: 28). In addition to the moving body, we would suggest that the costume innovations of modern dance – the loose tunic-like garments and the prosthetic scarf – were also important features of the cinematic effects that helped produce the impression of modern society's movement and flow.

With the exception of the Ballets Russes and some iconic ballet dresses, costumes have, however, largely been neglected as aspects of dance's history. It may seem obvious but they are unique objects that endure in the material culture of dance. Made and fashioned for performance from technologies and corporeal images available at any given historical period, they also possess singularity, and over time mold to a specific shape through wearing. For the dancer, they may play a transformative role, since a given movement style will be influenced by the garment. And in relation to the mise en scène, the costumes become potent, semi-autonomous, elements producing their own special effects, and affects such as desire or surprise, for the viewer.

Material objects

We do not have time here to elaborate on these many aspects of costume and moving images, although the preparation of a contemporary dancer when confronted with a small yellow silk dress meant radically adjusting her sense of the scale and energetic range of movements. Many of the Natural Movement costumes were made of hand-dyed silk, using natural dyes, with a crushed texture provided by tensile knotting during storage. The dress chosen for this project was that of a "Daffodil", a small yellow silk dress whose close-fitting tubular shape was draped by a matching two-metre-long orange-yellow scarf, and performed to the music of Debussy's *Danses Sacres et Profanes*. Its surface texture retained an irregular folding that clings to the body, while distorting the garment much like crinkled skin; this property also gave the costume an elasticity that was important to the dancer who moved inside it. These details of the costume's motion were significant to the work we wished to render as part of the bodily expression of the dance.

By way of comparison, in a large-scale collaboration between a costume museum and a digital laboratory, the Drexel Digital Museum has produced many "3D replicants" that simulate historic garments as an alternative method of displaying their collection. The digital costumes are complex ensembles, however; the avatar-like dresses move robotically, with their surface colours and textures somewhat flattened by the process of rendering. Aware of these limitations, the museum director Kathi Martin suggests that "to be fully appreciated, fashion needs to be experienced on a moving body" (2014: 3).[3] In iWeave, this notion of the moving body and its kinaesthetic qualities in relation to the surface rendering of the costume included delicate attention to tonal variation and the significance of folds in the material. For dance historians, the re-animation of the costume through

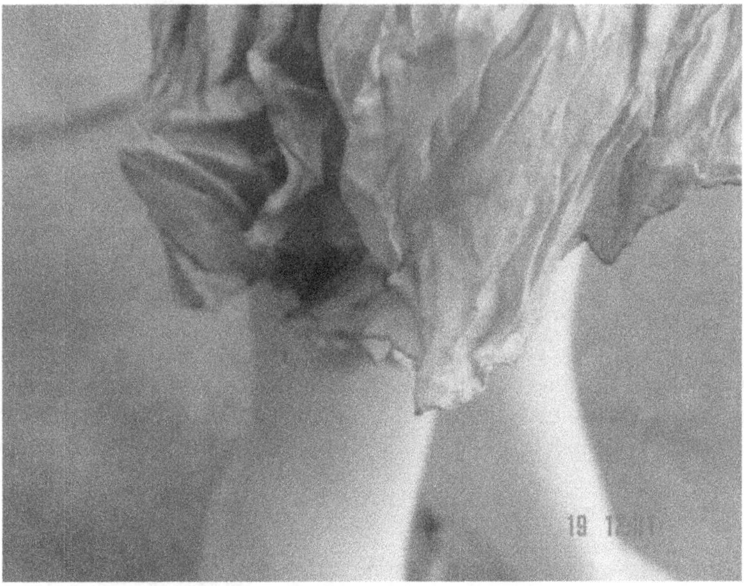

Figure 10.1 Detail of crinkled skirt for Daffodil dress, Natural Movement. Courtesy of © NRCD, University of Surrey

digital visualization ought also to communicate the dress's texture, colour, shape and weight as well as convey the movement quality of the dance to another user.

In archaeological, heritage and artistic practices, the 3D modelling of static objects is well advanced and movement animation is increasingly used to construct simulated environments, with interactive avatars and other participatory systems that reflect naturalized perceptions or world views.[4] There is also an ontology of the digital archive that leads choreographic objects to generate new perceptions of performance (Fensham 2013: 161). In this project, the three or four dimensions (including temporality) of the dancer's movements wearing the costume were, as we have indicated, not arbitrary but instead drew upon musical and movement repertoires that were derived from Natural Movement choreography. The aesthetic stylization refers not only to the costume therefore but also to the movement vocabulary and its specific kinaesthetic properties, which were critical components of the project, and notable since dancing does not have the regularity of pedestrian motion. Although this preliminary process of virtualization may have aimed to be a reconstruction, it was, as dance scholar Mark Franko notes, closer to radical reinvention: "Guided by the reinterpretation of a period's most characteristic aesthetic preoccupations [...] reinvention sacrifices the reproduction of a work to the replication of its most powerful intended effects"

(1989: 58). The rigours of dance research included the studied reproduction of poses from photographs in the archive, preparation of dance steps and phrases acquired in the contemporary pedagogy of Natural Movement dance styles, and the spatial analysis of its movement dynamics. Movement theorist Rudolf Laban provided a mode of analysis to understand the dancer's kinesphere and the effort-shape alignments of different poses. His theories also suggested which planes within the visualization of forms within the choreography could be used to structure appropriate movement sequences for the costume. The Natural Movement style particularly requires an elevated balance through the arches and heels of the feet, which enables lightness of movement and sudden turns and folds of the spine. Simultaneously, the dancer is animating an internal dynamics sometimes in response to imaginative suggestions of a natural element or archetype. At others, the suggestion is that the dancer herself becomes the vital force. Responding to these abstract ideas with both sensuous and regulated abandon, the intimate relationship between costume and embodiment begins to include expressive forms whose flow, shape and pattern were unfamiliar to the dancer. Both the historical research and the movement analysis are therefore components that interpret the tensile qualities of the costume for the camera; and key effects included its elasticity and thus the relative freedom it gave the dancer to shape the daffodil in motion.

Could 4D performance capture provide a means of preserving these dance artifacts digitally, while still providing a compelling experience for archive users to appreciate the dance? The V-Lab in Centre for Vision Speech and Signal Processing (CVSSP) at the University of Surrey has been active in developing performance capture technologies that can track human movement without body markers and from existing video filmed in "natural environments". When combined with high-resolution cameras that more expertly capture surface details, and a twelve-camera set-up that films in the round, the resulting images register significantly more than the joint movements of earlier systems of motion capture. The footage includes the surface shape dynamics of a moving body and a colour palette that registers the rhythms of clothing and hair. Since no markers need to be worn during filming, the dancer performs in her fragile costume within a blue-screen studio that consists of approximately six by four metres of usable space. The result is a 3D model that moves coherently over time (a so-called 4D capture), which when combined with video texture enables video-realistic replay of performance from any viewpoint the viewer specifies at the time of playback. Moreover, this 4D performance capture can be manipulated using other body software to animate the captured dancer in order to perform actions beyond those in the original capture.

Using the CVSSP Visual Media Lab (VML), six dance performances were captured, each one of which was approximately three minutes in length, featuring trained dancers performing rehearsed choreographies and wearing

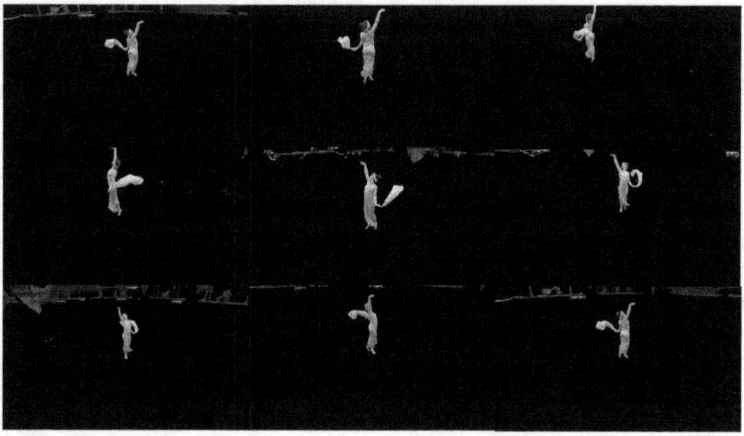

Figure 10.2 Eight camera frames of dancer in Daffodil dress with scarf. Courtesy of © CVSSP, University of Surrey

six different costumes.[5] From this video footage a sequence of 3D models was reconstructed preserving not only the movement of the performer but also surface detail on the clothing. One set of photo frames derived from eight of the cameras, show the turning of dancer Melina Scialom dressed in her yellow shift, waving her scarf.

This image is remarkably similar to the "instant whatevers" of Muybridge's dancing figure captured in stop-motion sequence, except that it exists as simultaneous multiples that can be folded into one another with the synthesizing power of the computer. This notion of simultaneity and multiplicity in the gestural virtuality of the image makes the two-dimensional or linear depiction of movement resemble the haeccity of material, and helps to replicate the moving body in a dress.

The rendering of these sequences into a unique 3D positioning of the dancer also required a grid-like frame that could set the dancer spinning across a simulated floor. In effect, a viewer could zoom in to see her costume more closely or turn their orientation to watch her dance from different angles. An unusual side-effect of changing the screen to grey scale, and watching the sequence in black and white, made the folds in the costume more visible, and heightened the affective response to the dancer's kinaesthetic rhythm. Another misalignment meant that gaps in the transition from one camera to another created the crinkly effects of an incomplete surface much as the coloured dye might have created on the original fabric. These coincidental gaps in the relay from body to data allowed the animated costume in this 4D footage to communicate traces of the costume's history in a ghostly way, with the movement qualities of the individual dancer repeating in an uncanny loop of Natural Movement choreography. Susan Kozel calls this "wavering", which as Sarah Whatley explains is an experience of the "autonomy of data in the projected image, [and which] paradoxically seems to capture something of the human condition, reminding us of the presence, not the absence, of the body" (Whatley 2012: 275). This kind of gestural stuttering can intensify the concentration and effect of the digital dancer, even if there is a loss of subjectivity. Perhaps less pleasing was the final rendering of the virtual dress, which remained an object at a distance, whose proximal surfaces could not reproduce the unique folds of the silk fabric. However, through the motion capture process, the costume had become re-visualized, and virtualized, as an embodied object.

Costume and choreography are, as we have suggested, intimately linked; one is designed with the other in mind, and in the last phase of iWeave, having produced a likeness in digital form of the dancing dress, we wanted to enable the "try-on" of the costume to function within the context of the dance for which it was intended. It is this phase of the project that proved most complex and raised the most issues for both sets of researchers. The aim was to enable the user to move with the costume by standing before a camera, and making a sequence of body poses that were echoed by the 3D captured performer (and viewed from any angle). A source of contention between the different researchers was the undesirability of allowing the user to simply "puppet" the costume, that is, to move the performer as, Franko suggests, "an outside within an outside" of precise and immediate control (1995: 153). Rather, we hoped that the user would generate through kinaesthetic empathy a responsiveness to the qualitative properties of the dancing avatar in her daffodil dress. Dee Reynolds describes the inner movement of a "dance's body" that might be experienced at the interface between the screen dancer and spectator, and one of the novel investigations of this project was to try to identify a means of translating this kinaesthetic experience into a defined system of pattern recognition in the computer (Reynolds and Reason 2012: 133).

To extend the interactive possibilities, a unique relationship was established between the surface rendering and the construction of a mesh skeleton, post-filming. This interior-exterior mesh dynamic became the device for structuring subsequent interactions based upon surface coordinates in the dancing dress. As a network of intersecting polygons, the mesh had a tactility that was unusual, based on shifting forces across a surface, not unlike Laban's conception of Platonic solids and their surfaces within the kinesphere, and again not limited to joint movements. The mesh dancing held open the possibility of the costume as a relational object, connecting the spaces of an exterior surface with the interiority of the organism.

To facilitate interaction with the mesh figure, concatenative synthesis was used. The essence of this idea is to identify points in the motion capture sequences that can be seamlessly stitched together, and identify an appropriate set of motion segments to concatenate the desired motion. Often the

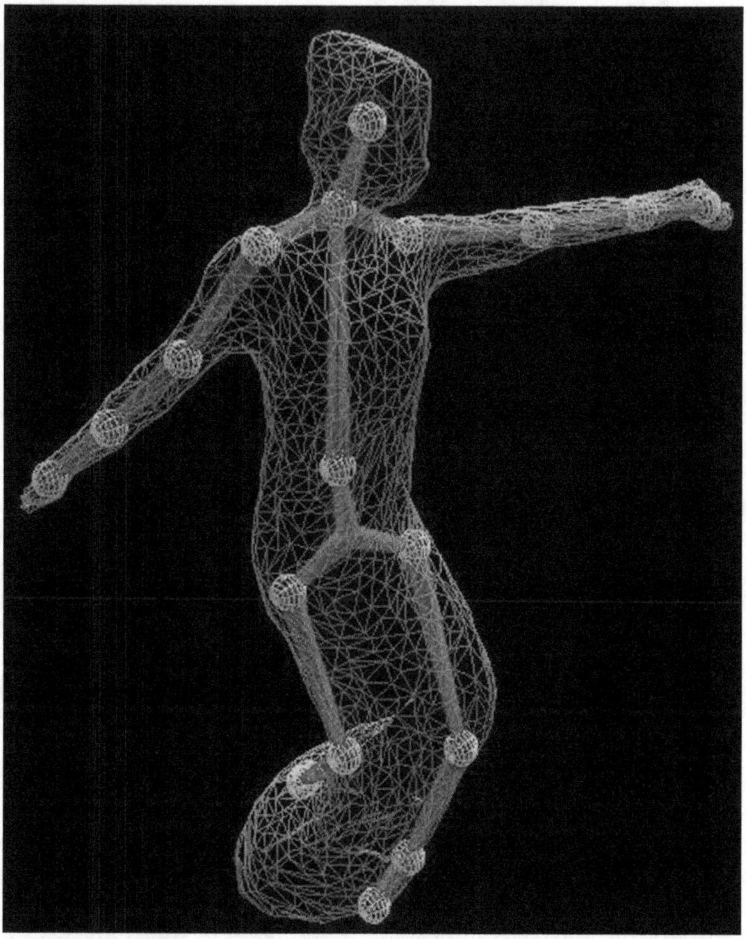

Figure 10.3 Mesh still from video of dance skeleton. Courtesy of © CVSSP, University of Surrey

motion segments are represented as a graph data structure, and moves made between nodes in the graph are used to generate video, a so-called "motion graph" (Kovar, Gleicher and Pighin 2002). This patterning produced a series of choreographic sequences in which the moving body could effectively produce an arc between set points defined by particular gestural combinations of the skeleton that could be combined in different ways depending upon an inputted starting pose.

Having produced a range of unique calculations, the use of the proprietary software package produced by Microsoft Kinect, with its fixed

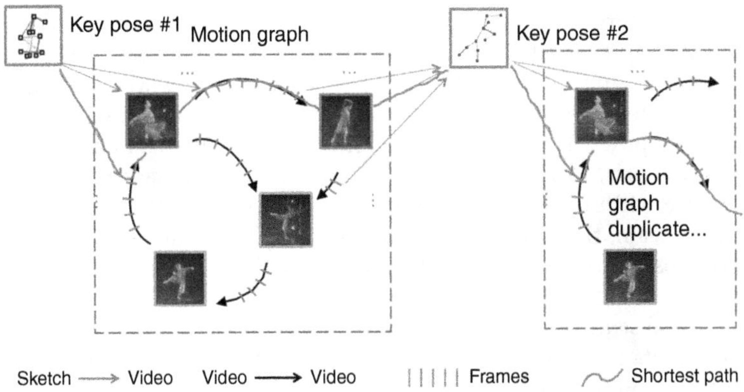

Figure 10.4 Motion graph with legend. Courtesy of © CVSSP, University of Surrey

camera, introduced another set of limitations. The Kinect has been hailed by computer designers and software engineers because it introduces the notion of "natural user interfaces" (NUI), particularly within the space of interactive gaming. This software technology is designed to enable users to have greater sensory feedback from their human-machine interactions, and thus can be said to increase emotional engagement by engaging players, or users, at an embodied kinaesthetic level (Shinkle 2008: 912). Its rapid uptake by games designers and its utilization in settings such as hospitals implies, according to Kenton O'Hara and his team at Cambridge Microsoft Research, that NUI design draws upon "gestures in everyday life, by identifying the physical movements used to manipulate and understand the world, [in order that] new interaction paradigms can be developed that will allow people to act and communicate in ways they are naturally disposed to" (O'Hara, Harper, Mentis, Sellen and Taylor 2013: 1). With its 3D camera, Microsoft Kinect technology is sophisticated enough to recognize faces, to identify bodily gestures and poses, but not yet to follow a body moving quickly or turning in a full rotation. Where it loses contact with the body of its operator, the system either freezes or makes up the transitional gaps with grey scale blurred markings. The "natural"-ness of the interaction, as O'Hara et al. point out, is also problematically limited to knowledge of objective representations rather than an understanding of how social users interact with the technology, and thus the NUI requires much more detailed analysis of social context and behaviours (2013: 11).

In our project, the Kinect was used to estimate the pose of the external user by linking their gestures (a set of joint angles) back to the skeleton inserted in the captured mesh. This was achieved by manually specifying a skeleton for a single video frame, and tracking the skeleton over time using a non-linear mesh tracking algorithm (Budd, Huang, Klaudiny and Hilton 2013). The user in the laboratory, the computer programmer, then planned to generate a sequence of poses that could retrospectively match points in the motion graph in real-time. Optimized route-planning algorithms were designed to determine the shortest path connecting the current video frame with the desired user poses. The resulting video fragments are then stitched together in real-time and, in response to these target poses, the dancer "performs" a sequence within her choreography of the dress. From the perspective of the dance team, the limitations to the user were the game-like instructions of body poses used as triggers for the dance to "flow" through movements within the original captured sequence. While functional in launching the dance, these sequences lacked the kinaesthetic empathy that we had seen in the live dance. Instead of the user being compelled to "wear" the costume by responding to the dance on screen (although this would be possible for someone improvising in a studio), the daffodil costume avatar became a cipher for the user's limited choice of movements.

To facilitate the user interaction with the costume, this software was incorporated into a portable demo (on a trolley TV stand) comprising a 50-inch flat screen plasma TV, a high-end graphics PC and a Microsoft Kinect installation.[6] Following user feedback, further refinements were then made to the system to improve responsiveness. A key issue was balancing a feeling of immediacy in the interaction against the gaming of the object, and this became an intensive point of difference in understanding between the computing engineers and the dance researchers. Prior to iWeave, the scientists had tested the 4D technology on sequences only a few tenths of a second in duration, for offline playback. However, iWeave required sequences of several minutes, with the ability to dynamically seek and skip between sections of playback with little or no notice due to the unpredictability of the user's pose in front of the Kinect sensor. This motivated some new algorithmic development to produce efficient data-caching strategies, and data representations, for real-time playback. The choices of the virtual dancing costume, not unlike the NUI discussion, were limited perhaps by the ways in which programing depends upon game structures, or binary responses to interject into the flow of a discursive structure. For example, in the sports training equipment TacTowers, a series of coloured lights prompts a psychomotor response from a pair of "players" (Fogtman, in Reynolds and Reason 2012: 310). Sensory interaction in more aesthetic computing technologies is not as simple as designing the prototype for interaction; not least because "naturalness is not bound up in a representation of our gesture and body movement". What is significant, however, is "how technologies reconfigure our relationship with the material and social world" (O'Hara et al. 2013: 9). Visualized in multiple dimensions, the digital costume object remained connected to a material and social world through the sensory qualities of the mesh and other surfaces at play in the rendering process.

However, the "touchless" interaction design of the proprietary software and the logics of the gaming structure lacked the flexibility of a "naturalized" movement.

In iWeave, this visual animation of the costume remained a rich source of movement research, and the specific materiality of the costumes presented a challenge to the computing engineers; for instance, the cameras captured performances containing significantly more flowing cloth, and thus dynamic patterning, than had previously been attempted with the technology. Not only was the dance vocabulary more complex than pedestrian movement, the clothing was also not form-fitting; rather, it was an aesthetic prosthesis capable of producing the effects of flow, air, or lightness, as it had been for the Natural Movement dancers. For V-Lab, this new dance data provided complex sets of video material with which the engineers could continue to improve temporal surface alignment in ways that would allow skeletal motion to be tracked and analysed in a more fluid way. Their research on these questions continues with the study of sports people and real-time hospital settings; and the question of fluidity remains one goal of improving technological performance.

Conclusion

A key outcome for the NRCD is the archival benefit of having access to high definition 3D capture of an historic costume presented within a reconstructed typical choreography, with the potential to broaden the impact of the physical costume archive, which is accessible only for highly specialized purposes. The dress can be seen dancing again, and users can play with this virtual costume through the limited pattern of movement choreographed in this research. They could also potentially manipulate it still further to perform other dances within a spectrum of recorded gestures. The virtual costume has therefore digitized a heritage value by animating the past for the present.

If, however, the criterion of this research had been pure reconstruction, or the likeness of "recognition" in a natural user interface, then the project might have failed. In the range of costumes digitized for the Drexel Digital Museum, the complete rendition of the actual into the virtual was also only partially successful – some costume details, such as pleats, were too hard to capture, and remain static and lacking in the weight of movement. If, however, as with iWeave, reproducibility by verisimilitude is not the goal, then watching the playful attributes of the daffodil dancer floating her arms and pattering her feet acknowledges that movement from an "outside to an outside" can incorporate the experience of multiple dimensions. The costume avatar became more or less "uncanny", a sensory doppelganger of the dress, whose trace effects are neither those of an original nor a future enactment, merely a reinvention. Her gestures did evoke a kinaesthetic empathy, particularly in her shadowy self as she invited the viewer to watch her again and again, reaching through empty space to feel the quality of a flow made up of tiny triangulations of energy. Her rhythm lives in a virtualization of the mesh, as well as its tactile surface – its non-linear composition of the costume's body. The transmission of this accessible movement content to a new user, through kinaesthetic empathy, was, however, stilted by reductive notions of how the "natural" or the "touchless" and their correlation to a user interface should be designed around a gaming paradigm. From the perspective of the dance scholar, we imagined that the resulting transference of the virtual costume would be closer to the cut-out of an historical costume in an exhibition or fun-fair display. In these silhouettes you see your own head and arms "clothed" in a virtual outfit and this distorted composition prompts you to move in unexpected ways. The proprietary software of the Kinect device may invite arm movement from an inexperienced or structured pattern of movement. However, the knowing corporeality of the dancer would add extra vitality.

Recognizing that expert knowledge and practised technologies are required in the field of dance history and archive management as much as they are in the field of high-end computing was an important moment in the research process. As it turns out, iWeave confronted multiple problems within the digitization of objects, computer vision, motion capture and the interactivity of screens. In the less than six months of this project's life cycle, we aimed to close some of the gaps between those areas of heritage understood as collection items, heritage as performance, and corporeal history, by experimenting with the interactive display of costumes as embodied virtual objects. The costume as materiality therefore became the stuff of invention, opening itself to an interweaving study of the daffodil dress in motion.

Notes

1. *Bridging the Gaps* was based at the University of Surrey, and for this component of that EPSRC scheme, Fensham was principal investigator and Collomosse co-investigator in a project called *Visualizing Motion in 3+D*, funded £65,000 in 2012. The research assistants were Nicolás Salazar Sutil and Qizhi Yu and it was also supported by library and archival staff in the National Resource Centre for Dance, Helen Roberts and Sharon Maxwell.
2. See: www.dance-archives.ac.uk
3. The Drexel Museum collaborated with the Digital Clothing Suite at Seoul National University to simulate historic fashion and visualize the physical parameters of selected costume materials. See: http://digimuse.cis.drexel.edu/home.html
4. Digital cultural heritage is an extensive field of contemporary study and research, e.g. *AVIE* is a 360-degree stereoscopic immersive interactive visualization system developed in association with Professor Sarah Kenderdine and installed at the ZKM Centre for Art and Media, Germany; City University of Hong Kong; Rensselaer Polytechnic University, New York; and Museum Victoria, Melbourne.

5. The dance researchers were Melina Scialom and Bernardette Limon; the research assistant Qizhi Yu contributed the technical expertise to develop the rendering of textured graphics and align them with the Kinect software.
6. The iWeave system was demoed to visiting companies and the EPSRC as a model for successful cross-disciplinary collaboration, and during the Surrey Science Circus (September–October 2012) iWeave was exhibited in the Lewis Elton gallery for three weeks, where numerous visitors tried out the iWeave demonstrator.

11
Gestural Materialities and the Worn *Dispositif*

Johannes Birringer

Moving with a body/moving through data

How deeply can you move through or with data? Or, as Nicolás Salazar Sutil asked in an early outline for this book on *digital movement*, how deeply can you live within data? While it would be presumptuous, though obviously not without political implications in an era of paranoid security, to make any unscrupulous claims about how living is correlated to digital data and those sensing technologies believed to offer potentials of motion capture/analysis, it is true that the performing arts – music and dance in particular – have been attracted to sensing technologies ever since gestural expression was linked to the notion of a "controller" or actuator enabling the playing of an instrument that can generate data in real-time – and in extension to performing *as* an instrument in real-time. If it were true, however, that performance might be intrinsically connected to forms of controlling movement and bodily/gestural activity – while also measuring such activity to convert its data into useful information for various commercial, scientific or artistic purposes – then some fundamental questions could be raised regarding the need or desire for capture (or control).

Initially I had planned to write about wearable technologies, but I feel an urgency to reflect on control systems – the *dispositif* or arrangement of such control in contemporary and earlier modern performance environments – and the affective experience or digital affectivity of such "control" if the latter were re-articulated by the performer.[1] From the outset, then, I wondered whether performers worried about controllers and control systems. Why do we hear so little about this matter? A question might be posed to a dancer, namely whether she felt her gestures (as data) transmuted something in the audiovisual/kinematic/choreosonic environment, or whether she was muted? Whether the sensors *worn on the body* changed, enhanced or reduced her movement techniques and sensations, and enabled new expression and physical responsiveness?

My own experience in the theatre tells me that performers do not live deeply within data at all. Many of those who have worked with sensors (myself included) complain that calibrational difficulties make the work exasperating and the aesthetic gain minimal, although the sense of limitation may be different when working, say, with Microsoft's Kinect camera system or other newly developed systems like Reactor for Awareness in Motion (RAM) (prototyped by Yoko Ando at the Yamaguchi Centre for Arts and Media [YCAM], Japan's leading experimental media arts centre), which promise better reciprocity and real-time feedback. The limitations we take for granted. The (re)compositional and interweaving possibilities in the singular, real-time event of the dance still arouse curiosity, as does the elusive feedback and the matter of enjoyment disguised by interactivity (as Mladen Dolar noted, we delegate our *jouissance* to the machine [2008: 136]). I have lived with machines and software in dance now for twenty years, looking for room to manoeuvre, for a way out of the tools' modalities to constrain, regulate and command.

In a discussion forum on dance-as-philosophy, André Lepecki ventured to call choreography itself a disciplined "apparatus of capture" troubled by movement's imperceptibility (escaping the partitioning of the sensible, those forms of sensing and sense-making that are recognizable):

> To see choreography as an apparatus – moreover, to see it as an apparatus that captures dance only to distribute its significations and mobilizations, its gestures and affects, within fields of light and fields of words that are strictly codified – is to delimit those hegemonic modes of aesthetically perceiving and theoretically accounting for dance's evolutions in time. (Lepecki 2007: 120)

This argument follows Deleuze in assuming that such an apparatus always foregrounds perception as tied to modes of power, which distribute and assign to things – matter – either visibility or invisibility, significance or insignificance. When reading this, I asked myself how it might relate to gestural interfaces for digital media, especially in cases where the transmutation of data, from/within the gesture, is imperceptible and therefore renders the gesture potentially insignificant or intransitive.

On the surface, questions of control were, for the most part, negligible in the history of modern and contemporary dance. Beyond the traditions of instrumental music, the idea that the dancer's body could be explicitly understood as an instrument was perhaps first introduced into the discourse of modern dance by Mary Wigman and other pioneers of German Ausdruckstanz, who saw the *Tänzermensch* as someone who chooses the body and its movements as one's instrument of knowledge, expression and thought, as one's own vision of corporeality and sensorial/corporeal technics. Watching Valeska Gert's gestural eccentricities and misalignments, performed in her cabaret acts, reveals a significant displacement of the control protocols that resided in the 19th century codified balletic traditions. Anita Berber, performing her dances of ecstasy mostly nude and under the influence of hallucinogens, cared very little about the capture of choreography. Codification of technique and notation systems, from Feuillet to Stepanov, gained an influential twist, during the interwar years, through Laban's choreutics and expanded analytics of steps (*pas*), arm movements (*ports de bras*), spatial directions and organizing principles – here modern dance (and Wigman had just opened her own school) strove towards its own professional legitimization, gradually wanting to elaborate on the effort, weight, flow, and able-to-be-sensed vectors of movement/gestural body language. Laban remained indebted to the idea of a harmonic system even as he expanded the "scales" and "spatial tone rows", if we bear in mind the musical analogies that Laban, a close contemporary of Schoenberg, worked with at the time of his choreological research (cf. Louppe 1994: 117–18).

The correlation between dancing body and musical instrument remains problematic as the creators of early modern dance in the West (absolute dance, as they called it) tended to divest movement from any already established codes (ballet technique, musical or verbal code), or the instrumentalizing and rationalizing biomechanics of the kind Vsevolod Meyerhold devised at his Petersburg laboratory, to have a clearing for a new poetics of invention. Wigman, who attended Émile Jaques-Dalcroze's Hellerau workshops on eurythmics and vitalities of inner musicality, performed her early solos in silence, and many of the great modern choreographers from Duncan to Hawkins, Graham and Cunningham (and the later Judson Church avant-garde) avoided following any score given in advance. Laurence Louppe, in her eloquent *Poetics of Contemporary Dance,* calls this the "extreme nearness" of modern dance to itself: "The dancer has nothing exterior or supplementary to the matter/material of the self with which to build a signifying universe, an intelligible imaginary" (2010: 23). Louppe uses the term "instrument" in the Wigmanian sense, suggesting that "contemporary techniques, no matter how scientific, no matter how long it takes to acquire them, are before anything else the instruments of a knowledge leading the dancer to this singularity" (23).

In her chapter on "Choreographic Works: Forms of Appearance" (202–29) Louppe does eventually address the "supplementary" objects conjoined with choreographic practice – and in postrevolutionary Russia the constructivist artists Varvara Stepanova or Liubov Popova, working alongside Meyerhold, had been eager experimenters designing technical apparatuses and devices, as did Oskar Schlemmer with his sculptural costumes, appendages and figurines for the Bauhaus dances. It is these material phenomena and interfaces that interest me here, since tools and appendages are amplificatory choreographic and scenographic elements that may become fundamental for the plasticity of articulation of movement and movement relations to space. When I first visited the Festspielhaus Hellerau in the early 1990s, after it had been reopened to the public following the departure of the Soviet army (the site had been occupied as a military facility since the end of World War II), memories of Adolphe Appia's path-breaking stage and lighting designs for dance were evoked in the vast, empty architecture of a hall that once housed Dalcroze and Appia's staging of Gluck's *Orpheus and Eurydice* in 1912, in a bare hall illuminated by elaborate special lighting (indirect, behind translucent linen) and equipped with modular elements of staircases, steps that provided upward, downward and sideward motion possibilities for the dancers and chorus.[2]

The set, and the lighting in this case, provide the multidimensional arrangement for the movement choreography which Appia understood to be rhythmically embodied in a light-producing, radiating space full of pulsating energies that were partly in-formed by his animated modular (moveable) architecture of the staircases for the choral passages. Illuminating convergences could thus be found between dancer Loïe Fuller and her use of light and cloth, Appia's designs for rhythmic movement, the constructivist stage machine designers around Meyerhold, Tretyakov and Eisenstein, and the Bauhaus designers and visual artists (Schlemmer, Kandinsky, Moholy-Nagy) in these early decades of the 20th century. Appia's spectacular lighting

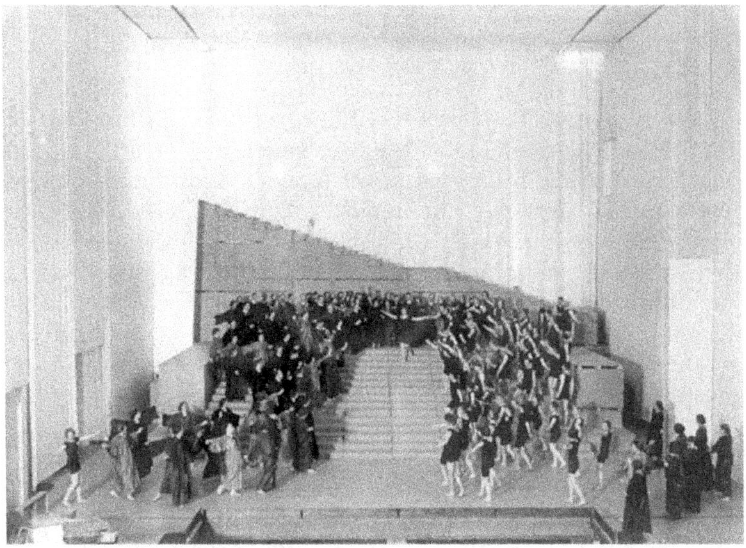

Figure 11.1 Archival photo of Dalcroze and Appia's 1912 *Orpheus and Eurydice* production at Hellerau Festspielhaus showing movement chorus distributed like a swarm network (envisioned by Biloria) in multidimensional space. Photo reprinted with permission © Richard Beacham

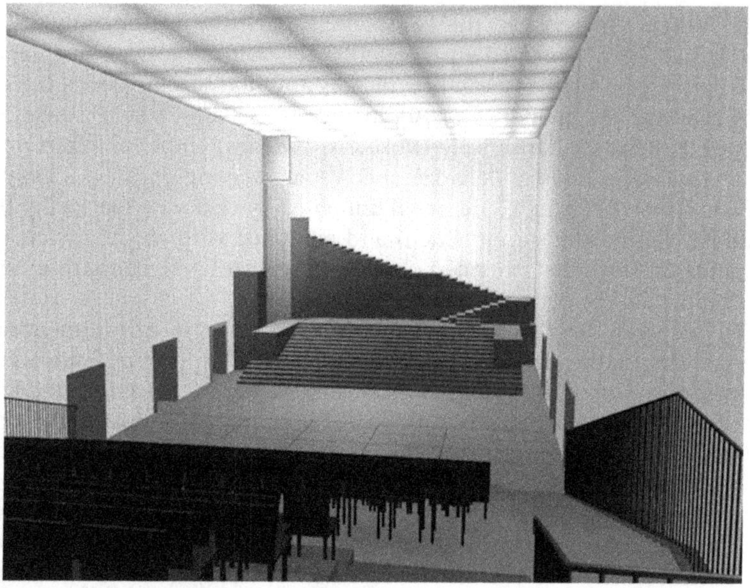

Figure 11.2 Richard Beacham's computer 3D model/reconstruction of Appia's mise en scène of Rhythmic Space for the 1912 *Orpheus* production at Hellerau Festspielhaus, created at the King's College Visualization Lab, based on the original design by Appia using modular units. Photo reprinted with permission © Richard Beacham

and spatial designs for Dalcroze's eurythmic performance of *Orpheus* at the Hellerau Festspielhaus also anticipate later kinetic and op art which made moving light a medium for "visions of motion" and virtual volumes, as Moholy-Nagy had suggested with his moving light/film sculptures that seem inspired by Étienne-Jules Marey's earlier graph-writing machines.[3] Dalcroze called the union of music and movement, of internal experience and external expression, *plastique animé* – a remarkable notion which projects the tactile, sensorial understanding of enveloping spatiality or animate design we now find in the architectural work of Greg Lynn, Diller + Scofidio, Peter Eisenman, Lars Spuybroek, and the Hyperbody Group for Digitally Driven Architecture.[4]

Nimish Biloria, who directs Hyperbody and with whom the DAP-Lab collaborates on the EU-funded *METABODY* project, describes his interest in proactive environments in a way that resonates with Appia's rhythmic spaces, yet his work investigates spatial affordances through *computational algorithms* which model ambiguous topologies of movement, mutations of the densities of space that can generate different mental and physical associations and, thus, behaviours.[5] The algorithms precede (anticipate) the physical 3D prototypes and interpenetrations of bodies: computer processing simulates

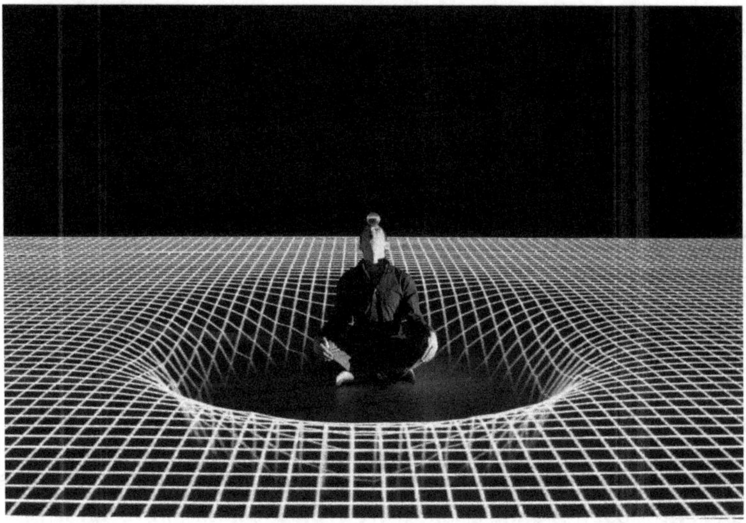

Figure 11.3 Cie. Adrien M/Clair B – *Cinématique*, Digital Media Performance, Bains Numérique Festival in Enghien-les-Bains, 2009. Photo courtesy of © Adrien Mondot

how vectorial forces can deform grid lines of a spatial field through temporal unfoldings and dilations.

One of the *METABODY* leitmotifs of investigations is this idea of dis-alignment – dis-aligning perception from the centrality of linear vision or any dominant perceptual paradigm, and generating boundary perceptions at the frontier of the intelligible – which would relate to the deformations of a stable (geometric) grid or architectural framing. Eisenman's *Diagram Diaries* hint at such virtual permutations modelled by the computer. Seeing virtual modelling projected around two onstage performers in *Cinématique* (Cie. Adrien M/Claire B, 2009) is partly confusing, however, since the performers are not modelling behaviour (or are they?). They act as if they were not in a projected and moveable space but a real space – literally stumbling into the digital warps and cracks opening up. Yet the *faux* movement is perceived to be an effect of the "shifting ground", and the bodies become parapractic.

If we now look at an archival photo of the 1912 *Orpheus* performance in Appia's rhythmicized space (Figure 11.1), we can imagine how the spatial organization affected the sensory economy of the body, how the dancers, so to speak, concretely had to navigate or adapt to the depth and verticality of the staircases, responding to spatial densities (height, width, breadth, etc.) and wearing them like an envelope, moving with them. I will return to this idea of a tactile-kinetic envelope near the end of the essay when discussing Rauschenberg and Klüver's event *9 Evenings: Theatre and Engineering* (1966) and Chunky Move's intra-active dance pieces *Glow* (2006) and *Mortal Engine* (2008).

Thinking of wearable space or the mutual embedding of body and space through *sensorial dilation*, the question of alignment or dis-alignment then becomes critical, since ambiguous and difficult spatial environments tend to destabilize proprioception and intra-actional behaviour just as immersive multimedia environments can become nervous systems where anything can happen and where there are no privileged points of intensity or clearly defined places (centre, foreground, etc.) for directional impulses.[6] Ambiguity, at the same time, is not what the quantitative logic of predefined parameters for motion tracking is set up for: the tracking system inevitably implies a readable syntax of body movement, and it needs to calibrate positions and postures. The term "nervous system" for a sensitive, interactive *dispositif* is owed to David Rokeby, who in the 1980s was one of the first spatial sound composers working with a computational environment to explore movement through a space using sensors. Rokeby's *Very Nervous System* (VNS), first installed in 1983, was quite unobtrusive as an apparatus; infra-red sensors or video cameras observe the room into which visitors enter, and the system is inactive until recipients move around the system's radius of action. The sensing system then records the motions, analyses them digitally and responds by emitting (via small loudspeakers) sequences of sound, timbres of musical instruments or everyday noises. Recipients explore the room (and thus the system) as they begin to realize their movements generate particular sounds: they can play the system (or the system plays them).[7] Rokeby kept modifying VNS over the years, later also adding video projections (e.g. *Reflexions*, 1983; *Silicon Remembers Carbon*, 1993–5) and verbal languages (speech-synthesized voices describing objects that visitors were invited to place in front of the video eye in *Giver of Names*, 1991). The system is to a large extent paradigmatic of most responsive/interactive installations and their functional parameterization (how motion detection translates such data into sound synthesis or graphic image behaviour, whether projected onto screens, walls, floors or in an immersive all-around manner). The camera vision systems must necessarily register motion, position, amount of movement and action behaviour within the parameters; most other commonly used software systems (Max/Msp/Jitter, EyeCon, Isadora, EyesWeb, MotionComposer, Gesture Follower, etc.) and wearable sensor systems act in the same way, "capturing" movement and gestures, then translating the input data into real-time (or delayed) media output – sound, images, 3D projections, animated avatars, text – and the various recombinatory loops and generative processes during which data perform their sonic and graphic emanations.

Working with controllers and conducting research in motion analysis based on computational data is a fairly recent phenomenon in the dance and technology community (since the early 1990s) that owes its existence to particular circumstances and cross-disciplinary developments (scientific motion studies and investigations of biomechanical or psychomotor laws characterizing human movement; kinematics and human factors research; neuroscience and cognition studies; software development in Human-Computer Interaction) and an overlap between the music technology and dance experimenters. Researchers working in these fields had become concerned with what Antonio Camurri and his team at the Genoa InfoMus Lab, for example, call "extraction" of expressive, non-verbal features of gestural and full-body communication (Camurri and Volpe 2004: 460). More concretely, music and dance experimenters came to share a scientific interest in the integration of extraction of movement cues into multimodal interfaces for performers or users, enhancing physical participation and expressive articulation in the human-machine interaction processes, controlling dynamic and interactional data displays, as well as understanding, evaluating and classifying information from gesture tracking and recognition techniques.[8] In the following pages, I want to reflect on the issue of capture and what significance – in the context of motion responsive technologies – a capturing of motion holds for performance artists and designers who must have been aware of a longer history of practices that engage media, animation, sound and image diffusion integrally in live performance.

Capture/control *dispositifs*

Aside from Taylorist time-and-motion analysis and studies on worker movement and efficiency (see Salazar Sutil, this volume), capturing or recording movement in the modern era begins with early chronophotography and the initial developments of the cinema (Lumière Brothers, Edison, Marey, Muybridge). Seeking to transpose bodily movement to the mobility of projection (images passing through a projector), Muybridge's cinematic stills form a special case of serial recomposition still highly influential for today's students of animation. Cameras were pointed at bodies, landscapes, urban space, trains, factories, and actors. Lest it be forgotten, cameras in the hands of Dziga Vertov or Alexander Medvedkin were factographic tools used not to capture an accurate image of the world but to transform it, activating the spectator to provoke revolutionary change. If we now consider the relations between corporeal and incorporeal materialism, with the historical role of the agency of the camera in mind, we could begin an archaeology of technology (and technological *dispositifs*) in choreographies of movement by looking at the spatial organization of bodies, and the utility of bodies (as it has now surfaced quite clearly in interactive performance that limits and controls action while pretending that the interactor is a controller). It is a remarkable historical detail to discover that Medvedkin's *kino-poezd* (ciné-train), consisting of three carriages transformed into film-production studios and projection room, travelled all across the Soviet Union to agitate and document labour, generating silent films almost in real-time – "film today and show it tomorrow" (Stark 2012: 129) – and thus participating in a productivist political project as well as an operational inclusion of the recipients just as Eisenstein had imagined with his "montage of attractions" focused on the audience as "material" (Raunig 2010: 53).

Gerald Raunig points out that Meyerhold and Eisenstein performed for workers (sometimes inside the factories) to show and develop particular competences that gestured at a precise scientific investigation of the audience to aid the socialist construction, and we can understand this to have been part of the logic of collective production/reception. The Lumière Brothers, a few years earlier (in 1895), had filmed "Workers Leaving the Factory" and demonstrated the precise choreography (organization) of bodies in front of the camera – framing and editing movement in a way that we can now read (following Kracauer) as an ideological (cinematic) production of worker conditioning (Taylorism predating today's ubiquitous surveillance systems) which in the subsequent silent film era became a dominant mode of representation.[9] In early experimental film and slapstick comedies (e.g. Buster Keaton's), the body is manipulated through artificial techniques (i.e. slow or stop motion, different speeds), which helped dis-play false movement/gesture and thus, we could argue, also enabled the performance of a contradictory effect (on the recipient). Movement production and performance became a funny slapstick contortion rather than sober functioning, or proper biomechanical automation (as in an assembly line).

Is this historical cinematic *dispositif* not a clear precursor of today's interactive operationality, and the contortionist aesthetics of catching sounds or jumping to perform your digital double (closed-circuit playing with your image)? The *dispositif* is based on rules and perceptional alignments that link instrumental constellations with various material stagings of processual activation. These alignments also link the framing (spatio-temporal limitations) and parameters with the cause-and-effect technical elements of the user interface (microphone, sensors, cameras, keyboard, mouse, joystick, buttons, sliders, projectors, screens, loudspeakers, etc.), which in turn control input and output media. Who would have thought that the rules of the game are so clear, that recipients intuitively want to "master" the functionality of the system and offer their labour?

Let me clarify both the notion of the *dispositif* and the question of who controls what, or indeed what controls whom. Unlike theatre theory, which after Brecht speaks of the "apparatus" when it looks critically at the operations of the stage/illusions, film theory since the 1970s has preferred the term *dispositif*, the French word meaning "disposition" or "arrangement" (Baudry 1978: 23). Philosophers of media and social/political theory such a Foucault, Deleuze, Guattari, Lyotard, Flusser, and Zielinski became hugely interested in the notion of the *dispositif* in the 1970s and 1980s, utilizing it as a conceptual category for examining environments (material, technological, medial) of regulating, strategic frameworks that are configured in certain ways making it possible for certain types of phenomena to occur. Foucault tended to emphasize the regulatory and panoptic formations that produce power, knowledge, and subjectivity. Deleuze and Guattari became more interested in the drifting and disjuncture between heterogeneous elements in a multilinear collective assemblage or *dispositif.* In other words, these authors acknowledge that arrangements are precarious and cannot always control outcomes; the lines that compose a multilinear ensemble – referred to as *agencements collectifs d'énonciation* or "collective enunciations" – can change direction or become unbalanced and forked. When human and technological processes are intimately intertwined or cannot be easily differentiated, the component materials, forces, energies, rules and conventions, and lines of communication are not stable, their contours are not fixed but subject to a series of variables (Deleuze 1992b: 159). Although factories were closing and new media art galleries opening up (Steyerl 2012: 99) it took a while for interactive art to catch up with its aesthetic self-analysis, distracting from the control apparatus and emphasizing a postvisual affective phenomenology of sensory deterritorialization (while Deleuze's academic stock seemed always on the rise).

In the music technology field, whenever "mapping" is invoked it tends to be concerned with precise controller functions. Kia Ng, in an article for *Organised Sound,* prototypically describes wanting to facilitate an intuitive and non-intrusive interactive multimedia performance interface that offers the users or performers *real-time control* of multimedia events using their physical movements. The mapping framework he proposes is intended to be a dynamic real-time performance tool, able to sense and track activities and changes:

> Physical movement, gesture and expression play an important role in stage performances, irrespective of the mode of human communications: verbal or non-verbal (audio or non-audio), or the language used. With the advances in electronic and computing technology, there has been increasing interest in new musical instrument design... to augment traditional instruments... with new capabilities, for example triggering digital sound and visual output. (Ng 2002: 191)

In music contexts, such controllers were often keyboard interfaces (to pitch), with each key assigned to a particular pitch (on the MIDI scale of 1 to 127), or any other potential modulation or filtering function on the scale (accented notes, sound effects, distortions, any data – audio or visual – prepared in the programming environment). This one-to-one mapping, as well as one-to-many mapping and many-to-one mapping designs, and the "triggering" of data responses that in music and dance performances/installations produce sonic effects and video projection outputs, are so familiar that we do not need to dwell on them. However, it became clear over the years that the action-reaction model required increasingly complex programming to allow for more interesting, complex behaviour (beyond slapstick) either on the part of the performer who had become the controller (e.g. wearing a sensor interface or being motion-tracked via camera sensing system), or on the part of the computational system (including servers, networks, GPS, etc.) able to provide dynamic, generative, and even autonomous agencies or audiovisual formations. One assumption of contemporary hybrid interactive architectures is their complex multimodal operation – involving "tangible interfaces" that afford various sensory modes (auditory, visual, tactile, olfactory, gustatory).

Tangible matter always mattered in performance, not just in relation to instruments and hardware but also through material processes and techniques of enactment. The sculptural Bauhaus stage workshop experiments or the Russian constructivist performances are important examples of this kind of gestural materiality we studied in the DAP-Lab.[10] The sculptural or biomechanical gestures projected a particular *gestalt* – image shapes, properties and assemblages – hinting at new forms of work-design/choreographic objects. These gestures could touch upon the early 20th century industrial and aesthetic abstraction of human motion, as well as concrete abstraction of object motion. The kinetic *dispositif* for *Lichtspiel* suggests that László Moholy-Nagy's emphasis on coalescence of sensory communication and sculptural gesture was complementary to Oskar Schlemmer's. In their experiments with motion, light, space, and *Faktur* (Moholy-Nagy's term for surface aspects and the perceptible effects of the performative treatment of a material), both artists exploit material presence, not as mere productivism (for industrial usage and union of purpose and material) but in the technogenetic and transductive sense Erin Manning has demanded in her perhaps shortsighted argument that technological systems tend to "operate prosthetically and are often attached to the human body". She adds: "[Technological systems] operate on the basis of more-than, 'enhancing' a dancing body's capacity to create space-times of experience" (2009: 63). But this enhancement is of course often imperceptible to the dancer, and thus it cannot be experienced as enhancing. Unlike the performers in *Cinématique,* most dancers in interactive stage performances do not see/look at digital projections on the screens behind them. I always wondered how they could possibly be affected.

Open scores and gesture followers

The sounding of performance gesture/controller tools and the atmospheric and aural transformation of an entire space were demonstrated most memorably in the collaborative event Robert Rauschenberg and engineer Billy Klüver staged in 1966 at the 69th Regiment Armory in New York (*9 Evenings: Theatre and Engineering*), working with Steve Paxton, Deborah and Alex Hay, John Cage, Yvonne Rainer, Lucinda Childs, Robert Whitman, David Tudor, and Öyvind Fahlström. How do we discern the "controllers" and material gestures in the *dispositif* of *9 Evenings*? And how do gestures matter in the most tangible sense, if the *dispositif* were to tend towards the incorporeal effects of extraction? How can gestures be affectively transformative? The tactile dimension of sound, and the tactility of sonic gesture, arguably form a core idea for Rauschenberg's experiment with *Open Score*. Rauschenberg's spatial design and dramaturgy at the same time illuminate the regime of "optical tests", which Walter Benjamin had so brilliantly described in his commentary on film (2003 [1939]: 259), while undermining it through a cunning transformation of the visible coordinate system. I want to suggest that the notion of "materiality" can be disentangled from materials (instrument, sensor, wearable) to the extent that "digital movement" and digital materiality allow a perspective on engineered objects that combine material and im-material, generating processes of dilation between these domains through gestural constraint, which the trope "tactile sound" can illuminate. Movement of the body and its material wearable becomes coupled, invisibly, with the intangibility of sound, intimating what we can perhaps call, after Benjamin, "audible tests". This testing of the audience, of course, is a provocation in the tradition of acousmatic sound art where distinct sounds cannot be located (perceived) in a source that causes them.

Rauschenberg's *Open Score* performance began with Frank Stella playing a tennis match with Mimi Kanarek on a real tennis court set up on the large floor – a sports parameter for theatricalized re-engineering of nonaesthetic gestures (hitting the ball). The racquets were wired to amplify the sounds of each hit. The audience hears a "plonk" sound, while at each sound occurrence one of the lights in the Armory switched off. In other words, Rauschenberg and Klüver had designed an interactive tennis match in which gesture controlled both sound and light. As the game proceeded, the hall increasingly fell into darkness. The gestures grew invisible. An infrared closed-circuit TV projection system was then turned on to throw images onto large suspended screens: a cast of nearly 500 extras. People hired from the streets for a small fee had come on in the dark to perform various gestures invisibly.

Algorithmically speaking, they performed instructions for simple tasks given to them ("touch someone who is not touching you"; "touch two places on your body where you are ticklish"; "hug someone quickly then move on to someone else"; "men take off jackets, replace them, repeat", etc.).[11] These gestures could not be seen: the audience could sense their presence but not perceive anything except spectral images of the crowd swarm on the screens, blurry ghosts that were captured but eluded signification. Their gestures survive only in the score. They existed, back then, only as tactile images inferred from, though not logically related to, the previously enacted "tennis game" of controllers, the sublimely silly and tedious "plonk" that dimmed the light in the space. *Open Score* ended when "the people" had vanished in the dark,

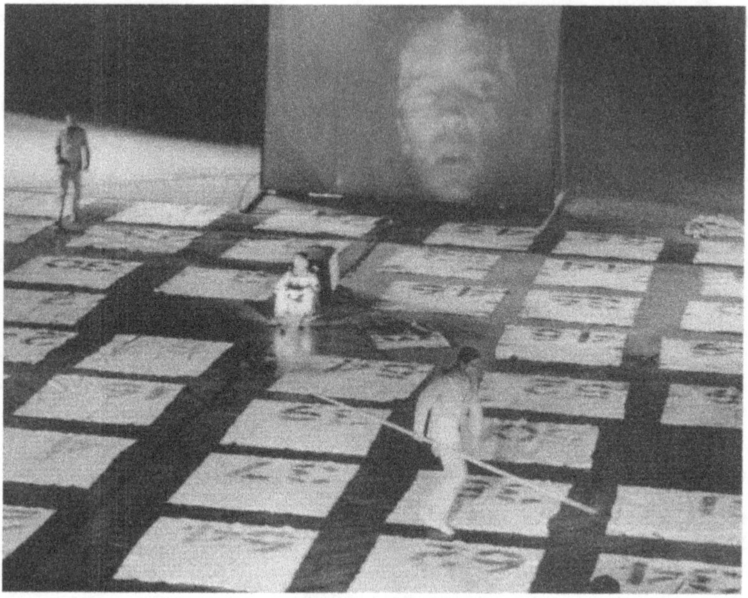

Figure 11.4 Robert Rauschenberg, *Open Score* (1966). Performance presented as part of *9 Evenings: Theatre and Engineering*, 69th Regiment Armory, New York, NY, October 14–23. Still from the factual footage shot in 16 mm film by Alfons Schilling. Courtesy of © Julie Martin (Experiments in Art and Technology) and the Daniel Langlois Foundation for Art, Science, and Technology

and a single spotlight picked out Rauschenberg carrying something obscure in a burlap sack. Inside was dancer Simone Forti, singing in a soft voice the tune of an old Spanish folk song, while Rauschenberg carefully pulled her about the empty space.

Behind the surface plonk, Klüver had enlisted nearly thirty engineers from the Bell Labs to work on the various artist-scientist collaborations devised for the Armory building. He had also quietly imported the special infrared closed-circuit TV projection system (not allowed for general use in the US, only for the military) to be deployed for *Open Score*. The engineering operations and circuits for the projects were drawn up into diagrams by Herb Schneider in order to visualize the coordination of the elements of action, film, video, sculptural objects, sound, and electromagnetic waves. Klüver had put much effort into the prototyping of a new Theatre Electronic Environmental Modulator (TEEM) to be used at the event. TEEM was designed as a central control panel to operate all the remote functions, composed of more than 250 elements including decoders, encoders, power amplifiers, power relays, tone control units, FM receivers, FM transmitters, photocells, speakers, programme drums, preamplifiers, Speaker Distribution

Matrix, Proportional Control System, and a specially developed wireless system that set up a "networked" environment in the Armory. Such a wireless system, using transmitters and FM receptors, was novel at the time, untested by the artists. What they envisaged was an interactive performance using new materials (such as the electrodes worn by Alex Hay, which transmitted muscle activity, heartbeat, eye movement and brain waves into sound) and "physical things", as Steve Paxton called the enormous plastic inflatable tunnels made out of polyethylene. Schneider noted that TEEM and the AMP equipment functioned like a proto-computer, demonstrating to the artists the fundamental principles and logic of computer science: programming, data storing, shifts between one media form and another, random logic, combinatorics, and the wired and wireless system architecture (cf. Morris 2006: 55–63). He also admitted that there was no guarantee that anything would work, since most of the engineers had trouble understanding the artists' visions.

The overarching *dispositif* of TEEM allowed for the flexible insertion of specific devices, such as Robert Kieronski's Vochrome and various other objects built for David Tudor's *Bandoneon! (a combine)*. Klüver envisioned Tudor turning the cavernous Armory into a musical instrument for immersive acoustic experience. Yet in spite of the carefully engineered patch-boards and controllers, this "self-composing" system with its multiple microphones and loudspeakers struggled with the Armory's extended reverberation time and ambient noise. Moreover, when "played" by Tudor the instrument became literally uncontrollable. When activated, its many evolving sonic and visual conditions were affected continuously by feedback, reverb, and signal processing, making the performance quite unstable. While Tudor, Cage or Paxton probably welcomed this insurgent behaviour of the effects of the devised *dispositif,* other artists were more baffled by delays and technical failures. There were glitches (so fashionable in today's sonic arts) and, ironically, TEEM did not work well during the first night of *Open Score,* forcing the engineers to manually unpatch the lights meant to go out during the electrified racquet game (without anyone in the audience noticing the difference to the second night, when TEEM worked perfectly).

This paradox of a controlled environment generating simultaneous, equivalent processes (causal, emergent, dissociated, acausal) is still a major techno-aesthetic problem today, fifty years on. Alex Hay, collaborating with Paxton and Rauschenberg as performers, pushed the performance medium to the same limits we still encounter today in interactive sensor-driven performance. The biofeedback apparatus worn by Hay simultaneously emphasized the body's materiality and its abstraction. The body's interior workings were translated into immaterial aural and visual effects, while other, exterior actions were carried out physically: 100 six-foot squares of cloth which Hay had laid out in a modular pattern were moved around in an arithmetic progression and placed centrally.

The imperceptible inner organic rhythms were here conjoined with external "natural" movement of a formally coherent action, even if this action appeared nonsensical. Hay's description of his work's unrelated, simultaneous processes is purely syntactical: they are equal in time. He locates the organic body (his own) in the spatial centre, and this body remains still: his involuntary internal system movements are comparable to Alvin Lucier's brain waves activating sound amplifiers in *Music for Solo Performer* (1965). In Lucier's case, however, the cones of speakers set into motion by his alpha waves were the actual performers – the composer himself remained a silent and stationary presence. What Louppe calls the "extreme nearness" of dance here reaches a new dimension. Hay absents all movement and gestural displacement while making an invitation to be seen or heard. This suggests autonomous nervous system activity is composed of internal "gestures" (mediated through worn biosensors) that might be experienced through a kind of synaesthetic resonation. Biofeedback, I suggest, is the limit case of dis-alignment. We do not know what it is we hear.

Paxton's *Physical Things* – along with Cage's *Variations VII* the only audience-participatory piece – performs a similar syntactic manoeuvre: the transparent, plastic inflatable tubes play with inside and outside, confounding the textures of plastic membrane and skin. The visitors entered into a translucent intestine-like space, a resonating "blurred" architecture:

> Wading throughout the warrens of Steve Paxton and engineer Dick Wolff's *Physical Things*, the *9 Evenings* audience was also made to confront ruptures in interactivity and transmission. Spectators palpated the tunnels' translucent plastic skin, then entered a magnetic potlatch of sound picked up on handheld receivers. Bodily sensation and receiving process overlaid each other... *Physical Things* mapped not only the space of the Armory but the commercial airwaves that girded it. During the first night, the work also entailed infamously long delays. The transmission to the modified transistor radios was weak, resulting in less aural incident than intended. As one critic complained, "There was nothing to throb over." Yet Paxton himself opposed such climactic thrills. Rather, the work was to unfurl in a slow series of haptic discoveries... The intrusion of dead air and delay enhanced this halting process, as the synaesthetic turned to an awareness of mediated reception. Unlike the brassy showmanship of much kinetic art, these works inhabited a space of fissures and temporal lags. It was in this sense that Klüver explicitly positioned *9 Evenings* against the immediacy of "flashing lights and psychedelic effects." (Morris 2006: 33)

The haptic discoveries point to the biophysical imagination about skin conductivity underlying such installations. One could even interpret Hay's "grass field" as an analogue metaphor for the many numbered pixelated (electrodermal) skin patches made of felt material. Hay's instructions say that "all properties and dress will have the color identity of the skin of the performers", while engineer Robby Robinson who developed the body-worn electrodes notes that "it was like preparing a man to go into outer space with the sensors attached to his body and the radio transmitters and amplifiers scattered over his body so they would not interfere with his movements in performing his tasks" (Morris 2006: 14). Both *Grass Field* and *Physical Things* conjured images of metabolisms and electromagnetic energy systems. These early interactional performances were modelling complexity, and they understood organisms to be complex control systems that are vulnerable to perturbations (cf. Birringer 2009: 294 ff.).

Dis-alignment/haute couture

This notion of perturbation (and the Cagean ideology of chance and indeterminacy) has been resituated, forty years later, in increasingly overdetermined capture systems of contemporary digital interface architectures. Contemporary systems can envelop the performer in totalizing projection environments, reducing movement choices while tuning the pre-programmed assets of the system to the gestures the dancer wants (to follow). Chunky Move's *Glow,* witnessed at the Hellerau Festspielhaus during the 2007 CYNETarts Festival, and their later *Mortal Engine* (2008, seen during a reprise at London's Southbank Centre in 2012) in fact struck me as immediately invested in "flashing lights and psychedelic effects", curiously entrapping the performers in the surround graphic and laser projections devised by Chunky Move's software engineer Frieder Weiss and laser designer Robin Fox. Both performances display an overwhelming suturing effect, an overlapping of the graphics and light effects onto the performed action that makes it hard in fact to remember Gideon Obarzanek's choreography. The temporal enframing here is largely owed to the software's gesture following, its spectacular technography that "writes" or inscribes its particle physics behaviour as fluid projections onto the floor canvas – the stage on which the dancer writhes.

The *dispositif* is surprisingly simple: a single overhead camera can observe all the action on the stage below; the LCD projector is mounted beside the camera and projects the liquid graphic mutations in real-time (with extremely low latency) downward onto the body of the performer and the floor. Both *Mortal Engine* and *Glow* use the Kalypso software program developed by Weiss for these data-intensive dances: it captures the contours of the performer and feeds the information into programmed visualizations such as lines, curves, contouring silhouettes, blobs, blotches, smears or harder polygonal geometries (the system does not "see" the body, but calculates the changes in the pixel field and structures this continuing input information via subtraction into the computer according to the possibilities of the

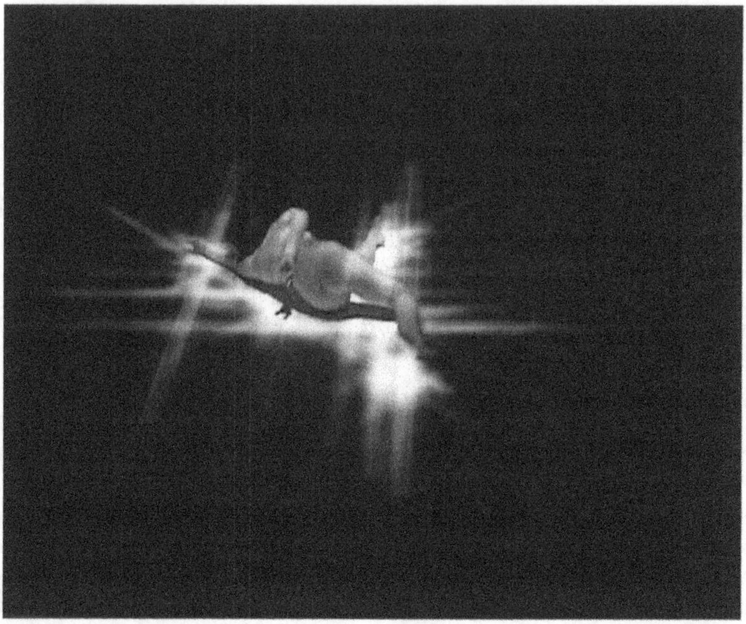

Figure 11.5 *Mortal Engine,* Chunky Move, choreographed by Gideon Obarzanek, interactive system design by Frieder Weiss. Queen Elizabeth Hall, Southbank Centre, London, 2012. Photo courtesy of © Frieder Weiss

Cartesian grid – from the material image to numerical grid). The effect is visceral and stunning.[12]

At the same time, the system here clearly acts as the controller manipulating the flow of images that appear to be generated by the immediate motor-sensory linearity of action, giving the dancer (Kristy Ayre and Sara Black, alternating in *Glow,* and both performing in *Mortal Engine*) the sensation that her gestures or postures can push the envelope, and that she can feel the "resistance of the instrument". This is how Kwastek describes the relationship between a virtuoso and the apparatus, thus of course addressing trained performers in staged processual interactions rather than untrained visitors to installations within which the operational rules are unknown to the recipient (2013: 172). In *Glow,* the dancer knows the rules. Her crouched, flattened body, her splayed form on the ground, becomes entwined with the appearing and disappearing graphic animations dropping over her, which she cannot see but which she experiences kinaesthetically on a level, perhaps, of blindness. What I described as the work's stunning visceralness is largely effected by the powerful rhythmic quality of the light (the graphic animations are the only light appearing – always intensely mobile and dynamic, creating constantly mutating shapes and

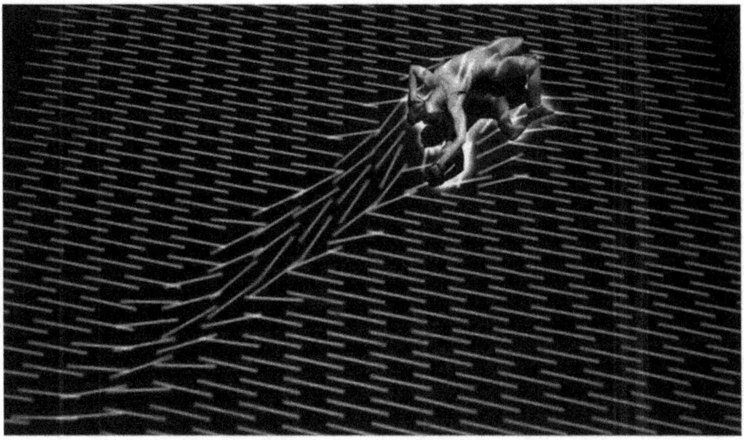

Figure 11.6 Sara Black in *Glow* (2007), co-created by Chunky Move and Frieder Weiss. Videostill courtesy of © Frieder Weiss

contours around the moving body), which traces the gestures. It must electrify the somatic consciousness of the dancer who acts in the moment, wholly aware of her corporeal relation to Kalypso's contouring digital accoutrements. She might as well dance with her eyes closed, intuiting her floor work as a continuous exfoliation into the shards, lines and curves of light radiating outwards and coalescing around her. Her gestures are inconsequential as sounding gestures since in this *dispositif* they are not tracked (sound design is by Luke Smiles), which one can read as a compromise. The system does not expect choreosonic gestures by the dancer to be capable of actuating or controlling sound correlative to the complex visual animations. In this case, then, the affective connection is largely tactile-kinetic.

Seated on raised platforms either side of the stage, we look down upon each movement and each tilting of the body generating an animated calligraphic gesture, a response, an echo, a whispered word from the shadowy lover Kalypso (why not call our control/capture systems our lovers?) Halfway through, Sara Black becomes quite audible in her ecstasy. She slides and slides, spinning around, stuttering words of love in the unknown dialects of the possessed dervishes. Suddenly all is white, and each gentle motion of her arm and elbow draws purple outlines of her coffin below her. She literally draws her envelope, wearing down her lover's embrace – the Draculan sucker who will climb into the coffin with her when the sun rises. Now it is night again, we are in the dark, only thin stripes flicker across, lines like knives stab and cut, and the dancer continues, happily enslaved by her pulsating underground. The electronic music vibrates in our ears, but our hearts melt, we are swayed to stay on this island, like Odysseus, drawn to the mimicry of the perfectly coupled re-union of dance and technology.

Near the end, the dancer jumps up, and lets herself fall, again and again, creating powerful dark blotches on the white floor. Her body leaves the silent black blob of a silhouette on the surface, and as she leaves and retreats to the side, anxiously, she separates from the loved silhouette, but then the dark blobs slowly begin to move, follow her across the space, to finally catch up with her. She is reunited with the impression of her body, embraced by digital movement, silhouette and person become one again. Distance and unity have been calculated and recalibrated.

I am not sure whether I want to end with these images of a worn-out marriage. The surveillance *dispositif*, Edward Snowden revealed, loves us and keeps track. InfoMus Lab's EyesWeb software, which I recently experienced in Genoa, has been trained to develop ever more sophisticated high-level parameters (though they are hardly clearly defined concepts) for physical motion extraction and the analysis of movement features including kinetic energy, "objective" and measurable features (i.e. velocity); perceptive and semantic aspects; emotional and social signal analysis, expressive features, etc. What interests us in the *METABODY* project are material gestures that can be processed. And yet, these material gestures remain illegible or cannot be reduced (and transposed) to a tracking syntax that remains indebted to biometric parameters and mapping coefficients. The old skeletal model of the human body is perhaps also due for revision. All the markers for motion capture have been attached to the wrong places. "Perhaps now is the time to do away with pictures of things which encourage our pleasure centers before trying to destroy them," Kara Walker once proposed in an exhibition (*American Primitives*, 2001). I was reminded of *Glow*'s silhouettes when I saw her retrospective *My Complement, My Enemy, My Oppressor, My Love* at the Whitney Museum. It then occurred to me that the visually stunning coupling of gesture to interactive systems is always also an involuntary political comment on ideological alignments, on how bodies are enframed or incorporated, insufficiently resistant to mortal engines.

In conclusion, the couplings to system controllers – with Kinect cameras now becoming the latest fashion – can be considered suspect if mapping operations are not questioned or if sensory affect is taken for granted and not probed in regard to meaning, form and pattern of visual, kinetic, tactile or auditory awareness generated by the "reactor" (Yoko Ando's term). The description of Ando's RAM reads like a diagram for an incubator: "via Motioner or 'Microsoft Kinect™', a computer detects the movements of dancers, collects and converts various relevant information, and visualizes them in the context of the dancer's body. Processing everything in real-time, it allows dancers to decide their next movement, create their own 'rule', based on visual, aural and haptic information."[13] But this bioreactor or chemical reactor – is it not built to contain/control reactions? The

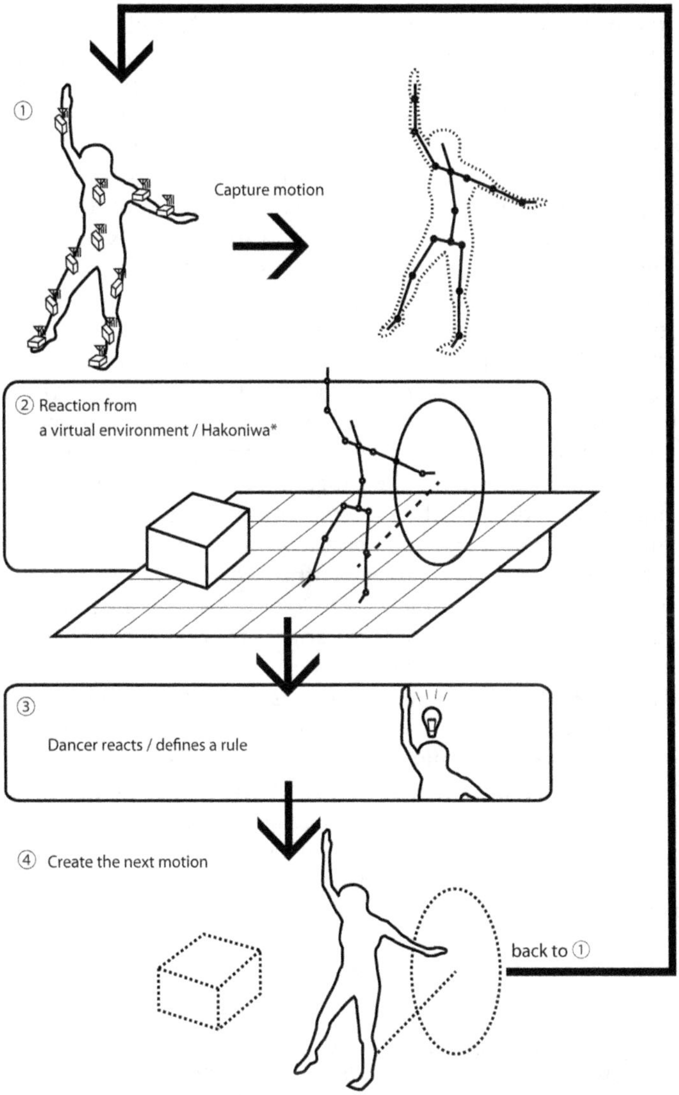

Figure 11.7 RAM's conceptual diagram: via Motioner or Microsoft Kinect™, a computer detects the movements of dancers, collects and converts various relevant information, and visualizes them in the context of the dancer's body. Processing everything in real time, it allows dancers to decide their next movement, create their own "rule", based on visual, aural and haptic information. Courtesy of © YCAM InterLab

Hakoniwa*: small physical laboratories that react to dancers' movements as an environment.

question then occurs what is "relevant information"? And how does the programmer decide how to visualize the information? Having watched Ando rehearsing with the reactions of the virtual environment (graphic objects projected onto three large screens at YCAM), my attention was drawn to *her* reactions, the rules she created, and I could not really fathom how her inwardly/outwardly directed movement meant to create an "intelligible imaginary" (Louppe 2010). The tactile-sonic experience of RAM is modest, as the dancer either remains aware of sound triggers or, as in *Glow,* has no control over sound assets. Regarding *Glow,* asymmetries and dis-alignments are hard to achieve: the body does not deform or drift off. It tends to want to con-form, even if blind, to the rules it knows and the system it remembers. I suspect the choreographic impulse behind *Glow* is a structured improvisation (largely horizontal floorwork) prepared to be "fitting" the graphic animation. Or to describe it from the point of view of the programmer of the gesture and contour-maker: based on recognition of the contours of the figure (cf. the RAM diagram), the programmer is fitting the bespoke projected images according to the cut-out pattern (mathematically based on measurements of the figure) and specifications of the dancer. We are in a world similar to Jean-Paul Gaultier's cutting-edge design (on view, beautifully orchestrated, at the Barbican Art Gallery, April–August 2014). Except that Gaultier's costumes are often amazingly bizarre psychosexual fantasies, freakishly punk and ambivalent driftworks.

The dancer's anatomy, contoured, remains her destiny in the dance performance, unless we imagine not noticing the silhouette. At the same time, the affective experience of the moving silhouette on the dancer and her own gestural motion awareness, like the perceptional-cognitive processing that Ando pulls out of RAM to drive each of her single executable movement options, will not be known to the system. Fortunately, the system does not know what pattern of thought, awareness or emotion underlies the dancer's movement.

Notes

1. I first began to write on the *dispositif* during the 2010 Live.Media + Performance Lab which I co-directed with Mark Coniglio at EMPAC (Experimental Media and Performing Arts Center, Rensselaer Polytechnic Institute, Troy, NY). The blog was posted at: http://empaclivemediaperformancelab.blogspot.com/ (accessed 18 March 2015). I wish to thank the participants, as well as all the members of the DAP-Lab ensemble, especially our designer/co-director Michèle Danjoux, for their contributions to the research/performance projects conducted over the past years, in particular the various versions of UKIYO (Moveable World). Danjoux's writings on her design of wearables have been published, for example, in "The Sound of Movement Wearables" (Birringer and Danjoux 2013) and "Choreography and Sounding Wearables", *Scene* 2:1/2 (2014), 197–220. The core concepts of her research, featured also in our new production, *for the time being*

[Victory over the Sun] (Sadler's Wells 2014), will be delineated in her PhD thesis *Design in Motion: Choreosonic Wearables in Performance.*

2. Regarding these scenographic experiments for early modern dance, and the implications for our discussion of equipments and the *dispositif,* see Brandstetter and Wiens (2010), especially Richard Beacham, "Thinking with Things, Speaking with Spaces. The Enduring Legacy and Lessons of Appia's 'Expressive Elements' in the Digital Age", pp. 38–59, and Michael Takeo Magruder, "(Re)Configurations of Space and Movement. Performative Extensions of Appia in *Second Life*", pp. 60–75. See also Beacham 2006a and 2006b.
3. Erin Manning writes poignantly about Marey's graphic "movement machines" and photographic experiments in *Relationscapes* (2009: 83–111), noting that Marey devised new techniques for perception that "in turn became new kinds of sensing machines" (83) able to capture and compose durational rhythms through successive, overlapping images assembled/spread over the photographic plate. For my interpretations of Schlemmer's and Moholy-Nagy's designs, see Birringer 2013.
4. Regarding the astonishing freedom of spatial imagination in the vast, open Hellerau Festspielhaus architecture, which early on made me think of movement choreography as being enveloped by a transparent "costume" without front or back, without a determined way to be seen or felt, see my reflections on the 1995 *Parsifal* opera installation we created there at the 9th Dresden Festival of Contemporary Music (Birringer 1998: 46–50, 348–51).
5. For Biloria and Hyperbody (Department of Architectural Engineering and Technology, TU Delft), see: http://www.hyperbody.nl/research/projects/metabody/ (accessed 18 March 2015). The *METABODY* project was initiated in Madrid (July 2013) by a collaborative network of arts organizations and performance companies engaged in a radical rethinking of perception and movement away from the mechanistic and rationalistic tradition towards a (digital) embodiment that puts emergent differentials of bodies and affects in the forefront of its concerns. The *METABODY* project is coordinated by Jaime del Val (Asociación Transdisciplinar Reverso) and encompasses seventeen partners including DAP-Lab, STEIM, Palindrome, K-Danse, Fabrica de Movimentos, InfoMus Lab, Kouros, the Hyperbody Research Group, Trans-Media-Akademie Hellerau, and others (http://www.metabody.eu).
6. The idea of sensorial dilation, evoked for me in my study of Appia's spatial modulation, deserves further explication for which there is not enough time here; I learnt the expression in a workshop on dilated bodies conducted by Eugenio Barba back in the 1980s. Karen Barad's notion of "intra-action" deserves mention here as it provocatively riffs on the tired interactivity which has become such a technological vaudeville. Barad suggests that unstable (rather than formatted) "phenomena are produced through agential intra-actions of multiple apparatuses of bodily production. Agential intra-actions are specific causal material enactments that may or may not involve 'humans'" (2003: 817); they dilate – generating differential boundaries – the capturing instrumentations set up to work in particular ways. She implies, therefore, that apparatuses are not necessarily closed but open-ended practices.
7. Katja Kwastek, visiting the work's installation at the Lentos Kunstmuseum in 2009, patiently describes the interactional behaviour of audiences in detail, noting the "material minimalism" of the space and how visitors slowly explore "the effects of different movements" and how sounds were triggered and how

they could "catch" them (Kwastek 2013: 234–40). Kwastek is concerned with an empirical approach to interactivity and relationality and thus offers many detailed observations (including audience feedback, interviews) of user behaviour in interactive installations. Unlike Popper (2007: 271), who disarmingly claims that VNS is not a control system, she is aware that interaction/interactivity is always limiting, yet seeks to distinguish between *interactivity* as a characteristic of certain sensing systems (bearing the potential for interaction), and *interaction* as the actual feedback process itself. What has to be understood as limiting, in her view, is interactivity, even if this characteristic makes interactivity interesting. She also proposes that, instead of perpetuating an earlier 1990s rhetoric circling around expectations of interactivity as implementation of an ideal of collaborative authorship, or the critique of unfulfilled expectations, we acknowledge its foundation on programmable control as a distinguishing feature.

8. See the special issue on "Choreographic Documentation" edited by Scott deLahunta and Sarah Whatley for *International Journal of Performance Arts & Digital Media* 9:1 (2013), and especially Bertha Bermúdez Pascual's essay on Emio Greco | PC, "(Capturing) intention: The life of an interdisciplinary research project", pp. 61–81. The Forsythe Company's collaboration with Ohio State University (OSU) on the creation of *Synchronous Objects* is documented online: http://synchronousobjects.osu.edu/ (accessed 18 March 2015). McGregor's research with cognitive scientists was recently exhibited in *Thinking with the Body* at London's Wellcome Collection (19 September–27 October 2013).
9. During a workshop in Croatia, dramaturge Goran Sergej Pristaš pointed me to Marko Kostanić's "The Choreographic Unconscious: Dance and Suspense", included in the programme for BADco's performance of *Semi-Interpretations, or how to explain contemporary dance to an undead hare* (Zagreb 2012). BADco's political and dramaturgical interventions into the relations of performance and image are relevant here as they seek to recover and deconstruct early modern choreographic "scores" and enframings, noting how bodies are slowed down when the camera moves and how our attention is conditioned. We had a prolonged discussion about the "responsibility for things seen" and the "actionable image", i.e. the question of whether images can have agency. I am redirecting the question here to the role of agency or control of sensing systems in generative real-time capture environments where motion is both analysed (by the system) and where the interactor can feel/realize and improvise dis-alignment from an affective environment.
10. The DAP-Lab's research on the Bauhaus dances and futurist/constructivist performance inspired the choreosonic composition of *UKIYO [Moveable Worlds]*, 2009–10, and *for the time being [Victory over the Sun]*, 2012–14, both presented at London's Sadler's Wells. For filmic extracts that highlight the role of the costumes and accoutrements in the performances, see: http://people.brunel.ac.uk/dap/Ukiyo_Sadlerswells_movie.html; https://www.youtube.com/watch?v=NTcm-UFk0x8 (Part 1); http://youtu.be/oXBW4oWyK0k (Part 2) (all accessed 18 March 2015).
11. The MIT List Visual Arts Center offered its 2006 exhibition *9 Evenings Reconsidered: Art, Theatre, and Engineering, 1966* as a critical homage to the original event, featuring the records of 1966 to focus on a ground breaking link in the history of performance, art, and technology. The excellent catalogue of the exhibition was edited by curator Catherine Morris. My reference is to Rauschenberg's original

typescript "Instructions and List of Cues to the Group of 500 People" (Morris 2006, pp. 36 ff.).

12. Naturally, when I first watched *Glow* in 2007, I was sceptical about the hype that Chunky Move had caused regarding its interactive system. The CYNETart 2007 November newsletter announced that "in most conventional works using light/video projections, the positions and movement sequences of the dancer need to be cued precisely to the space and temporal structure of the video playback. The role of the dancer is reduced to the difficult task of making each performance an exact copy of the original. In *Glow* the machine vision of the software observes and analyses the performer and reacts to her movement. It thus releases the dancer from restrictiveness and monotony." My impression was exactly the opposite, namely that the downward projection encapsulated the moving body entirely, restricting the dancer to the precise programming feedback/mutations of the largely horizontal motions she was asked to elicit from the gesture follower to maintain the topological logic of continuous geometric variations.
13. The diagram is shown here: http://interlab.ycam.jp/en/projects/ram/about_ram (accessed 18 March 2015). Addressing the "Motioner", the website describes the outfit cheerfully as "an inertial motion capture system which is relatively low cost and fairly accurate, light weight, low stress and low latency".

Part V

Motion Technology, Interaction, Performance

12
Communication through Haptic Interaction in Digital Performance

Doros Polydorou, Tychonas Michailidis and Jamie Bullock

Introduction

This chapter aims to contribute towards the theory and techniques of creating an immersive stage environment for choreographic performances and installations. A number of digital performances are known to employ audio-visual material with the intention of creating relationships between the real-time projected material and the performers. However, such approaches often cause problems in practical terms. Dancers are expected to form a relationship with other dancers on stage and also with projected imagery. Due to the nature of the performance space and the performer-audience relationship, the dancers cannot experience visual elements in the same way as the audience. The performers tend to be in the midst of the action while the audience is situated in the optimal position to experience the composition of the performers, the projected visuals, the audio and the scenography of the space. While dancers are able to interact "live" on stage with ever-changing visual and audible aspects of the performance, their sense of interaction with the technology or other performers may be diminished due to a lack of feedback gained from visual or acoustic cues. The performers may not be in line of sight or audio might not be adequate during the heat of the performance. This chapter proposes ways to overcome this problem, by offering different input stimulations through haptics that enhance the relationship between performers, or between performers and technology, adding an additional layer of spatial and emotional awareness.

The chapter begins with an introduction to the relationships between aural and tactile feedback and a discussion of gesture in haptic interaction. It continues by constructing a theoretical approach to technologically mediated embodied performances with reference to the choreographic installation *Ukiyo [Moveable Worlds]* and the participatory installation *Whisper[s]*. The former example concentrates on the interactional qualities between one performer and an artificial intelligence system, while the latter utilizes networked technologies, sensors and motors to initiate technologically mediated relationships between participants. In the following section we explain how current technologies can be utilized to create a reactive space involving haptic technologies. Finally, we draw the various components of our research together by proposing an analytical framework for thinking about different scenarios and modes in which these technologies can be used.

Aural and tactile feedback relationship

The relationship between tactile and aural feedback is often difficult to illustrate. This is because both feedback modalities are a by-product of the same action-reaction mechanism, making it difficult to separate them. Action refers to the energy or energies that have been applied and transferred from one agent to the other. As a result of the applied energy, the "reaction" produces audible and tactile results. Such audio-tactile reaction is perceived as feedback, allowing the initiator to make adjustments. For instance, this could include tapping the fingers on the table or hitting a bell with a hammer. Both actions provide aural and tactile feedback gained from a single action. The "volume" of feedback received, aural and tactile, depends greatly on the applied energy.[1] The more powerful the hit, the louder it will sound and the more the hammer and bell will vibrate. What is important here is that both feedback modalities are supplementary to each other, and it is impossible to separate them from activities in everyday life. Apart from the physical laws that govern the universe, a fundamental aspect of this multimodal integration (Calvert and Thesen 2004) is due to the formation of habit over the years regarding the relationship between aural and tactile (Mäki-Patola 2005). Consider again the example of hitting a bell with a hammer. It would be "unnatural" to experience the received aural feedback three seconds after hitting the bell. Our perceptual understanding of audio-tactile relationships is closely related to the habit formation of latency that exists between the two. We have established a habit relationship of coupling the aural and tactile feedback in a particular way. In instrumental performance, the introduction of latency in the audio-tactile feedback relationship changes learned kinaesthetic-acoustic mappings and disrupts the inner prediction of the instrument's responsiveness. That is, the behaviour of the instrument changes unexpectedly for the performer and needs to be relearned (Mäki-Patola 2005).

Looking at the available technologies and how we interact with them, such audio-tactile feedback relationships are often *designed* and introduced artificially into products. The majority of handheld touchscreen devices provide audible feedback in relation to user interaction in order to simulate acoustic feedback that occurs naturally in physical interaction with "real world" objects. The incorporation of such feedback modalities can be a key factor in differentiating a successful product. For example, when using the onscreen keyboard on a touchscreen device, there is insufficient auditory and tactile feedback through contact with the physical device. An auditory cue indicates whether a key has successfully been "pressed", as found with the traditional typewriter or with computer keyboards. The addition of auditory feedback to the user therefore serves to enhance and confirm the actions of the user.

In performing arts, and in particular in instrumental music performances, the aural-tactile relationship forms the basis of learning and performing an instrument. The instrument reacts to the energy that it receives from the performer by producing both aural and haptic feedback. Through instrumental practice, the performer is able to learn and internalize these responses. Interestingly, a threshold exists on the amount of latency with which the performer can cope when separating the aural from the tactile feedback. Exceeding such a threshold exposes the performer to a problematic performing environment. A study by Dahl and Bresin (2001) shows that 40–55 milliseconds could be the "break-point" where the performer notices latency and needs to adjust to other strategies for sensory input information in order to continue playing. The study shows that different musical training can produce different thresholds for such reliance on aural and tactile coordination. The familiarity with different performance settings, such as an orchestra or jazz ensemble, suggests there are variable levels of confidence for the performer in terms of the relationship between tactile and aural feedback (Dahl and Bresin 2001). Orchestral performers were able to play in-tempo with larger delays present between physical input and resultant sound due to the necessity of an orchestral timing where long distances between performers often require them to adjust their tempo.

Research by Sawchuk et al. showed a variable break-point of latency depending on the instrument used and piece performed (Sawchuk et al. 2003). For instance, they noticed that the performer could tolerate latency at 25 milliseconds when playing a synthesized accordion, but at up to 100 milliseconds when using a synthesized piano. Furthermore, they observed that performers felt more confident in a large performing hall where the reverberation of the hall is greater, rather than in small halls with tighter reverb.

The importance of such audio-tactile relationships becomes apparent in particular when examining digital controllers and their use in live electronics performance. The majority of digital controllers available to performers do not provide sufficient tactile feedback (if any at all) to enable expressive nuances in the performance to be felt. For example, it is difficult for performers to know the angle of an expression pedal without visual cues because its range of movement is small (around 30°) and insufficient kinaesthetic information can be obtained through foot movement within this range to determine accurately the relative position of the pedal. Furthermore the movement of a digital controller may be mapped nonlinearly to the controlled output (audio or visual) and it is generally very difficult to infer the nature of this mapping through the physical state of the controller. It is therefore necessary to provide additional haptic cues through vibrotactile feedback, supplementing the information that the performer gains from other sensory modalities. A study conducted by Michailidis and Bullock (2011) examined whether incorporating haptic feedback in music performances can improve the overall control, perception and musicality of live electronics by instrumental performers. The results suggested that adding vibrotactile feedback to a sensor-based controller can significantly improve a performer's understanding of relationships between the controller and resulting sound produced. Additionally, the use of haptics suggested new musical possibilities not previously considered by the performers when using the controller without the vibrotactile feedback.

Gesturo-haptic interaction

The most common uses of haptic feedback involve direct (often linear) mapping between input values (e.g. from physical controllers such as foot pedals or sensors) and output values controlling the magnitude of the vibrotactile vibration. That is, the specific volume of vibrations felt on the body has meaning *in itself*. For example, data received from a gyroscope sensor measuring the angle of rotation could be used to control in a linear fashion the amount of vibration received. However, an alternative, somewhat more advanced approach is to allow *multiple* values over time, in multiple dimensions, to carry meaning. Rotman (2002) describes this approach as "gesturo-haptic". Movement data can be captured through combinations of motion capture technologies (gyroscopes, accelerometers, range and depth-sensing cameras) and stored on physical media as time series. This data can later be "played back" through mapped vibrotactile feedback devices, effecting a re-embodiment of the disembodied movement. A simple example would be using an accelerometer to measure the movement of a hand back and forth. This data could be replayed later through a combination of two vibrotactile motors on the front and back of the hand, one motor "pushing" back, the other "pushing" forwards.

Movement in itself does not equate to gesture. A gesture has a beginning and end point and is often (although not always) intended to express meaning or communicate information (Cadoz and Wanderley 2000). In the context of digital arts performance, gesture is certainly multifaceted, serving a range of functions from abstract and expressive gesture in dance to "functional" gesture in composed sensor-based control-processing relationships in interactive music. Once a gesture is captured, it becomes dislocated from its original embodied form. That is, once the gesture is stored numerically, it is deprived of its original meaning (if any) and becomes only a representation of movement. As Rotman writes:

> [H]aptic gestures do not communicate as such; they are not signs in de Saussure's or Peirce's sense though they may become so retrospectively in that they come to signify (if that is the term) their own happening; their meaning is the fact and consequence of their having occurred. (Rotman 2002: 102)

There is thus an opportunity for the digital artist, in the process of mediation between movement capture and haptic gesture, to manipulate creatively the gestural time series in order to bring about a particular haptic sensation for the performer or to synthesize entirely artificial haptic gestures, which may have significance only within the digital virtual space of the work in question. This process becomes particularly interesting in the case of having two or more performers, with one performer originating the motion (which is captured), and the other receiving the captured haptic gesture. A system like this could be used bi-directionally, to simulate the "throw and catch" of a virtual ball, for example. However, any such correspondence between received haptic gesture and meaning (I am receiving a ball) would need to be learned, or at least inferred, within the given context. As Rotman warns:

> [T]he gesturo-haptic is a medium with features opposed to the alphabetic: by enacting and executing its "messages" it is evidently governed by a participatory rather than functional logic, having to do primarily with action, deeds not words and inscriptions, with *faire* rather than *dire*, with performative and interactive states before constative and descriptive statements. (Rotman 2002: 103)

Technological embodiment

In his paper *After Choreography*, Birringer (2008) emphasizes that real-time digital performances mostly involve some sort of projectional activity. Performances follow traditional scenographic structures, in general, where the audience is situated in the optimal position to receive primarily audio/visual stimulations. At the same time there is a shift in emphasis away from the physical body towards the inter-relationship of the digital elements *with* the physical bodies. Birringer states:

> [T]his marks the enactment of movement that has less to do with steps, phrases or placements of the limbs than with gestural or postural articulations of motion "mapped" onto image, sound and light movement. (Birringer 2008: 119)

One example of such a technique can involve a real-time video feed which is filtered through software and then projected back to an on-stage screen. This technique involves having a dedicated camera, aimed and oriented at a very specific part of the set. This results in a camera which is set up for a specific task, most likely for a short period of time, which will probably not be utilized for the rest of the performance. If that projection image provides visual feedback to the performer, then the static position of the camera means that the performer must be within the range-view of the camera and must also notice the projected image. From the performer's perspective this can be an overwhelming situation in the context of an already-complicated performance, and presents an unnecessary cognitive burden.

Maurice Merleau-Ponty views the phenomenal body as our primary access to reality and further subdivides it into two notions: the body schema and body image (Merleau-Ponty 2002). In his book *Bodies in Code*, Hansen distinguishes between the two:

> The body image characterizes and is generated from a primary visual apprehension of the body as an external object, the body schema from what, with autopoietic theory, we have called the operational perspective of the embodied organism. (Hansen 2006: 39)

In other words the body image refers to how the body is *represented* whereas the body schema refers to the *organism*, which is caused by movement and subsequently causes it. It is clear that when a performer is offered visual stimuli from a video loop, the visualized body image drives the body schema. As described by Polydorou:

> In the first generation interactive media works, where a performer or a member of the audience interacts with a direct one-to-one relation with the projected image, the visualized body image drives the body schema. The system will only respond interactively after it is being offered an analog signal, usually in the form of movement from the interactor. All subsequent movement, even though originating from the body schema, is a result from the immediate visual or audio stimulation derived from the body image. (Polydorou 2011: 58) (See Figure 12.1)

Talking about 1st Generation Interactivity, Birringer mentions that this "layering of indirect and remote manipulation of the kinaesthetic and the kinaesonic can be highly complex, especially if the kinaesthetic qualities of the projected images contradict the human body motion observable in the phenomenal body" (Birringer 2008: 119). Birringer continues to warn that these digital articulations created by the body but "at a distance from the body" can induce new associations between sensations and sense

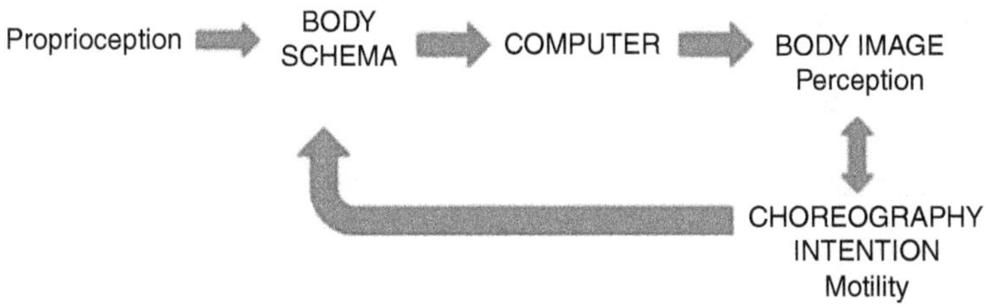

Figure 12.1 1st Generation Interactivity – Interacting with body image. Reprinted courtesy of © Doros Polydorou

perceptions that cannot be assumed to belong in the "intuitive" vocabulary of a dancer's physical thinking and kinaesthetic experience (Birringer 2008).

In order to avoid this situation, the technological stimulation that aids the choreographic intention must not be sourced by purely looking at the body image, but rather by an all-rounded perception, which originates from the performance space itself. Polydorou describes such a scenario in his thesis:

> At the start of an interactive dance, a connection is established between the performer and the space, through the interfacing technology. We have the visualized space, the computer which is hosting the rules imposed by the designer and the performing body which becomes the mediator of the two. The dancer perceives the space shifting and along with the motility derived from the choreography/intention as well as the dancer's proprioception, the body schema becomes an embodied potential. [...] [T]he body schema is transferred through the embodied agent directly to the computer. As the body schema is digitized, the space architect needs to interpret the data in ways which make the space transform and not only become an extension of the dancer but also possibly a dance partner with its own initiative which equals that of the dancer. (Polydorou 2011: 60) (See Figure 12.2)

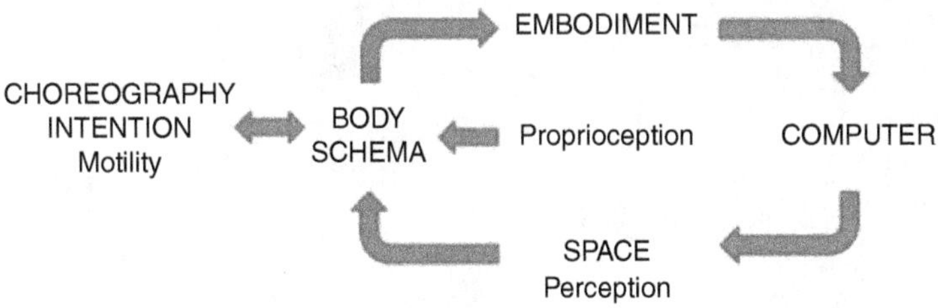

Figure 12.2 2nd Generation Interactivity – Interacting with body schema. Photo reprinted courtesy of © Doros Polydorou

Polydorou continues by comparing the 1st Generation Interactivity (Figure 12.1), where the live video feedback loop is prominent, with the 2nd Generation Interactivity (Figure 12.2), where the visual representation of the body image is absent and the focus is primarily on abstract audio/visual cues that are generated as a consequence of the dancer's movement. He explains that:

> The interactive audio-visual feedback is more subtle, taking more time to develop, and it does not directly relate to the body image. It is perceived more subtly, therefore it is being "absorbed" by the body schema at a slower rate, allowing the dance to get *influenced* by not *directed* by it. (Polydorou 2011: 58)

Hansen refers to this phenomenon as a disconnection of the fundamentally visual body image, resulting in the viewer (or performer) technically being enabled to utilize the excess of the body schema over the body image to increase her agency as an embodied being (Hansen 2006: 20).

For the section "Creation Scene" in *Ukiyo [Moveable Worlds]*, performed in 2010 and directed by Birringer (2009/2010), Polydorou created a system in which the performer and machine would work together to shape and evolve a real-time generative 3D world.[2] The "Creation Scene", as a scene in a choreographic installation, aims to immerse both performers and audiences in a dreamlike fragmented world that stimulates the perceptions and emotions of the viewers by offering them an ethereal journey through a world generated right in front of them.

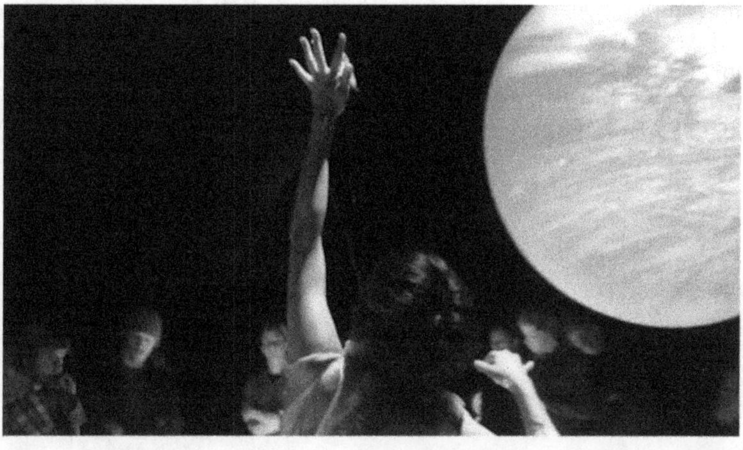

Figure 12.3 Katsura Isobe performing in Sadler's Wells (2012). Courtesy of © Johannes Birringer

During the "Creation Scene", the performer becomes immersed into her own virtual space, working along with the system, and aims to shape the Hiroshige-inspired landscape, turning the dark industrial space into a peaceful green landscape with falling leaves or an icy, eerie and mysterious forest with volcanic rocks and walking trees. Matching the concept of *Ukiyo [Moveable Worlds]* and the mixed realities, the "Creation Scene" exists in two realms at the same time. The dancer performs her choreography around a weather balloon, while an evolving world is projected onto it. Using an inflatable weather balloon, suspended above human height in the air, the image engulfs the sphere adding extra texture and dimension to the image, while the dancer softly moves around it acknowledging its existence. The dancer is wired with sensors to an interface controlling the virtual realm, and so her actions acquire a double meaning. Not only are they viewed by the audience as performing acts (the touch of a hanging leaf), but also they have an effect on the evolution of the virtual space. Starting from a flat concrete surface, the space is slowly transformed into a beautiful landscape, with mountains, grass, trees and lakes and populated by the eerie walking tree and mysterious black figures.

"The place," Polydorou describes, "like an extension of her wishes flows onto her dress and then the dress extends all around her, creating a fortress of solitude. She can invoke trees, she can make the grass grow, she can call for the sun or she can call for rain" (Polydorou 2011: 80). As the three-dimensional world passes through the different evolutionary stages, the dancer and artificial intelligence system work together to shape the malleable environment according to their impulses at any specific moment in time. Birringer described Katsura Isobe as a performer with an "acute sense of bodily self-awareness and alertness, that listens through all her sensory channels, perceives through her entire body, with movements that feel animal-like, suspended somewhere between the rhythmic and the arrhythmic as she navigated real and virtual spaces" (Birringer and Danjoux 2013: 234). The world is beautifully rendered on the suspended weather balloon, right in the middle of Isobe's "magic circle" of interaction.

Upon reflection, the communicative limitations of the *Ukiyo* project are apparent. Even though there was an artificial intelligence system developed, which was working both independently and in collaboration with the dancer, the communication infrastructure was heavily one-sided. Through the performer's technological embodiment (sensors and a camera vision system) the system could identify where the performer was located in the space, identify her gestures and movements, and react accordingly. However, there was no way for the system to communicate back to the performer, for example, that the tree she planted was fully grown or that it had suddenly started to rain. The only way for the performer to obtain this information was to turn towards the projected image and view the result. In doing so the camera, which was programmed in advance, would capture that movement from the dancer and trigger another section of the world which was not intended. This not only caused loss of agency and reduced the sense of belonging in the digital space but, most importantly, it forced the dancer to rethink the kinaesthetic experience.

The unhealthy divide

Before investigating the on-stage links between multiple performers, it is beneficial to examine another performance/installation stemming from a phenomenological perspective. In her book *Closer*, Kozel (2007) claims that phenomenology can help to bridge the unhealthy divide between solitary and shared experiences. By working alongside Thecla Schiphorst, Kozel aimed to analyse the "conditioned reactions and physiological behaviours of performers and audience members". In their installation *Whisper[s]: wearable body architectures* they aimed to represent the participants as "common and shared signals" in a "multiple networked system" (Schiphorst and Kozel 2002). According to Kozel, the design of their system encouraged the audience member to attend to his or her physiological data or affective corporeal state and to send it to another as a "poetic amalgamation of sound, visualisations and haptics". The two goals of the project were to "amplify the poetic capability of our mobile devices and their convergence with our bodies, both in what they convey and how they are worn" (Kozel 2007: 238). By identifying a number of characteristics that they believe are the glue of human exchange – invisible layers of emotion, physicality, vitality, imagination, gesture and attention – Kozel and Schiphorst created different configurations of sensors, actuators and networking protocols in order to access and transmit this data.

The *Whisper[s]* project had three slightly different iterations, all of them in a participatory installation format inviting members of the public to wander around the space while wearing garments embedded with small wireless computers. In the first iteration, participants could access their own breath and heart data through simple gestures and transfer those, in the form of real-time visualizations, onto video "pools" which were spread around the installation space. In the second, more playful iteration, both men and women were asked to put on skirts embedded with small fans and vibrators. Wanting to move away from a primarily ocular-centric feedback system, Kozel and Schiphorst embedded garter belts that sensed the muscle contractions of one person, which were then used to trigger the fans and the vibrators in the skirts of another. Kozel explains in *Closer*: "We wanted to escape the visual in order to enhance the kinaesthetic and tactile, to draw people into different qualities of awareness that did not privilege vision" (Kozel 2007: 314).

While reflecting on the procedure and the reactions of the participants, Kozel identified three distinct and very interesting relationships. It is important to note that *Whisper[s]* aimed to investigate the participation of a public audience with little past experience in using networked haptic and responsive technologies. It was therefore expected that most participants would spend a big portion of their initial exposure to the system in a mode of self-reflection and experimentation. The first relationship, revealed as soon as the participants put on the garments, was between oneself and one's own visualized physiological data. When this relationship became apparent, participants shifted their attention and intention to an exploratory mode, which slowly revealed the possibilities of being embodied through technological mediation in a digital networked space. The second relationship that was revealed was between self and others. Kozel (2007: 316) explains that some people were fundamentally uncomfortable with the procedure and chose to remain engaged with exploring their own data. Perhaps the most interesting relationship was the last one, self-to-system, where participants realized they were part of a holistic system, one that has been spawned into existence purely because they had entered the space. Furthermore, the space became a composition of a community of bodies with unmanifested potentialities that invited experimentation and self-reflection.

As this chapter concentrates on performers, there are some fundamental differences that need to be taken into consideration. As is evident in both *Ukiyo [Moveable Worlds]* and *Whisper[s]*, there is an urgent need to shift away from a purely visual feedback. In addition, Kozel states about her project: "The decision to focus on tactile and haptic outputs was born of an awareness that the visualization of body data out of the first version of *Whisper[s]* somehow limited the gestural and imaginative interaction" (Kozel 2007: 314). The one-to-one mapped interactions (Figure 12.1) greatly impacted upon the kinaesthetic and proprioceptive intelligence, as intention needed to be directed in controlling instruments. According to Birringer (2008):

> In such environmental practice there can be no set piece (choreographic), nor can one speak of improvisation, since the interactive potentials are shaped by particular aesthetic and mathematical principles requiring that performers adopt specific physical techniques to play the instruments of the medium and learn new proprioceptive and sensory process. (119)

It is therefore imperative that technologies do not limit kinaesthetic processes but rather become part of, and offer stimulations to, the mechanism that inspires and creates the intention of the dancer.

Polydorou identified three techniques that drove the intention of the dancer in *Ukiyo [Moveable Worlds]*. Firstly, it was the sense of being part of a narrative. Secondly, it was a sense of agency both in the long term (form the island) and short term (change the colour of the sky, plant, trees, etc.). Thirdly, and perhaps more importantly for this case, it was the reassurance of spatial belonging through technological embodiment with the virtual space (Polydorou 2011: 85). Kozel described the participants entering her networked space as hesitant, careful, always listening and aware, like they were "discovering the self anew, entering into a gestural dance with one's body in order to access things that were intimately familiar but strange at the same time" (Kozel 2007: 291).

Performers on the other hand need to be confident, without any doubts regarding whether the technology will work or not. In the case where the technology becomes an "instrument" for the dancer, such an instrument should be in a position to provide direct and sufficient feedback to establish confidence by providing confirmation about the performer's actions. Such an approach will form a technological "habit" that would potentially allow the experience of the "instrument" to become an extension of the body. Merleau-Ponty, in *Phenomenology of Perception* (2002), discussed how we learn to use tools in our everyday life. He introduced the importance of habit and skills and how they impact at a perceptual level on the body schema when it comes to manipulating objects. Through the process of acquiring skills the body schema changes, thus altering further the potentialities of carrying out an action. To a further extent, through this process, unfamiliar objects become familiar tools that are incorporated and act as an augmentation of the phenomenal body. This depends on the way that motor skills are involved in developing a body schema and how feedback enables the development of that motor skill.

The senses address the body schema to determine external properties of the environment and objects which are "near" or "far away", "large" or "small", "heavy" or "light", "high" or "low", and so on. Interestingly, the elements that allow one's body to distinguish the variation of external objects and the environment have evolving and interchangeable properties. In music performances, for example, such distinction is prominent to the instrumentalist where the performer's body schema evolves through years of practising and performing. The way that an instrumentalist experiences such external influences, such as how the instrument is registered on the body schema, depends greatly on haptic and tactile feedback. The body schema cannot be evolved through visual and aural feedback alone. An instrumentalist cannot develop any performative aspects without physically experiencing the instrument. Similarly the relationship between dancers and technology should provide some physical experience if possible to enable an expressive link between the two. Even though the performer controls and manipulates the properties of the system, the absence of haptic feedback creates a gap that limits the development of a new body schema. Hansen, who expanded the ideas of Merleau-Ponty into the domain of new media, argues that technologies can change or enhance our sensory experiences, consequently affecting our experience of embodiment. By envisioning a world with a fluid interpenetration of the virtual and the physical realms, he prioritizes touch, the perception of spatial depth and motor activity, claiming that the body can synthesize and complement our other senses, resulting in a holistic experience. Moving towards indirect interfaces (including optical, ultrasonic sensors or machine vision), creators of such performance systems generally prioritize the development of software techniques over physical interaction techniques, and the performer's sensual and cognitive needs are often not addressed.

A potential solution

Our aim is to provide and suggest a tool with which dancers and choreographers can further develop and explore the implications of vibrotactile feedback. We propose the use of vibrotactile feedback as a means to provide a creative multidirectional communicative channel between digital technologies and the dancers. In addition, the vibrotactile feedback, through the technology, can be utilized as a creative communicating link between the dancers themselves. At this stage, due to the wider implications of and approaches possible with vibrotactile feedback, we avoid suggesting any dogmatic "solutions". Instead we have explored the use of vibrotactile feedback through the examples provided in this chapter. Firstly we studied the performer-technology relationship and secondly the performer-performer relationship through the technology. The performer can benefit from the introduction of vibrotactile feedback as a means of communication while immersed in a technology-oriented environment. The technology and the overall system can provide, through vibrotactile feedback, meaningful information to the performer about current states and functions otherwise limited to visual feedback. The placement of the motors on the body has not been fully investigated creatively in a performance setting and can be a topic for further research.

In terms of hardware, we defined our needs based on the following three criteria:

- The device should be wearable and light in order to have minimum impact on the performer.
- The device must be wireless to allow free movement of the performer on stage.
- The device should be able to provide Pulse Width Modulation (PWM) in order to drive the motors.

Since the technology is not absolute, our approach is focused on how it can be best suited for the criteria above.[3]

Haptic interaction grid

Based on our research and artistic practice incorporating vibrotactile feedback into music and dance with digital technology, we have devised an analytical framework for haptic interaction in interactive performance. The framework consists of three primary interaction types, which we term "modes", and two secondary interaction types, which we term "scenarios". We illustrate this below through practical use-case examples.

Modes

The performer-technology mode concerns interaction between the human performer and the technology "system" (including all control-processing, sound generation, visual and haptic elements) within a digital and physical space. For example, we have used Microsoft Kinect to digitally capture movement of the human body to provide a virtual skeletal image in space representing the performer in a virtual world. For the purposes of this example we placed the performer in a virtual 3D dome.

When the skeletal image of the performer comes into contact with the boundaries of the virtual dome, a vibrotactile signal informs the performer about his/her actions (Figure 12.4). Even though such a relationship seems straightforward and relatively simple, it is nonetheless powerful as it provides informative feedback and thus confidence about the performer's actions as well as awareness of the performer's physical presence within the virtual world. In another example, the boundaries of the dome can change and interact with the performer depending on the type of movement and gestures initiated by the performer. The Inertial Measurement Unit (IMU) capabilities of the x-IMU (described above) can additionally provide information about the acceleration and position of the performer that can be further enhanced in the creative relationship between the performer and the technology. The shape of the dome may be a result of the movement and gesture of the performer during the performance.

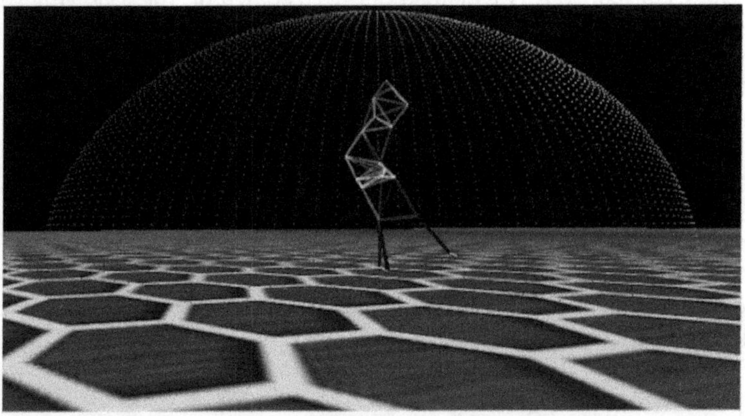

Figure 12.4 The performer performs within a virtual space and "feels" the boundaries of the dome through vibrotactile feedback. Courtesy of © Doros Polydorou

The performer-performer mode concerns the interaction between two (or more) performers mediated by a haptic communication channel. For example, two performers are able to interact through vibrotactile feedback and, in a similar setting to the performer-technology relationship, data captured from gestures and other actions carried out by one performer are transformed into vibrations communicated to the second performer. In this mode it is possible to experience different levels of density of the vibrating feedback. The received feedback is continuous and changes according to the movement and gestures of the other performer. Again it is important to mention the position of the motors on the body. For example, movement and the acceleration of the left hand are sensed as vibrations on the leg of the other performer. In addition, such creative communicative links between performers may be utilized in network performances. From the dome example above, a performer can determine the shape of the dome controlling the space "available" for the other performer to exist.

Scenarios

As interactivity subtypes for our proposed haptic interactivity modes, we propose three different scenarios of haptic embodiment that create a creative and corporeal link between two performers and the technology that is in use:

1. Co-exist: Performers are in a state of collaborative play. Their bodies move in unison around the space feeling a constant stream of vibration that symbolizes the union. As they move apart from each other, the intensity of the vibration decreases, and as they get closer it increases.
2. Intersection: Performers are in a state of competitive play. They compete for "ownership" of their own personal space (identified by the body tracking mechanism). Vibrations intensify as their personal space gets invaded.
3. Intimacy: In the third and final scenario, the performers are lying close to each other in an intimate position, having their body parts connected through technology. As the hand of one performer approaches the body of the other, vibrations on the second performer body intensify, creating a greater anticipation of the touch.

The previously described haptic interaction performance modes and scenarios can be combined to form an "interaction grid" that can be used as a tool for describing or conceptualizing an interactive process. This could be useful, for example, in analysing existing works or parts of works, or composing or choreographing new pieces. For example, in Figure 12.5, we show a grid with four states:

performer-technology / intimacy
performer-performer / co-exist

performer-performer / play + performer-technology / play
performer-performer / intimacy

These states could be viewed as a quasi-score indicating, for example, the progression of interactivity through the course of a musical work, or within a section of a musical work. Alternatively, this could be used as a tool in the description and analysis of existing works.

Conclusion

In this chapter we highlight the importance of sensorial feedback needed by performers in order to become more immersed in a virtual playground and share the experience of technology not only with the audience but among themselves. Based on the theories proposed by Mark Hansen and Merleau-Ponty, we suggested that attention should be shifted by enabling a creative approach and connection among dancers within the virtual space. This will allow performers to have an immersive experience of digital space and interactions. Furthermore, through vibrotactile feedback, the performers become more aware of the space and the virtual elements that are presented to the audience, but they are also aware of how other performers are interacting with the space and the technology. Three different scenarios have been chosen for experimentation and we have noted that the use of vibrotactile feedback, like any other technology, is subject to the creative intentions of the choreographer and the artist involved. Our framework for thinking about haptic interaction in the context of live performance is designed to assist in both the analysis of existing works and in the creative processes of producing new works with performers and interactive media. Vibrotactile feedback is not necessarily appropriate for all technology-related choreographic pieces, and neither does it imply any creative outcomes on its own. Due to the tactile properties, it should be seen as being like an instrument where experimentation and practice are essential. We propose

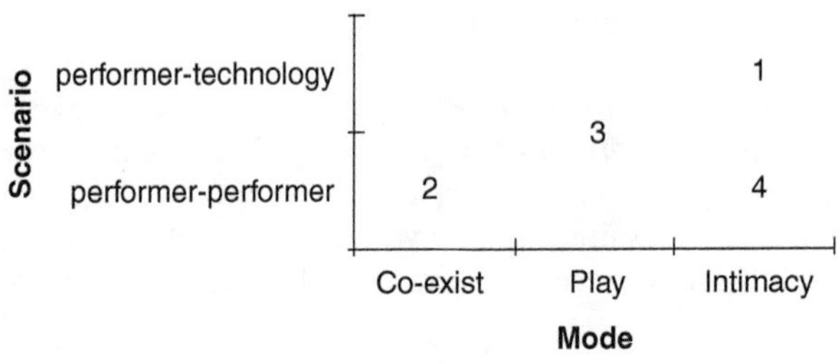

Figure 12.5 Haptic interaction grid showing four states distributed across combinations of mode and scenario. Courtesy of © Doros Polydorou

the use of vibrotactile feedback as a means of creative and corporeal linkage between the performers and the technology as well as between the performers themselves.

Notes

1. Other factors such as material, shape and temperature between two agents should also be considered.
2. Part of the performance can be found here: https://www.youtube.com/watch?v=mnCtnH1NdY0 (accessed 24 September 2014).
3. Even though many devices met our criteria to differing extents we decided to use X-IO Technologies' x-IMU (www.x-io.co.uk). The device is capable of providing up to 4 PWM outputs and can be connected to a laptop wirelessly via Bluetooth. The vibrating motor (www.precisionmicrodrives.com) is attached securely on top of the device with a plastic tape. The device weighs a bit over 50 grams including the motor, the battery and the plastic housing. It is easily attached around different parts of the body with a Velcro strap. In addition to providing vibrotactile feedback via PWM, the x-IMU contains a gyroscope, accelerometer and magnetometer, which can be used to gain positional and movement information. The data is managed through a bespoke software, IMU2OSC, created by one of the authors (http://birminghamconservatoire.github.io/IMU2OSC/ [accessed 17 March 2013]) that bridges the functionality of the device with OSC to allow flexibility and diversity in its use. The IMU2OSC software is able to provide data to and from the x-IMU. In addition, it allows network capabilities where a host computer is used as a bridge between many computers expanding further the creative possibilities.

13
I-CARE-US: Flying Robots and Human-Robot Interaction in Digital Performance

Fernando Nabais

Introduction

The impulse to create artificial creatures has been manifested throughout the history of human civilization in social, religious and artistic expressions. The ancient Greek story of Galatea – a statue brought to life by the goddess Aphrodite – or the Jewish legend of the Golem, a speechless anthropoid made of clay by humans, are just some of the examples of this human fascination with the creation of life and its representation. This pursuit has materialized in both physical and virtual forms. From the millenary shadow play through to techniques such as the famous Pepper's Ghost,[1] human beings have attempted to give life to the unseen. These constructs served many purposes, for example, to impress and sometimes to enhance the narrative power of storytelling, or else to induce fear in audiences, or merely to entertain a crowd.

Alongside these attempts at creating life in virtual forms, humankind has also tried to recreate it in tangible, physical form. Steve Dixon (2007) traces the "long historical lineage" of robot artworks and performances back to millenary forms of automata. These mechanical anthropomorphic figures capable of movement date back to the third century BC. Dixon mentions Joseph Needham's description of "a vast array of mechanical figures, animals, birds, and fishes constructed in ancient China, as well as mechanized chariots and flying automaton that he dates circa 380 BC" (Dixon: 282). Each new stage of technological development in human history generated new attempts at creating ever more sophisticated forms and concepts of artificial beings. The technology communities have devoted their efforts to solving many mechanical and control problems involved in developing machines that present effective movement and behaviour. With permanently increasing computer power, these machines are also gaining cognitive abilities capable of making decisions and collaborating with humans.

In parallel, "robotic artists with their work have been questioning our perception of science and technology as well as its influence on society, staging robots in typical human analogies and situations" (Ghedini and Bergamasco 2010: 4). In areas like social robotics, this mutual influence of science and art is increasingly evident and beneficial.

Robotic art/robotics in art

In his 1997 article "Foundation and Development of Robotic Art", Eduardo Kac marks the beginning of robotics in art with four works from the 1960s. Although acknowledging previous work on robots in the 1950s, oriented to entertainment and scientific research, Kac highlights the "pivotal" artistic works of Nam June Paik and Shuya Abe's *Robot K-456* (1964), Tom Shannon's *Squat* (1966), and Edward Ihnatowicz's *The Senster* (1969–70) because "they acquire a particular meaning when reconsidered today, since seen together they also configure a triangle of new aesthetic issues that has continually informed the main directions in robotic art" (Kac 1997: 61). For Kac, these works established three creative directions in which robotic art has developed: remote control, cybernetic entities, and autonomous behaviour.

Kac's ontological trilogy offered an important initial framework for robotic art pieces, from his own 1989 work with Ed Bennett, the telerobot Ornitorrinco, to the biomachines of Ken Rinaldo or Hiroshi Ishiguro's Geminoid F, starring in the theatre play *Sayonara* in 2010. However, as the field of robotics has evolved, so too have the possibilities available. Even the structural concept of what a robot might be has been questioned and extended to new expectations by the users and the audiences.

While trying to establish a chronological outline for robotic art, Kac points out that "[o]ne of the most problematic issues of robotics in art is the very definition of what a robot is" (1997: 60). Fiammetta Ghedini and Massimo Bergamasco reinterpret Kac to explain that "robots are an assemblage of mythological and folkloristic traditions, popular science-fiction imagery, often detached by the operational definition of robots as found in scientific research and industrial applications" (2010: 3). For Louis-Philippe Demers, machines are "it selves" media. In his resumé he defines himself as "a multidisciplinary artist using machines as media" (V2, 2012). Probably the only common point in most analyses and artworks involving robots is, as Kac points out, the precedence of behaviour to the detriment of form (1997: 60). Demers reinforces this idea by explaining that "[c]ontemporary robotic art brings a new aesthetic dimension that prefers modelling of behaviour over a representative form or a mimetic static object" (2010:2).

In relation to the performance context, David V. Lu's *Ontology of Robot Theatre* (2012) aims to categorize and classify robots based on the different algorithms that can exist behind the curtain to create these performances. Lu establishes a continuum of robot actors from a motionless device on stage that simply beeps or turns on a light, to a completely autonomous robot capable of interpreting a character in *Hamlet*, making decisions about its performance and interacting with other actors. Lu defines a bi-dimensional matrix with autonomy as one of the axes, going from human to hybrid and algorithmic, while the other axis, control, has values of open-loop, closed-loop or free. These classifications can accommodate the inclusion of all sorts of mechanical devices that have been incorporated in stage representation, from *puppeteered* robots to autonomous intelligent agents.

Demers (2010: 6) classifies the forms of robots on stage from agentic (intentional) to automatic (functional), raising the question, "at what point does a machine performer graduate from an automaton into an agent?" That graduation process has been facilitated by contemporary robotics development, where advances in sensory technology and artificial intelligence have been creating completely autonomous reasoning machines. The availability of a multitude of robotic development platforms, which solve and abstract their functional control and allow the development of layers of intelligent behaviour, has oriented the work of many artists and roboticists, validating Kac's and Demers' analyses of "behaviour modelling over form".

Flying robotic performers

From the large range of locomotion and movement strategies that robotics researchers have created, a particular type of robotic device has gained enormous popularity in recent years. These are Unmanned Aerial Vehicles (UAVs), commonly known as drones. They exist in a variety of flying shapes and their applications have been rapidly diversified: from the inevitable military missions, to providers of previously inconceivable cinema shots; for surveillance or simple physical manipulation and entertainment.

Of the many forms flying robots can assume, one in particular became an important field of research and a commercial success in the gadget markets: quadrocopters.[2] These small flying devices with four propellers achieve impressive stability due to the use of high-precision mechanisms and miniaturized computational power that runs very effective software-control algorithms, transforming them into highly flexible and easily controllable drones. An example is the AR.Drone by Parrot, whose enormous commercial success changed the company's strategic orientation from a producer of Bluetooth car equipment and smartphone speakers into a 500-employee enterprise, focusing intensively on their UAV products.

The DIYdrones community (DIYdrones 2013) became very active also, and projects like Arducopter, among others, made it possible for anyone to construct their own customized drone. Their application in artistic contexts started to appear recently and this section will focus on several pieces that include or are based specifically on aerial robots

Emerging from the research of the Institute for Dynamic Systems and Control at the Swiss Federal Institute of Technology (ETH) in Zurich, *Flight Assembled Architecture* (D'Andrea 2012) claims to be the first installation to be built by flying machines. Among other research subjects, this institute has dedicated considerable effort to investigating precision flight control of quadrocopters. Videos from their Flying Machine Arena[3] show the prowess of these machines. Equipped with pads, they are seen perfectly juggling multiple balls, or else performing in a choreographic fashion (e.g. *Dancin' Drones: Rhythmic flight of multiple quadrocopters*, 2010). Using gesture control with Microsoft Kinect, modular flying vehicles are capable of assembling, coordinating, and taking flight.

Building on these technological and scientific achievements, ETH professor Raffaello D'Andrea, together with architects Gramazio & Kohler, conceived *Flight Assembled Architecture*. This dynamic installation, exhibited at the FRAC Centre, Orléans, France (2011–12), creates an architectural vision of a 600-metre-high "vertical village" for 30,000 inhabitants, modelled in a 1:100 scale, which is assembled over several days by flying machines. One and a half thousand prefabricated polystyrene foam modules are precisely transported and placed by these aerial robots with the intention of addressing "radical new ways of thinking and materializing architecture as a physical process of dynamic formation" (D'Andrea 2012).

Two other recent works have merged technological research institutes and companies with artists to give starring roles to flying machines in artistic creations. *Meet your Creator*[4] was a robotic dance performance presented at the Saatchi & Saatchi New Directors' Showcase 2012 in Cannes. This live theatrical performance included 16 moving head spotlights, and 16 flying robots each equipped with LEDs and motorized mirrors that reflect and divert light beams on the stage. These flying robots created floating light sculptures while dancing (GRASP 2012).

The Ars Electronica Futurelab and Ascending Technologies GmbH, a company dedicated to the development of sophisticated flying machines, presented an artistic event performed by a swarm of quadrocopters at the Linzer Klangwolke (Linz Cloud of Sound) in 2012. In this open-air multimedia musical event, 50 quadrocopters swarmed up and around, changing colour and appearing almost like stars in the night sky. This magically lit, computer-controlled swarm symbolized the virtual network and digital communities of our time and made impressive movements collectively for a tri-dimensional effect. This particular event was the first presentation of this new approach to the visual representation of moving images. The author's intention is to extend the control to a much larger swarm of flying vehicles where each one represents a *spaxel*, i.e. a pixel in space, which is the "novel concept and paradigm [of] a new kind of spatial display" (Hörtner 2012: 1). Like pixels, each of these spaxels represents a controlled colour element. However, we could think of spaxels as pixels that got free from their static bi-dimensionality to gain a physical presence and the capacity to move freely in the real tri-dimensional space around us.

This project establishes a close parallel with *Flyfire* (2010), a project initiated by the SENSEable City Laboratory in collaboration with ARES Lab (Aerospace Robotics and Embedded Systems Laboratory). Its concept-videos present a swarm of pixels in space, configuring a "free-form display" capable of displaying a raster image of Mona Lisa or a vector display of a 3D face (Flyfire 2010). Besides static images, the real-time adaptable motion of the pixels allows for a spatially animated viewing experience. In the case of this "display", the illusion of movement is not only created from the colour transition of each pixel to form each new image – as in traditional 2D displays – but also from the fact that each of these pixels moves physically in space. To form a new image, each pixel assumes a new position in space at the same time as it changes its colour. Like the work of ARS Electronica Futurelab, "*Flyfire* serves as an initial step to explore and imagine the possibilities of this free-form display: a swarm of pixels in a space" (Flyfire 2010).

As flying assemblers of physical objects, sculptors of light or re-inventors of the concept of the pixel, all the previous works focus on the precise control of flying vehicles. In some of the cases this relies upon the ability of remote pilots, in others on algorithms running over sophisticated motion capture systems that manage to locate these vehicles in tri-dimensional space with high precision. Despite these differences they all share one common characteristic: the astounding technological prowess of these flying devices are centre stage and the human element is simply a spectator of this aerial robotic ballet.

Flying robots and human performers

In the previously mentioned projects aerial robots are the stars and sole performers, but humans are also starting to star alongside these mechanical actors and dancers. The coexistence of human and flying robotic performers raises many issues besides control and synchronicity, and thus presents another level of challenge. At the same time, it represents a rich territory in which to study and research a set of multidisciplinary and interdisciplinary aspects related to performance studies and human-robot interaction, among many other fields at stake in these types of productions.

A very adequate project to establish the transition from the previous section is *[VOILES / SAILS] Self-Assembling Intelligent Lighter-than-air Structures*, by Nicolas Reeves and Éric Poncet (2005).[5] This project represents an important example of science and art cooperation materialized in aerial robots. The type of flying vehicle involved in this work is a helium-filled balloon with locomotion capabilities, commonly known as a blimp or "aerobot". *[VOILES / SAILS]* is composed of four "aerobots" (one 180 cm prototype and three 170 cm beta versions), equipped with sensors, ducted fans, wireless communication and a 40G fully functional UNIX computer. The piece started as an installation in which a swarm of flying automata evolved in simulations of "assemblages and behaviours derived from the observation of animal societies (coral reefs, anthills) that can collectively produce high performance structures with extremely limited abilities at the individual level" (Reeves et al. 2005).

In these initial presentations the system was designed so that the interaction changed or evolved according to data sampled from the environment: noise levels, temperature and lighting, among others. Besides the system reaction to environmental conditions, the installation set-up considered interaction strategies with the visitors through a variety of sensors and actuators. In recent presentations, this started to include scripted interactions with actors emerging out of the observation of the audience's reactions in the many presentations over the years. These observations permitted the design of a more suitable sensory system and more effective behaviours leading to the development of hybrid choreographies between human and robotic performers. The aerobots' expressiveness through video projection on their surfaces from inside their shells; the stability they present to a certain extent of physical contact; the array of sensing modalities, and the emergent intelligent behaviours generated: all provide a great deal of potential that has been explored in these more recent iterations of the work.

Adding a contemporary edge to a classical play, the 2009 production of William Shakespeare's *A Midsummer Night's Dream* by Texas A&M[6] included seven flying robot "fairies" (one autonomous, military-grade flying robot and five smaller remote-control devices) performing along with human actors (Murphy et al. 2010: 44). In this rendition of the play, the author's intentions were to reinvent its format, blending classical and modern elements while, at the same time, grasping an invaluable "opportunity for interdisciplinary and experiential learning in the fields of acting, directing, scenography, computer science, computer engineering and electrical engineering" (Guerin 2010). These aerial machines, originally utilized for robot-assisted search and rescue missions, were cast as on-stage co-actors, participating in the action. They even had a stand-alone scene.

Central to the abovementioned works is the concern to overcome what roboticist Heather Knight calls the "the gulf between props and characters" (Knight 2011). Technical director Robin Murphy's description of the audience's attitude towards the flying robots whenever they crashed provides an interesting insight into the importance of the human performer behaviour in the construction of viewer perception. Murphy highlights that when a robot crashed in the middle of the audience, it was often treated roughly as a toy or as a simple object by the spectators, who threw it back carelessly to the technicians. However, when the audience had previously observed an actor treating the robot gently as a pet or a baby, that behaviour was replicated by the spectators. Murphy concludes that "the audience, over the course of a performance, changed their perception of the robots from Props to Social Actors" (Murphy et al.:152). This conclusion confirms what Heather Knight refers to as the "power of interaction partners to shape robot attributions" (ibid.).

In 2011, the American dance company Pilobolus cooperated with the Distributed Robotics Laboratory of the MIT Computer Science and Artificial Intelligence Laboratory (CSAILS 2011) to create *Seraph*, a stage work where a human performer develops a friendship with a remotely controlled flying device. With the backdrop of a Schubert piano trio – an unusual soundtrack for this type of stage work, especially considering the ever present motorized sound of the UAVs – the human and the robotic performer explored their differences and similarities co-evolving in very close proximity, creating visceral, yet sometimes humorous situations. The lighting design played an important part in expressing the mood and affective states of the performers. Through the LEDs in the quadrocopters and a simple flashlight used by the human dancer, the dialogue between the two became clear to the audience, adding another expressive dimension on top of their movements. The drama increased when a second quadrocopter invaded the stage, showing its jealousy towards this new human rival, going so far as to expel him from the stage. The three performers engaged in a highly physical and dynamic performance that led to a climactic finale where the tension between the three culminated in a harmonious ending, expressed in a celebration of friendship through choreography.

Like the previous work, *Copter*, by Niva Kov, establishes a duet between a remote-controlled helicopter and a dancer. It was a Place Prize for Dance 2012 commission and premiered on 20 September 2012 at The Place, London. The creator explains that "the piece is aiming to question our relationship with objects in a sometimes playful, sometimes unsettling way" (Place Prize 2013). As in *Seraph*, Kov elects this aerial device as her dancing partner. In *Copter*, the helicopter becomes the agent of an off-stage pilot, capable of performing constant changes in the power dynamic, expressing intentions and emotions as expected of a performer. The flying object with which Kov creates a relationship is a commercially available helicopter, popular among teenagers. Maybe because of this familiarity, its narrative carries a sense of "childhood fantasies viewed through the lens of adulthood, and fantasies of being an adult during childhood" (ibid.). It is interesting to note that many critics found the pilot's performance and his capacity to imbue the helicopter with personality to be the central focus of the piece, to the detriment of the human dancer's efforts to explore possible movement interplays with the flying object (Cloud-Dance 2012).

All of the above works have relied almost exclusively on talented independent pilots who remotely confer intentionality and affect to the flying characters. However, the work of the Japanese collective Rhizomatiks and the dance group Eleventh Play dares to conjoin the movement of the human performers with that of the autonomous flying robots on stage (Rhizomatiks 2014). Although maintaining a manual safeguard system to land the quadrocopter in case of a system failure, three-dimensionally accurate movement is obtained through a high precision motion-tracking system, as in the case of the previously mentioned project *Meet your Creator*. In order to be tracked by the high-speed video cameras, the UAVs are augmented with reflective markers that allow the system to permanently know their position, rotation and inclination, allowing for a very precise control of their movements. The human performers also carry the same markers so that the system can track their position and movements as well, allowing them to control the flying robots. In the work of Rhizomatiks and Eleventh Play, the relationship between three quadrocopters and three human performers evolves in a highly coordinated and collective choreography that attempts to express the "beauty and difficulty which come from resistance to gravity and inertia" (ibid.). Coming from a strong computer-graphics background, the technical members of the creative team were inspired by physical laws such as gravity, inertia and air resistance, which in the virtual world are merely simulated and not constraining. These differences are highlighted when the human performers are gradually replaced on stage by a video projection that augments geometrically the movement of the drones. This piece is the starting point for a line of work that the author intends to update in the future.

Although all the works mentioned in this section also require precise control of the flying vehicles, in these cases the movement is not just functional, a means of assembling or controlling something. The presence of the human element contributes to constructing in the audience a different layer of perception of these movements, assigning them intentions and emotions besides functionality. Alongside their human partners, the robots in these pieces become artificial creatures and ascend to the role of characters, contributing to the narrative by expressing their emotions and driving the drama.

I-CARE-US: learning to fly

This inquiry also aims to situate the author's research and creative work in a more historical perspective. In collaboration with choreographer Stephan Jürgens I have been exploring since 2011 the aesthetic potential of physical computing in its relation to the human body, while, at the same time, extracting insights from digital performance to question and enhance this hybrid relation. This continued collaboration materialized in the performance *I-CARE-US*, which was premiered in Lisbon in May 2014 (I-CARE-US 2012). This section will elaborate on some of the ideas, objectives and conclusions of this stage work.

An important dramaturgic reference for this work is necessarily the famous myth of Icarus. In fact, it presents myriad metaphors that can be established about the relationship between humankind and technology, as well as the many dystopian futures it can create. These scenarios became part of the public perception of drones, especially given their highly mediatized military applications. The *I-CARE-US* narrative revolves around fragments taken from different versions of the myth that converge to question and explore the many relational dimensions of these flying devices with human beings.

The use of aerial remote-controlled vehicles could also be seen as the hawking or falconry[7] of the digital age. From a hunting strategy to a popular sport and status symbol among the nobles of medieval Europe, the use of birds of prey by mankind has been traced back to 2000 BC. A flying predator leaves the hand of his master in a surgical strike to capture a prey. Through training, this bird becomes almost an extension of his master's killing will, by carrying out his lethal intentions. In shamanic rituals, the transcendental embodiment of a bird through altered states or religious ecstasy induces insights of the animal's own aerial vision. Through communication with the animal spirit the bird's vision becomes the shaman's vision. In a modern version of these practices, UAVs become the bird of prey in many combat and surveillance operations. At the command of their masters they fly away, either to kill or to capture the most valuable prey: information. Equipped with modern sensing technology, they have the capacity to analyse all sorts of data in challenging scenarios, while maintaining the safety of their human operators. For these agents of someone else's intentions, the "prey" can now be all of us. How do we feel when we are followed and analysed by such a predator? Do we think of the object directly or the person behind it when we become the subject of its attention?

In the beginning of *I-CARE-US* the audience becomes the prey observed by this creature that, although remotely controlled by a human, presents features of an autonomous being. By directly confronting the audience with the flying robot behaviour, our intention is to elicit a perceptive duality in the spectator. The presence of the pilot on stage in full view of the audience provides evidence of the remote control of the vehicle. Does the audience member maintain this perception while being directly confronted with this mechanical object? Do they detect the subtle relation between the flying robot behaviours and the human performer/operator's movements and expressions? Or in such a situation of stress and attention focus, do they attribute autonomous intentions to this unfamiliar species, even if momentarily?

Beside their autonomous activities, robots can also act as communication mediators between humans. As telepresence devices, drones become the "animal spirit" inhabited by the user, a kind of a technological shaman. *I-CARE-US* is an exercise in theatre robotics that tries to confront a theatre audience with a not-so-distant future where UAVs will be part of our domestic airspace. An important part of the communication between humans is non-verbal. What we affectively convey through our postures, gestures and gaze, among many other aspects of physical expression, can sometimes say much more than words. Human-robot interaction research has been dedicating special attention to these aspects of communication (Dautenhahn 2007). The way robots express their intentions and attitudes is crucial for their acceptance by their users. In many of the shapes that robots have acquired, these aspects are significantly mimicked from the human form. However, robots have also taken many non-anthropomorphic shapes that maximize their range of applications and their kinetic efficiency. These forms can have inherent constraints that minimize expressive potential. Nonetheless, human-robot interaction researchers have been studying strategies to overcome these limitations through movement, light and sound, among other aspects (Bethel and Murphy 2008). These strategies can also be very relevant in their theatrical applications by providing these robots with enhanced performative capabilities, allowing an expressiveness that can convey personality and intention.

One final source of historical inspiration for us is the work of Italian futurist artist and First World War aviator Fedele Azari, who provides an inspiring reference baseline with his *Futurist Aerial Theatre Manifesto* (Azari 1970 [1919]). Assuming that the airplane becomes an extension of the pilot's body, Azari declares in his text that flight can be an artistic expression of states of the pilot's mind. He even proposes a taxonomy of flight manoeuvres associated with affective states:

> In fact, looping denotes happiness and spins denote impatience or irritation, while the repeated alternation of wings from right to left indicates light-heartedness, and long falls, like dead leaves, give a sense of nostalgia or fatigue.

Although the vehicle used in *I-CARE-US* is very different in shape from Azari's airplane – in particular, because it has no wings – these manoeuvres can easily be transferred and some of these empirical conclusions may be implemented by the quadrocopter.

In *I-CARE-US*, the flight behaviour results from a mixed mode of direct remote control, in conjunction with pre-defined manoeuvres that can be triggered at any given moment. To allow a more accurate control of the UAV and provide a more expressive interface for the pilot, a specific software application was developed. This application included a set of features that enhance creative freedom and open many aesthetic possibilities at different levels. It became possible to record flight data and to interface with different UID (user interface devices) such as joysticks, route video and audio signals to external equipment, among others. The possibility of recording flight data allowed the showcasing of creative visualizations of the UAV movement through video projection, providing a second representational layer to the creature's choreography. The system also enabled the addition of the

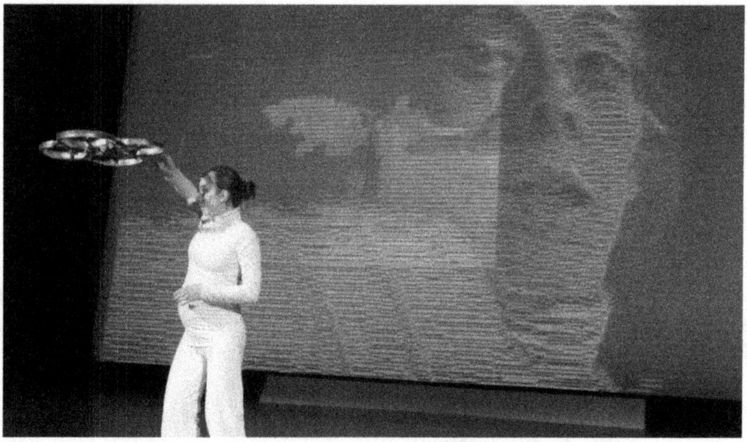

Figure 13.1 *I-CARE-US* (2013), performance for dancers and quadrotors by Fernando Nabais and Stephan Jurgens. Photo courtesy of © Fernando Nabais

recorded data, allowing the creation of a database of flying behaviours that became the dramaturgical vocabulary of the robotic creature. The possibility of routing the video signal from the UAV cameras to another computer running a real-time video-editing application enabled the creation of a rich imagery of the creature's inner vision. Equipped with two video cameras, one pointing forward and the other pointing to the ground, the UAV captures its surroundings and provides a remote vision to the pilot. By capturing the spectators and the human performer in real-time and projecting this video signal on stage, *I-CARE-US* creates a sensation of privacy disclosure that may induce a feeling of discomfort in the audience.

The exhibition of this flying point of view finds a strong analogy in another futurist movement, Aeropittura. The futurist fascination with the machine and particularly with airplanes contributed to the changing of painting from a central perspective to a terrestrial free perspective embodying dynamism and velocity: "We Futurists declare that [...] [t]he shifting perspectives of flight constitute an absolutely new reality, which has nothing in common with reality as traditionally constituted by terrestrial perspectives," they said in their manifesto (Balla et al. 2009 [1929]: 283).

Aeropittura was an important source of inspiration for the way video projection is explored in *I-CARE-US*, considering that it provides the possibility of previously inconceivable cinematic perspectives. This kinetic videography can also be understood as the creature's inner vision, enhanced by specific motion graphics work that merges evocations of First World War aerial reconnaissance images with the present point-cloud representation derived from depth-sensing sensors, such as Microsoft Kinect.

Interspecies dialogue

Initial experiments in the process of human-machine acquaintanceship were established without a priori familiarity between the human performer and the quadrocopter pilot. Very little was said to each of the performers about the other and what was required from each one of them. The guidelines were simply to go on stage and explore the interaction possibilities that the system presented as if the stage were a habitat to be shared by these creatures. Nothing was said to the dancer about the characteristics and capabilities of the quadrocopter, nor did he know anything about the aptitude of the pilot and the system. The idea was to allow a process of mutual discovery that could lead to empirical impressions and possibilities for dialogue.

What happened was that even without any special manoeuvres from the pilot, the movements that emerged from the stability disturbance provoked by the dancer were a rich expressive territory to be choreographically explored. The UAV stability is the result of sophisticated algorithms that fuse its set of sensors to make it hover, fly and respond to the pilot's commands. A change in the set of conditions that stabilize its flying behaviour will result in a crash – in case of a radical disturbance – or provoke a manoeuvre that will try to correct the perturbation.

Given the typical shape of its exposed propellers, direct physical contact with the quadrocopter is impossible without harming the human element and provoking an inevitable crash of the aerial device. However, when the drone is equipped with outer shells designed for indoor flight it is possible to establish body contact without necessarily causing the UAV to crash. The exploration of these limits between both performers – a kind of interspecies contact improvisation – was an important starting point for the development of choreographic ideas for *I-CARE-US*.

Embodiment and outer shell design

As in the case of human performers, appearance, alongside behaviour and context, is a crucial aspect of how the audience perceives and interprets a robotic actor's personality and intentionality. Unlike terrestrial robotic shapes, quadrocopters present an enormous challenge for costume designers due to the implications of form on aerodynamics, which can drastically affect its flight capabilities. It is not surprising that the most common expressive add-ons for an aerial robotic performer are artificial lights, normally through LEDs powered by the device's battery. It is important to highlight that two critical factors for shell design are the UAV payload – how much weight it can carry – and battery autonomy. The quadrocopters used in *I-CARE-US* have a payload of around 150 grams and a maximum autonomy of 12 minutes. These two characteristics are interdependent, meaning that a heavier load implies a reduced autonomy. The type of flying manoeuvres also have implications for the quadrocopter's autonomy and dynamic flight patterns, draining the battery much faster than a stable, steady flight.

Although light and colour can be effective in conveying affect and emotion (for a study of its codes and significances within the robotics community see Bethel and Murphy 2008; Harrison et al. 2012; Paiva et al. forthcoming), this is just a small part of the plethora of communication channels that can be called upon to enhance the quadrocopter's representational capabilities. Another aspect of emotion communication directly related to embodiment is pose. We get important emotional cues from the postures of our interlocutors. In the case of quadrocopters, because they have a rigid body, this aspect is very complicated to achieve. In the most logical biomimetic reference (i.e. birds), it is acknowledged that there are several clues for interpreting behaviour, such as flight pattern, posture, sound, but also through the position of feathers. While feathers play an essential role in terms of insulation, waterproofing, flight and camouflage, they are also an important aspect of behaviour display. Several species of birds use their feathers as visual displays to communicate very effectively their intentions and attitudes. Although roboticists are studying quadrocopter enhancements that focus on pure functional aspects – grips are a concrete example – it is our conviction that through the application of light smart materials[8] it is possible to develop shells that can hugely enhance the expressive potential of these devices, both for theatre purposes and for human-robot interaction.

Sing like a bird

Sound is naturally a very important component in *I-CARE-US*. A character with an audio presence such as a drone cannot be bypassed and requires inevitably a special approach in the way it gets integrated in the general sound design. This flying performer comes with its own striking voice that conjures strong impressions in the listener. In fact, the inherent sound of the four UAV engines is unavoidable, especially indoors, and provides an immediate component of the flying creature's personality. Complementing this perception, the sound of the UAV engines is a natural illustration of its movement, with its inevitable roar every time it has to execute a manoeuvre. It is not just a coincidence that we may find again in the words of Fedele Azari, distant in time and yet so close in context, the same questions and conclusions:

> [A] sense of characterization will be given by the type of airplane, the voice of the motor, and the diverse lines of flight. (Azari 1970 [1919]: 129)

In his manifesto, he then proceeds to establish an association in terms of genre between different planes and their types of engines and flight characteristics, going as far as to consider that some of them have feminine and masculine differentiating aspects. Although Azari's aerial theatre was naturally an outdoor spectacle, the inherent sound of the plane's engine was too important to be ignored by a futurist artist:

> The voice of the motor can be regulated to fullness or reduced, broken up into clearly defined and precise points, or modulated in scale from high to low, thus forming a very effective way to supplement music with expressive noise. (Azari: 129)

In cooperation with his friend Luigi Russolo, Azari invented a special type of wood system for the planes that could, in his words, "regulate the motor's sonority without modifying its potential" (129). In its computational formats, music has been a particular focus of research and creation in terms of systems and technologies that enhance its interactive potential. From the 1986 *Hyperinstruments* projects by Tod Machover,[9] to today's hugely successful video-game *Guitar Hero*, interactive music systems are a fertile territory that has been extensively explored. Software frameworks such as MAX/MSP or Pure Data became almost a standard tool for electronic musicians. Their ability to connect to a large variety of sensors and actuators (musical and others) allowed the development of many digital media masterpieces and a plethora of new original instruments.

In *I-CARE-US* the drone becomes an instrument also. The fact that the custom software application developed for *I-CARE-US* allows the routing of the drone's sensors data to other computers running audio software, such as MAX/MSP, enabled the possibility of creating a real-time interactive soundscape. The evolution of the drone on stage modulates components of the soundtrack in order to enhance its dramaturgical intentions. When at a certain point in the piece it lands and becomes handheld by the human performer, the drone becomes a musical instrument that allows the dancer to control and manipulate "its voice", as if the flying creature is reacting to the movements inflicted by the human character.

Although this performance intends to provide research results in the different aforementioned areas, it has been guided by the creative development of a stage performance, with its narrative and dramaturgy as a starting point to analyse and dissect aspects of robotic behaviour and human-robot interaction. The author hopes that *I-CARE-US*, and its associated research work, may contribute to the questioning of the unstoppable presence of drones in our society and to the improvement of their social capabilities, leading to a socially conscious evolution of their roles and applications.

Notes

1. "Pepper's Ghost uses a large pane of plate glass and carefully controlled illumination to allow audience members to view the reflections of hidden performers alongside performers who are seen directly on stage ... The effect became wildly

popular among Victorian audiences and was used in lectures, traditional full-length theatre pieces, novelty presentations and touring fairground exhibitions during the latter part of the nineteenth and the beginning of the twentieth centuries" (Reilly 2013: 198).

2. The terms quadrocopter, quadricopter, quadrotor, among others, are often indistinctly used when referring to the typology of a flying device with four engines and four propellers. In academic and industry communities it is common to use the term that defines their vehicle category: UAV, or even UAS – Unmanned Aerial System.
3. See: http://flyingmachinearena.org/videos/ (accessed 17 March 2015).
4. This project represents the collaboration between a large team of artists and engineers directed by the Marshmallow Laser Feast and Saatchi & Saatchi creatives Jonathan Santana and Xander Smith with the technology of KMel Robotics. Behind KMel Robotics are Alex Kushleyev and Daniel Mellinger, graduates from the University of Pennsylvania GRASP Lab (General Robotics, Automation, Sensing and Perception). Together with Professor Vijay Kumar they created the now famous swarm of quadrocopters performing the James Bond theme.
5. This project started in 2005 but had several iterations throughout the years and continues its research purposes at the time of publication of this book.
6. A collaboration between the Department of Computer Science and Engineering and the Department of Performance Studies at Texas A&M University, College Station, TX, USA.
7. *Hawking* or *falconry* is the hunting of wild quarry in its natural state and habitat by means of a trained bird of prey, such as a hawk or falcon.
8. Smart materials possess the capability to change one or more of their characteristics in reaction to changes in the environment. Some examples are Shape Memory Alloys (SMA) and Electroactive Polymers (EAP), which contract or expand according to changes in temperature or magnetic fields (both controllable by voltage).
9. See: http://opera.media.mit.edu/projects/hyperinstruments.html (accessed 17 March 2015).

14

Embodying Digital Creativity: Designing Computer Tools to Support Spontaneity and Creative Work in the Digital Arts

Nicholas Toothman, Tyler Martin and Michael Neff

Introduction

"An essential voice in creatively based thinking is the language of the body..." (Johnston 1986: 21). Depending on your background, claims such as this may seem obvious or surprising, but are they actually true and can they be validated? If true, can we use this insight to design computational tools that better support creativity? These questions drove an extended investigation of how increasing embodiment impacts upon creative processes when using computational tools for artistic tasks. Unlike Lise Amy Hansen's *Sync* tool (Chapter 7 in this volume), which aims to enhance creativity in dance and the movement arts, our work is focused on supporting creativity for common tasks in computer graphics and animation. During this work, we built software tools designed to support greater spontaneity and embodiment: features common in artistic practices like theatre, but poorly supported in most computational tools. We then studied how working with these tools impacted upon users' creativity. Preliminary results indicate that embodiment does appear to enhance certain factors related to creativity. This chapter will first summarize some of the relevant theoretical background and then describe our investigations into the role of embodiment in supporting creativity. It concludes with a discussion of lessons learned and next steps.

In developing a framework for creativity, Johnston (1986) breaks the creative process into a series of progressive stages, listed with their associated activity in brackets: pre-axis (incubation), early-axis (inspiration), middle-axis (perspiration), late-axis (finishing and polishing) and integration (into a personal and cultural context). It is the early phases of incubation and inspiration during which embodiment and spontaneity seem to have an especially strong role to play. These stages are also poorly supported by computer tools for graphics and animation. Indeed, the early development of animated scenes is dominated by creative exploration, but this work is normally done entirely outside of the realm of computational tools (Panko 2005, Kapijimpanga 2005, Burke 2007), by acting out the scene, seeking reference footage, writing or drawing thumbnail sketches. Computer tools are only used once the major decisions have been made, after much of the creative exploration is done. It is worth noting that the early period of heightened creativity will often require two out of the five days spent on a scene in a common animation cycle (Burke 2007).

Computer tools see limited use in the main creative phase in part because they are slow to use and limit spontaneity. Exploration through means such as acting is fully embodied, versus an almost entirely mental process when the animator begins to work in a computer keyframe system. The gestures of moving the mouse to adjust a control have no correspondence to the intended motion of the character. Along with offering advantages related to response time and aiming accuracy (Mazalek et al. 2011), embodied exploration allows the animator to access a much wider range of faculties such as psychological recall, muscle memory, proprioception, visual feedback and the felt effects of gravity. Through improvised movement, various possibilities are developed, examined, enhanced and/or rejected (Halprin 2002). Movement metaphors can express thoughts that are difficult to communicate otherwise: "What is very real and understandable in movement does not necessarily have an equivalent in words" (Blom and Chaplin 1988: ix). The sections that follow further elaborate on the role of spontaneity and embodiment in creative activity, survey relevant background material and describe our experiments to date building novel tools.

The case for embodiment and spontaneity

Techniques that encourage spontaneity and those that rely on embodiment are often the same or closely intertwined, with embodiment appearing to support spontaneity. As Shusterman (2008) and Dissanayake (1992) have pointed out, we have inherited a Western philosophical legacy, starting from Plato and reinforced by Descartes, that denigrates the role of the body while elevating the role of the mind. While they observe that this bias has been challenged by many philosophers in the last hundred years (with some going so far as to develop body-based philosophies of aesthetics), this legacy is no doubt reflected in the development of computer technology, where early artificial intelligence was developed largely without reference to the body and the dominant computer interface remains a keyboard and mouse. Most computer systems take a fundamentally disembodied view of their users.

As Richard Shusterman (2008) explains, this view is deeply limited, as "...the body constitutes an essential, fundamental dimension of our identity. It forms our primal perspective or mode of engagement with the world, determining (often unconsciously) our choice of ends and means..." (2–3). Theatre has often taken a deeply embodied approach to exploration. For instance, the renowned playwright and director Bertolt Brecht would not allow discussion in rehearsal. All ideas had to be tried – in the body. Director, movement educator and Grotowski disciple Jennifer Tarver (Tarver and Bligh 1999) instructs her students during workshops with the perhaps unintuitive phrase: "The body is smarter than the mind." For acting theorist and director Jerzy Grotowski the goal of a training method is to develop the plasticity of the body such that the actor perceives a stimulus and reacts, rather than perceive a stimulus, consciously think about it, and then react (Grotowski 1968). Such methods try to access a pre-conscious thought, the flow of ideas that comes before conscious deliberation. Important to creative engagement and inspiration, this idea will be further explored in the discussion of spontaneity. Johnston argues that in the pre-axis stage of the creative process, knowing is based on the "kinesthetic language of the body" (1986: 84).

Numerous facilities within our bodies are often ignored in our society. Beyond the five classical senses, neurophysiology identifies the somesthetic senses that "are often divided into exeroceptive (relating to stimuli outside the body and felt on the skin), proprioceptive (initiated within the body and concerned with the orientation of the body parts relative to one another and the orientation of the body parts in space), and visceral or interoceptive (deriving from internal organs and usually associated with pain)" (Shusterman 2008: 2).

Embodiment and spontaneity: support tools

Warren Lamb, a contributor to the development of Laban Movement Analysis, divided movement into two broad categories: gestures and posture changes (Lamb 1965). Gestures are movements of parts of the body – arms, legs, and head. Postural movements involve the entire body, including an engagement of the core. Lamb argues that effective, beautiful, expressive movement includes postural adjustments even while gesturing. Following this, tools that support embodiment will be those that naturally allow postural movement during use. As an added criterion, the movement of the body should also align with the attribute being controlled. For instance, a fully embodied pushing of a button is unlikely to have the same benefit as a system that more directly maps the timing of body movements to the timing of attributes in the application.

Dance professor Della Davidson suggests that "creativity is about the ability to play, and the ability to be spontaneous to your environment and to the tools you are working with" (2003). Spontaneity is tightly tied to creativity in the performing arts and psychologists Runco and Sakamoto suggest: "creativity may depend on spontaneity" (1999: 62). Jacques LeCoq claimed that the fundamental goal of his renowned school for actor training in Paris was creativity and that this relied extensively on play and spontaneity, arguing that "...through improvisation, [they] externalize what is latent within the students" (2002: 27). Grotowski similarly put spontaneity at the core of the creative work of an actor. Their embodied approaches reinforce the linkage between the body and spontaneity in creative work, and the role of avoiding deliberative thought in creative exploration.

Keith Johnstone, one of the founders of improvisational theatre, similarly supports the importance of spontaneity and warns about the danger of deliberative thought in inhibiting creativity. He writes: "Imagination is as effortless as perception, unless we think it might be 'wrong', which is what our education encourages us to believe. Then we experience ourselves as 'imagining', as 'thinking up an idea', but what we're really doing is faking up the sort of imagination we think we ought to have" (1981: 80). Tools that support spontaneity must allow the real-time exploration of ideas. This implies that they must A) be real-time and interactive, and B) allow control over a significant portion of the total creation in real-time such that the user has the experience of direct creation rather than contemplative reflection.

While most industry-standard tools for creating computer graphics, such as Maya or 3D Studio Max, rely on a traditional mouse and keyboard interface, perhaps extended with a stylus, research and art systems have experimented with more embodied and performative interfaces. Space here only allows us to touch on a few computer animation examples from a large body of work. Sturman presents an early survey on computer puppetry (1998). Oore et al. developed a system that allowed users to puppet a character by moving tubes in 3D space that were mapped to character bones (Oore et al. 2002a, 2002b). Dontcheva et al. developed an animation system that uses a motion capture system to track widgets in 3D space as input (2003). Those approaches rely on layering multiple performances to produce a final animation. Work by Laszlo et al. provides interactive control of physically simulated, planar characters (2000, 2005) using a mouse and keyboard. This approach provides spontaneity, with a still low embodiment interface. Igarashi et al. embed character poses in input space and automatically blend them as the user drags a mouse pointer through this space (Igarashi et al. 2005). Neff et al. tackle one of the limits of mouse-based input by developing methods that allow the two degrees of freedom of mouse input to be mapped to a much larger number of degrees of freedom in a human character model (2007). Other work has explored performance as an effective means to time-animated motion (Terra and Metoyer 2004, Baecker 1974). In the realm of painting, Keefe et al. (2001) created an embodied interface that allows users to paint 3D scenes while working in a Cave.

Shneiderman (2007) emphasizes the importance of rich history-keeping and tool design that is straightforward enough for novice users, yet powerful enough for experienced ones. Constraints play an important part in the creative process and serve to establish an artistic foundation. Task constraints, such as choice of medium and style, serve as loose guidelines for the artist's creative process (Stokes 2005). Work focused on embodied interfaces discusses how exertion can be considered an aid, rather than a hindrance, when using embodied interfaces. Expecting a certain amount of exertion can encourage one's positional awareness (Lyons et al. 2012).

Measuring creativity

Psychologists have developed a wide range of techniques for evaluating creativity.[1] The techniques are aimed at answering varied questions, such as "How creative is a particular individual?" "What personality traits correlate with creativity?" or "Can providing particular information increase creativity?" Our work is focused on a specific issue: how does the use of a particular tool alter a user's creativity? The process, user experience, and final output are all distinct components when doing creative work, and one alone should not be treated as a definitive authority (Cropley 2000). Plucker and Renzulli (1999) describes several approaches for measuring creativity. One of the most common and well-validated psychometric measures is the level of divergence in someone's thinking, as this has been shown to be a good proxy for creativity. A second technique is to examine the product of the creative process and determine how much creativity is embedded in it. This is generally done by external judges. This approach is based on the view that "People know creativity when they see it" (1999: 45), which has been at least partially validated and is widely used in practice. The Creativity Support Index (CSI) has been proposed as a standardized instrument for evaluating creativity in software (Carrol and Latulipe 2009). It is based on research that identifies key factors that support creative work: exploration, enjoyment, collaboration, expressiveness, immersion and results-worth-the-effort. With it, a subject weights the importance of these factors for her given task and also how effectively the tool supports them. The CSI then calculates a rating between 0 and 100, which can be reliably compared with scores for other tools.

Our first study uses the CSI. While prototyping the second study, subjects reported a tendency to repeat answers in the CSI, especially those used to weight components, after responding to the same set of questions multiple times after sequential tasks. We therefore extracted a shorter set of questions based on the CSI, eliminating questions on collaboration, which is not possible in our tool, and component weighting. This leaves five specific factors for our subjects to rate, as detailed in Table 14.1. We believe this approach provided a good indication of self-reported creativity levels with less risk of survey fatigue.

Experiments with a 3D animation interface

In our first exploration with more embodied interfaces, we created a system for quadruped animation dubbed "CAT" (Figure 14.1). The system used a tracked Wiimote to provide six DOF continuous input (position and orientation), along with button controls. Animation was produced in a layered approach where part of the cat's motion was specified and then additional details were added in subsequent layers. Controls included automated locomotion generation in which the user would draw a path and the system would generate the correct gait along the path; multi-joint controls, where the user could move a paw to control the cat's leg or adjust the height of the cat; and single joint controls that vary a single degree of freedom. We evaluated the system with ten participants across five different tasks, using the CSI to judge how well it supported creativity.[2]

The most enjoyable tasks according to CSI ratings were those that were well embodied and responded directly to the user's movement. The tools

Table 14.1 The five questions asked after each task

Measure	Question
Output Satisfaction	I was satisfied with what I created using this system
Exploration	It was easy for me to explore many different ideas, outcomes, or possibilities
Use Frequency	I would be happy using this system on a regular basis for this type of activity
Creativity	I was able to be very creative while doing this activity
Absorption	I became so absorbed in the activity that I no longer had to concentrate on using the controls

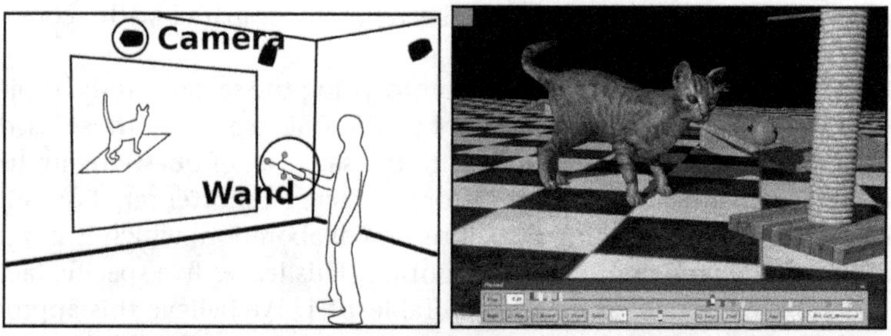

Figure 14.1 The physical workspace of CAT is shown on the left, and a screenshot of the produced animation is on the right. Courtesy of © Nick Toothman

that affected multiple joints were rated the most creative. This suggests that our mostly novice users require some amount of automation to convey their creative ideas. Survey responses indicated both that users had a significant preference for the automation of multiple joint control over single joint control and that CAT's automation impacted upon participants' creativity.

Embodied interfaces for sketching, editing and puppeteering

We developed a software platform called Sandbox (2013) that supports basic curve and polygon drawing. The platform also allows editing through the use of deformation lattices and animation by puppeteering a physically simulated marionette. The system supports two main forms of input: a mouse and keyboard, and a pair of tracked Wiimotes. The position, orientation and input buttons of the Wiimotes are all used as input. The input volume can be scaled so that the entire screen can be spanned with a single arm movement while the user is seated (small) or the user needs to walk to span the screen (large). All output is generated on a 2D plane so that the user does not need to deal with the added complexity of depth control.

Comparative studies were performed with the system using a within-subject design in which subjects would use a different interface to complete tasks during two separate visits to the lab. In one experiment, we compared the mouse and keyboard to a tracked Wiimote. In the second, we compared Wiimotes in a small volume where the subject was seated to a large volume where the subject was standing and needed to locomote to cover the screen.

The less embodied interface in the first experiment is a 3-button mouse and keyboard with an LCD monitor for display.[3] Directly mapping input movements to the large screen encourages users to move throughout the environment to span the screen. A 12-camera Vicon optical motion-tracking system is used to provide accurate position and orientation information for the Wiimote. For the 3D interface, Sandbox orthographically projects the device location onto a screen position, and displays a glyph at that cursor location. A scale factor is introduced that allows the input space to be made smaller than the screen space in case subjects cannot reach the top. This scale factor is further reduced to allow users to span the screen while seated for the less embodied interface in the second experiment.

We instructed participants to complete (in order) a drawing task, an editing task, and an animation task. No task was allowed to continue for more than 15 minutes. In the drawing task, users were instructed to draw emotionally expressive faces. Users could draw curves and shapes, scale, rotate and translate them. Editing was done by manipulating free-form deformation lattices (Figure 14.2) (Sederberg and Parry 1986). Each lattice contained a 3×3 grid of control points for a set of Bézier curves that deformed the underlying shapes as the control points were moved. The user could select and move arbitrary sets of control points. Users edited a template face to produce

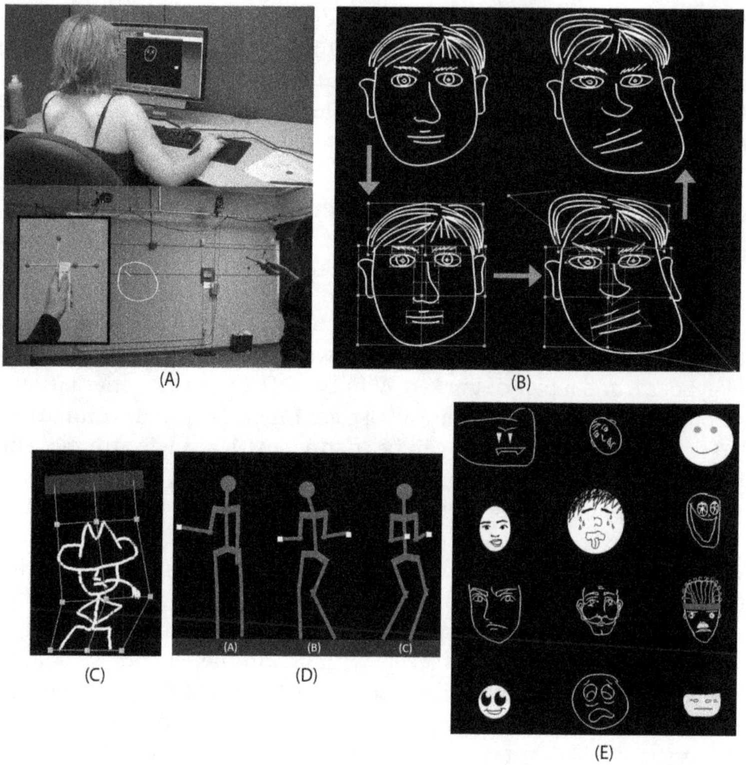

Figure 14.2 Sandbox. (A) The desktop interface (top) and Wiimote interface (bottom, with controller inset). (B) Free-form deformation lattices for editing. (C) The marionette used in Experiment 1. (D) The marionette used in Experiment 2. (E) Sample face drawings from Experiment 1. Courtesy of © Nick Toothman

emotional expressions. The first task was an animation exercise where users puppeteered a marionette. As discussed below, the controls differed slightly in the two experiments so that the first Wiimote interface was comparable to a mouse in its degrees of freedom. The stick figure marionette was physically simulated and collided with the screen edges for enriched interaction, but not with itself. Written and oral instructions on the interface controls were provided prior to the beginning of each task. Subjects were allowed to practise the controls and ask questions. At the end of each task, subjects rated the prompts in Table 14.1 on 5-point Likert scales, and also completed an exit survey rating their preference for each interface on each task.

Care was taken to minimize unwanted experimental effects. Subjects could not complete both sessions on the same day. The intersession interval encouraged learning decay between tasks and reduced fatigue effects.

In each experiment, half the subjects started with one interface and half with the other to minimize ordering effects. Prior to each task, subjects were presented with instructions on the interface controls and given time to practise to ease learning requirements. Finally, subjects were advised that about 45 minutes were allocated per session to avoid developing fatigue from extended use. These decisions are based on best practices in HCI research methods (Lazar et al. 2010).

Experiment 1: mouse and keyboard vs standing interfaces

Some unique qualities of the first experiment are described here, but it largely follows the aforementioned general method. It compares a standard mouse and keyboard interface with a more embodied Wiimote where the input volume is scaled so that the user must take steps to span the entire screen. Given the mouse's dominance in desktop computing, we found it an appropriate device for the 2D interface. A pen and tablet could have been configured for drawing and editing, but a mouse offered more comfortable control for the animation task, and we wished to use a single input system for each session. Figure 14.2 shows both interfaces in use.

The marionette system for this experiment was designed to use one Wiimote and only vertical and horizontal translation to maintain the same number of degrees of freedom with the mouse interface. The marionette was constructed using a deformation lattice with rope-like constraints between control points and a top handle that was moved by the user translating the mouse or Wiimote, as shown in Figure 14.2. Handle rotation was controlled by rotating the mouse scroll wheel or axially rotating the Wiimote.

Seven women and five men, aged 18 to 27, participated in the study as paid subjects. Six were randomly assigned to use the mouse in their first session, while the remaining six were assigned to use the Wiimote first. The intersession intervals ranged from 1 to 7 days. We chose a parametric test for our data as it has been shown to be effectively equal in power to non-parametric tests such as the Wilcoxon signed-rank test (Kaptein et al. 2010, Carifio and Perla 2008). The survey responses were analysed separately for the five creativity support factors. Table 14.2 summarizes the results of paired t-tests on the populations from these three tasks, including sample means, standard deviations, and Cohen's d effect size. We designated small, medium, and large effect sizes as 0.2, 0.5, and 0.8 respectively. Grouping performance by interface across all three tasks, all factors except for Exploration were rated significantly higher for the embodied interface. The mean ratings for the embodied interface were all higher than the paired mouse and keyboard interface. For the drawing task, the results indicate a statistically significant higher rating for the embodied interface on the Creativity factor,[4] but not significant differences for Output Satisfaction and Use Frequency. These three factors all have effect sizes between small and medium. The analysis for the editing task indicates a statistically significant preference

Table 14.2 Analysis of responses to survey questions for each task using paired t-tests. Survey responses were on the 5-point Likert-type scale from 1 to 5; 1 indicates strong disagreement with the question and 5 indicates strong agreement. μ is mean response, σ is standard deviation, effect size is Cohen's d. N = 12 for all analysis measures, so DOF = 11 for t-tests. Bold indicates significance ($p < 0.05$), italics indicates tendency ($p < 0.1$)

	Question	2D	3D	*t*-test	*d*
Drawing	Output Satisfaction	$\mu=3.25, \sigma=1.14$	$\mu=4.0, \sigma=0.74$	*$t=2.14, p=0.06$*	0.36
	Exploration	$\mu=3.5, \sigma=1.38$	$\mu=3.92, \sigma=0.79$	$t=1.16, p=0.27$	0.18
	Use Frequency	$\mu=3.0, \sigma=1.28$	$\mu=4.0, \sigma=0.85$	*$t=2.17, p=0.05$*	0.42
	Creativity	$\mu=3.16, \sigma=1.21$	$\mu=4.0, \sigma=0.74$	**$t=2.27, p<0.05$**	0.39
	Absorption	$\mu=2.92, \sigma=1.24$	$\mu=3.67, \sigma=0.78$	$t=1.52, p=0.16$	0.34
Editing	Output Satisfaction	$\mu=3.17, \sigma=1.27$	$\mu=4.17, \sigma=0.72$	*$t=2.03, p=0.07$*	0.44
	Exploration	$\mu=3.67, \sigma=1.27$	$\mu=4.17, \sigma=0.72$	$t=1.32, p=0.21$	0.24
	Use Frequency	$\mu=3.25, \sigma=0.97$	$\mu=4.08, \sigma=0.51$	**$t=2.42, p<0.05$**	0.48
	Creativity	$\mu=3.58, \sigma=0.9$	$\mu=4.08, \sigma=0.9$	$t=1.20, p=0.26$	0.27
	Absorption	$\mu=2.5, \sigma=1.24$	$\mu=3.83, \sigma=0.58$	**$t=3.22$, p <0.01**	0.57
Animation	Output Satisfaction	$\mu=3.67, \sigma=1.15$	$\mu=4.0, \sigma=0.74$	$t=1.08, p=0.30$	0.17
	Exploration	$\mu=3.92, \sigma=1.16$	$\mu=4.08, \sigma=0.29$	$t=0.48, p=0.64$	0.10
	Use Frequency	$\mu=3.5, \sigma=0.90$	$\mu=4.08, \sigma=0.51$	*$t=2.03, p=0.07$*	0.37
	Creativity	$\mu=3.42, \sigma=1.0$	$\mu=4.0, \sigma=0.60$	$t=1.73, p=0.11$	0.33
	Absorption	$\mu=3.42, \sigma=1.24$	$\mu=3.92, \sigma=0.79$	$t=1.15, p=0.27$	0.23
Composite	Output Satisfaction	$\mu=3.36, \sigma=0.89$	$\mu=4.06, \sigma=0.56$	**$t=2.31, p<0.05$**	0.42
	Exploration	$\mu=3.69, \sigma=0.84$	$\mu=4.06, \sigma=0.43$	$t=1.32, p=0.11$	0.27
	Use Frequency	$\mu=3.25, \sigma=0.86$	$\mu=4.06, \sigma=0.33$	**$t=3.05, p<0.01$**	0.53
	Creativity	$\mu=3.39, \sigma=0.85$	$\mu=4.03, \sigma=0.62$	**$t=2.10, p<0.05$**	0.40
	Absorption	$\mu=2.94, \sigma=0.98$	$\mu=3.81, \sigma=0.5$	**$t=2.73, p<0.05$**	0.49

for the 3D interface on the Use Frequency and Absorption and a tendential difference for Output Satisfaction. Output Satisfaction and Use Frequency have effect sizes slightly below medium, and Absorption has a large effect size. For animation, only Use Frequency has a tendential difference, while Use Frequency, Creativity, and Absorption have effect sizes between small and medium.

Finally, the exit survey responses were analysed as single-sample t-tests using a hypothetical mean of 3 to indicate equal preference for the interfaces, 1 representing preference for the mouse and 5 for the embodied interface. There were no significant preferences for either interface in any task, although editing,[5] animation,[6] and overall[7] preferences tended to be higher for the embodied interface.

Use Frequency showed a preference for the embodied interface over the mouse and keyboard. A possible explanation portrays the mouse as ubiquitous and less interesting than the more novel Wiimote. However, survey responses also described the embodied interface as "more intuitive" and especially praised the naturalness of rotation operations, suggesting these results are not entirely arising from novelty. This preference for regular use of the embodied interface occurred despite higher fatigue, suggesting that the characteristics unique to using an embodied interface, such as exertion, may contribute positively to the user's experience.

We thought the animation task would most clearly benefit from embodied interaction, but only Use Frequency showed a tendency for preference for the embodied interface on this task. High Absorption and Exploration ratings for both interfaces indicate that the task is quite engaging. Subject comments revealed important differences between the two interfaces, noting that the 3D interface "made it easier to imitate a walking motion by making it easier to rotate while moving the character", and that it was "particularly intuitive and engrossing. You felt like you were in more direct control of the character, having to move your body in similar patterns to accomplish the intended motion." Not all participants favoured embodiment, one finding the wrist rotation difficult and one preferring the familiarity of the mouse. Based on qualitative feedback, self-reported measures, and user preference, we preliminarily suggest a relationship between task complexity and the effect of embodied interfaces. For embodiment to have a meaningful impact on creativity, tasks may require a minimum level of complexity such as that which was present in the drawing and editing tasks, but not in the animation. Reducing the embodied interface to only provide x, y translation and axial rotation may also have limited some of the benefit that a more fully embodied version might provide.

Experiment 2: small volume seated vs large volume standing

While still preliminary, the results from our first experiment indicated that embodied interfaces better support factors related to creativity. To further understand the role of specific variations in embodiment, we designed and performed a comparative study in which the interface was the same in both tasks, but the input volume was scaled. In the small condition, the person could span the entire input space with an arm movement while seated. In the large condition, the user needed to walk to cover the whole input space, leading to greater body usage.

Subjects were first asked to draw lines connecting pairs of coloured points placed at arbitrary distances. The same set of points was used for every subject. This task allowed participants to familiarize themselves with the application. It also provided quantitative data on how the two interfaces performed. After this, the drawing and editing tasks were the same as in Experiment 1, except that the animation task was enhanced using two Wiimotes. One controller translated and rotated the marionette's overall pose by transforming the head and neck, and the other allowed the user to move one or two limbs, symmetrically or in parallel. Participants were instructed on how to use the controls, given practice time, and then invited to perform basic motions such as waving and gesturing. Once comfortable, the prompter asked them to make the marionette dance to a piece of music for 90 seconds. After this, participants used the marionette as an avatar and held a short dialogue with the prompter.

Eleven women and fourteen men (ages 18 to 39) participated in the study as paid subjects. They attended two sessions, completing each task (drawing, editing and interaction) with one interface per session. Thirteen were randomly assigned to begin with the seated interface, twelve with the standing interface. After all subjects had completed the experiment, the creativity of their output was judged.

Analysis on the performance measures indicated no significant difference between average time spent drawing lines[8] or residual error between lines and targets.[9] This indicated that people were equally capable of drawing lines with either interface. Participants rated their interface preference for each task on a scale from 1 (seated interface preferred) to 5 (standing interface preferred). Based on a one-sample t-test with a mean of 3 ("equal preference"), results show a significant preference for the standing interface for the drawing[10] and animation[11] tasks, and it was the overall preferred interface.[12]

As in the first experiment, survey responses were analysed with paired t-tests for each factor for each task, as well as a composite measure across all tasks for each factor. Only Q1 (Satisfaction) for the animation task had a significantly different rating, showing preference for the standing interface.[13] The mean responses for each factor across each task were all greater than the neutral value 3, except for Q5 (Absorption) on the standing animation task, indicating a favourable view in general of how the tools supported these creativity factors.

Task engagement or "flow", a term borrowed from psychology to describe when a person is highly involved in a task (Chen 2007), can indicate a high level of absorption that is an ingredient in creative work (Csikszentmihalyi 1997). The prompts for the drawing and editing tasks instructed subjects to notify the proctor when they no longer wished to create content. We counted the number of sessions where participants continued past the time limit as an objective measurement of task engagement. Long work times could indicate difficulties completing the task, but since people performed equally well with either interface in the performance task, we attributed differences to absorption. Participants continued past the time limit in 37 out of the 100 drawing and editing tasks.

There were 13 from seated sessions and 24 from standing. A Chi-square test between interface and task timeout revealed that the number of timeouts significantly differs between interfaces.[14] Tasks performed with the standing interface were more likely to continue until timeout.

The performance measure suggests that participants could work equally well with either interface, which is noteworthy given the quite different movements required to span the different input spaces. With regards to creativity, subjects appeared to create similarly creative outputs with each interface and only the Satisfaction factor on the animation task was rated significantly higher for the more embodied interface. However, participants were significantly more likely to work with the standing interface past the time limit. As performance on the interfaces is similar, this appears to indicate a greater level of immersion, a factor related to creativity. Participants also indicated a preference for the standing interface on the drawing and animation tasks, as well as an overall preference. In addition, the creativity factors were rated above neutral on average for every interface/task combination except for Absorption for the standing interface on the animation task (29 out of 30 tasks), suggesting both interfaces supported creativity. Taken together, this data is not sufficient to conclude that the large volume interface will have a specific impact on creative performance, but does suggest that people prefer the large volume, standing interface and appear to be more engaged while using it. Further investigation appears worthwhile.

Additional insight can be garnered from user feedback. After completing the seated animation task, one user remarked "My arm wasn't too tired... but it was just harder to animate someone who was standing up while I was sitting down", indicating that fatigue was not a factor, but that the larger, standing space felt easier, at least when it matched the task. Participant feedback suggests standing is a more natural choice for interactive tasks. One user noted that "Standing allowed me to be much more expressive, and the movements and gestures were much more natural. Sitting, it was more a matter of thinking about the gestures, rather than spontaneously doing them." Another subject, whose first session used the standing interface, reported after completing the animation task while seated: "I felt like I did not have to think about it as much when I did [the task while] standing, even though it was the first time using it then." These comments indicate that the large volume interface may bring about a different form of engagement, one more embodied and less based on conscious, deliberative thought. This is exactly the type of transformation sought in much actor training in order to encourage creativity, as discussed in the opening.

We attempted to identify the impact of embodiment on creativity by using experiments to compare traditional and embodied interfaces for creative tasks in digital media. Experiment 1 showed a significant preference for the more embodied interface across a range of factors, including self-reported creativity levels and task absorption. The picture in Experiment 2 is less clear, with limited differences on creativity factors, but evidence of greater engagement and a clear preference for the larger volume interface over the small volume. Perhaps most interestingly, participant comments suggest that their manner of engagement may be different with the change in input volume.

Discussion and paths forward

This chapter describes our initial explorations of the relationship between embodied interfaces and creativity support. There is a clear theoretical basis for expecting that increased embodiment can help support creativity. While still preliminary, our results indicate that embodied interfaces do offer value in enhancing creative tasks and further research is likely to be fruitful.

Thoroughly investigating these issues presents numerous challenges. First, a definition of embodiment is required. Interfaces lie on a spectrum from those that require little movement to those that engage the entire body. Second, particular types of activities must be identified as some tasks may benefit more from embodiment than others. While the comparative approach adopted here leads to clean experimental designs, it restricts the range of embodiment that can be supported as two interfaces must offer comparable affordances. This is particularly limiting when one interface is a mouse, with only two degrees of freedom, but in general there is a limit to the ability to create complex embodiments, optimized for a particular task (for example, acting out the movements of a cat). Instead of the comparative experimental approach, ethnographic techniques that seek to understand users' thought processes may prove useful. As a final note, the adoption of more embodied interfaces may have health repercussions as well as impacts on creativity. There is growing concern about the negative health effects of a sedentary lifestyle (Thorp et al. 2011). Given the importance of computation in modern work, more embodied ways to interact with computers may offer significant health gains.

Acknowledgements

Financial support for this work was provided in part through National Science Foundation grant 0856084.

Notes

1. For a survey of psychometric techniques, see Plucker 1999; for experimental techniques, see Runco and Sakamoto 1999; or for a broad overview, see Sternberg 2003.
2. Full details on the system and experiments can be found in Martin and Neff 2012.
3. The more embodied interface is a single motion-tracked Wiimote (two are used during the Experiment 2 animation task) with a projector with a 124 in. × 93 in. screen for display.
4. With tendential ($p < 0.1$).
5. $t(11) = 1.39$, $p < 0.1$, $\mu = 3.67$, $\sigma = 1.67$.
6. $t(11) = 1.76$, $p < 0.1$, $\mu = 3.83$, $\sigma = 1.64$.
7. $t(11) = 1.65$, $p < 0.1$, $\mu = 3.83$, $\sigma = 1.75$.
8. $t(24) = 1.19$, $p = 0.24$, $\mu = 0.08$, $\sigma = 0.33$.
9. $t(24) = 1.03$, $p = 0.31$, $\mu = 0.48$, $\sigma = 2.34$.
10. $t(24) = 2.72$, $p < 0.05$, $\mu = 3.88$.
11. $t(24) = 2.52$, $p < 0.05$, $\mu = 3.76$.
12. $t(24) = 2.34$, $p < 0.05$, $\mu = 3.72$.
13. $t(24) = 2.06$, $p < 0.05$.
14. $\chi2(1, N = 100) = 4.29$, $p < 0.05$, $\varphi = 0.23$, odds ratio 2.62.

Part VI

Digital Movement as Emergent Practice

15
Practising Choreography: Digital Movement as Emergent Praxis

Thecla Schiphorst and Tom Calvert

Setting the stage: choreography in a digitally mediated world

Choreography – the art of designing movement – has a rich creative history rooted in embodied practice and compositional knowledge. While human movement is necessarily expressed through the body, it is also increasingly shaped by our digitally mediated world. The relationship between choreography and technology has provided a growing number of examples of software systems that support movement composition. Yet these systems are varied, with differing goals and values, and general tools for choreography remain scarce (Fdili Alaoui et al. 2013). Simultaneously, contemporary choreography incorporates numerous compositional strategies that are applied to the design of movement; we can view these compositional techniques as being procedural in nature, which is inclusive of classical structures such as canon, call and response, imitation and inversion, chance procedures, generative and improvisational strategies, contact improvisation, somatic practice and the appropriation of everyday movement (Blom and Chaplin 1982). Long before the emergence of the digital computer, choreographers had used numerous procedural and algorithmic structures to create compositional strategies to structure movement in time and space. While human movement is the soft, messy design material of choreography and the felt fabric of expression, contemporary digital choreography continues to mediate, explore and appropriate this design material through computational means.

Today, as our contemporary digital sensing devices continue to expand the use and capture of movement data, human movement is increasingly implicated and appropriated by our digitally mediated world. For example, the "surrender" motif (two hands up to the sides of the body) has become the iconic gesture of Microsoft Kinect users. Mobile phones, smart tablets, depth cameras and touch devices extract, interpret, analyse and represent our movement data in increasingly complex ways. Yet, the question remains: how does choreographic practice and its inextricable foundation in the experience of lived bodily movement continue to inform technology design?

In 1991, when Merce Cunningham began to explore movement generation using the choreographic software LifeForms, an enthusiastic CNN reporter declared: "and now technology is finally coming to the rescue of choreographers" (Cunningham, quoted in Schiphorst 1997: 14). There is an implicit tension between computers and creative design processes that are both ideological and epistemological. In this chapter we explore the historical development of digital movement as emergent praxis, countering the CNN reporter by suggesting that choreographic knowledge can also innovate technology by informing, shaping and influencing its design paradigms. We review our own design history as lead members of the development team for one of the more influential digital tools for dance, LifeForms/DanceForms choreographic software,[1] through our collaboration with the choreographer Merce Cunningham, and we describe our ongoing research into digital tools for movement, meaning and interaction. We situate our research in the historical context of collaboration between two epistemologies of practice: technological and embodied. And finally, we review the development of LifeForms/DanceForms in relationship to other software systems that have explored digital choreography, setting the stage for future directions in software systems for dance in which choreography can inform technological design and maybe even come to the rescue of technology.

Choreography as design: practising collaboration with and for movement

All creative processes in the arts can be considered as practices of crafting, making and therefore designing. In choreography, movement is both the design material and the design outcome (Wilde et al. 2012). Incorporating technology into movement creation necessitates not only a formalization of choreographic knowledge but also the collaboration between different fields of creative design. The late 1980s were formative years in the evolution of the LifeForms/DanceForms choreographic software. We worked with a group of computer scientists and choreographers in the computer graphics lab at Simon Fraser University, influenced by concepts that explored the design process as a creative task implicated by technology (Calvert et al. 1991). At that time Herbert Simon (1981) had identified the emergence of a common ground between creative and computational processes:

> We have noticed a growing communication among intellectual disciplines that takes place around the computer... But surely the computer, as a piece of hardware, or even as a piece of software, has nothing to do directly with the matter. The ability to communicate across fields –

> the common ground – comes from the fact that all who use computers in complex ways are using computers to design or to participate in the process of design. (Simon 1981: 159)

The emergence of software tools for dance and choreography has also necessitated and been supported by cross-disciplinary collaborations. As Herbert Simon suggests, using computers in complex ways requires participation in the process of design. And just as choreography is a process of designing with and for movement, computational systems for choreography require that we explicitly formalize choreographic models by defining procedural descriptions of choreographic strategies. How does this influence the more ephemeral, somatic and intuitive processes of choreography, which are based in movement experience and practice, and more importantly how can we afford movement experience within our digital choreographic tools?

Developing computational models for choreography requires the intersection of divergent theoretical and practice-based knowledge that lies across disciplines. Emerging collaborations in digitally mediated choreography require expertise in both movement and computational modelling. Among this emerging set of interdisciplinary knowledge bases are Movement Analysis, Compositional Theory, Design Theory, Human-Computer Interaction, Artificial Intelligence, Machine Learning, Cognitive Science, Linguistics, Computer Games, Graphics and Animation, and Dance History. Even as early as 1988, Don Herbison-Evans, a computer scientist who innovated research for the digital representation of dance, recognized the immense value and challenges that interdisciplinary collaboration incited between these diverse disciplines:

> In perhaps no other application of computers is there such a disparity between the modes of thought of the users and of the system designers. The challenge to bridge this gulf is likely to benefit humanity as a whole... Computers can be humanized. How to do this may become clear from trying to "dance" with a computer. (Herbison-Evans 1988)

The humanization of technology has been a central goal of the field of Human-Computer Interaction since Herbison-Evans' call to collaboration in 1988. Yet the challenge to bridge the modes of thinking and practice that lie at the intersection of dance and technology remains fundamental to our research and creative practice today. We are still working to articulate a shared language between the somatic, experiential knowledge inherent in human movement and the computational formalization that is necessary to design technological systems for movement.

Choreographic thinking: algorithms, processes, procedures and techniques

Choreographic thinking is at once procedural and embodied. Choreography is a compositional process. And, like all compositional processes, its styles, techniques and procedures of transforming movement into choreographic structures are continually evolving. How has technology afforded, problematized or expanded choreographic thinking? And how has choreography challenged technological design strategies? Like musical composers, choreographers continually reinvent their procedural strategies for working with the material properties of movement. Consider Merce Cunningham's use of chance procedures, Anne Teresa de Keersmaeker's use of musical counterpoint and text, and Trisha Brown's use of gameplay rules (*RuleGame No 5*, 1964), return to gravity (*Leaning Duets* and *Falling Duets*) and mathematical structural accumulation (*Accumulating Pieces*, 1971). Choreographers construct and deconstruct the material of movement using procedural techniques that explore new ways of understanding, building and experiencing that movement. Cunningham acknowledged that his use of LifeForms/DanceForms was an extension of his use of chance procedures:

> My use of chance methods in finding continuity for dances [...] is a present mode of freeing my imagination from its own clichés and it is a marvellous adventure in attention. Our attention is, normally, highly selective and highly editorial. But try looking at events another way and the whole world of gesture, the whole physical world in fact, is as if jabbed by an electric current. (Cunningham 1955)

Indeed, many postmodern choreographic procedures were created to problematize the experience of designing movement, creating uncommon rules and structures and thereby defamiliarizing the experience of the performer and spectator (Loke and Robertson 2007, Carlson, Schiphorst, and Pasquier 2011). While choreography remains a living compositional form, dance scholar Scott deLahunta echoes Cunningham in questioning our own contemporary belief systems around current choreographic practice: "Perhaps our practices are outdated or can be improved. How can we doubt our own processes and question our own methods?" (deLahunta 2002: 136).

Doubting as constructive creative process is not an unfamiliar artistic strategy. Doubting is a reflective form of questioning: "How could this be different?" In our research, we are interested in extending deLahunta's question by asking how technology can form an equal partnership with the experience of movement in a way that helps us to uncover, doubt, question, inspire and expand our choreographic possibilities. This allows us to see movement differently and, as Cunningham has suggested, to discover something that we do not already know. In 1991, Cunningham described his first explorations with LifeForms in a CNN interview:

> One can *make* things with it [Life Forms], one doesn't have to put things in one already knows... one can make *discoveries*, and that interested me from the beginning. (Cunningham 1991)

How, then, could we represent or capture movement in such a way that it could be incorporated within a computing program and used for creative purposes? Even prior to the evolution of digital movement and software tools for choreography, dance theorist and artist Rudolf Laban in *Mastery of Movement* (1980 [1950]) distinguished between *measurable* and *classifiable* movement attributes. Laban defined the first category as "operative and objectively measureable" and the second category as "personal and classifiable". Even prior to the existence of digital movement capture technologies Laban anticipated our contemporary understanding of sensing technologies by describing functions of *capturing* measurable *data*, while simultaneously *representing* classifiable *experience*. He identified measurable aspects of movement qualities, with variables such as resistance, speed, direction, and control, and distinguished them from classifiable aspects of movement expression, such as levity, duration, expansion and fluency. The latter, he explained, refer to the lived embodied sensation of movement quality, an outward expression resulting from an inner intention to act (Laban 1980 [1950]: 73). According to Laban, these measurable aspects of movement can be correlated to the classifiable aspects. For example, the measurement of resistance correlates to the experience of levity and the measurement of direction correlates to the experience of expansion or plasticity (Laban 1980 [1950]). However, movement sensation, expression and meaning are contextual. What comes before and after an action influences both experience and interpretation. Laban illustrated that human movement can be observed and measured; it can be classified and as such can be understood meaningfully.

While many questions are still unanswered concerning the semantics and classification of movement, the relationship between observation, classification and experience remains critical to developing software tools for choreography (Cuykendall et al. 2015). Can classification limit or expand our experience of movement? How is experience altered through the digital choreographic processes that articulate them? Can we translate the experience of movement itself into digital formats? Laban suggested that through intentional practice we can represent, translate *and* experience a growing "mastery of movement" by articulating and attending to the subtleties of movement qualities, its relationship to space, environment (Laban 1980 [1955]).

Considering human movement in a digital world: a historical note

In his book *Changes: Notes on Choreography* (1968), Merce Cunningham described the potential of computers as creative tools, insofar as they could be applied to choreography and to the description of a spatial configuration of human stick figures that would visualize the dance:

> I think a possible direction now [in 1968] would be to make an electric notation [...] that is three dimensional [...] it can be stick figures or whatever, but they move in space so you can see the details of the dance; and you can stop it or slow it down [...] [it] would indicate where in space each person is, the shape of the movement, its timing. (Cunningham 1968, our insertions)

Cunningham's description corresponded to and was perhaps incited by Michael Noll's 1967 article in *Dance Magazine* titled "Choreography and Computers", which described the Bell Lab engineers' research into early scanned raster displays for computer graphics applied to dance. This technology enabled stereoscopic three-dimensional representation of computer art and animations (e.g. a dancing stick figure), which supported perception of spatial orientation and relative positioning on a two-dimensional screen. Even before this early form of visual representation of the dancing body, computers were used in the creation of a choreography. In 1964, Jeanne Beaman and Paul Le Vasseur at the University of Pittsburgh used a computer to generate random, performable dance sequences. They input 20 different time variations, 20 different spatial directions, and 20 different types of movements into a mainframe computer, which printed out 70 dances in text-based instructions over a period of four minutes. This compositional material was based on aleatoric composition in which the computer selected random combinations of material. The printed output gave directions such as: "Three medium beats, half-turn clockwise, arc backward", or "Six slow beats, jump, diagonally back" (Cybernetic Serendipity 1967). This exploration of aleatoric and procedural transformations was also incorporated into sound composition in the 1950s and 1960s. Cunningham and his long-time collaborator, composer John Cage, had already been integrating concepts of random movement and sound generation in the 1950s, in particular using the *I Ching* as an instrument to generate arbitrary compositional structures. From these early beginnings in aleatory compositional techniques and in visual representations of choreographic space and time, creative collaboration between choreographers, technologists and engineers stimulated the evolution of software for dance. As computer graphics became more sophisticated, there were many innovative combinations of computer-generated imagery with live performance, with the images most often designed as an augmentation or visualization of movement on the stage.

However, while there have been numerous systems that explore a specific choreographer's compositional techniques, relatively few tools with general applicability have resulted (Fdili Alaoui et al. 2013). Based on the literature that describes choreographic software tools there have been certain categories/taxonomies introduced by dance researchers and scholars, including deLahunta's emphasis on the use of choreographic technologies (deLahunta 2002); Carlson's stages of choreographic process as a lens to view choreographic representation (Carlson, Schiphorst, and Shaw 2011); and Fdili Alaoui's focus on goals and values of choreographic support tools (Fdili Alaoui et al. 2013). In the following sections, we will explore digital tools for choreography, contextualizing the development of LifeForms/DanceForms (Calvert et al. 1991), and consider the critical values and goals of ongoing research. Choreography remains a living art form and choreographic techniques continue to respond to, express and become shaped by the world in which they exist. As such, choreography mediates and continues to be mediated by contemporary tools for creativity.

Notation, notation, notation: an early use of computers in dance

Although dance is considered by many historians to be the oldest of the arts, preceding music, architecture, painting, sculpture, drama and literature, it has been very slow to develop any agreed system of representation (Czompo 1974). It was not until the early to mid 20th century that three notation systems emerged which were generally accepted – Labanotation (Laban 1924), Benesh Notation (Benesh 1955), and Eshkol-Wachman Notation (Eshkol 1955). Although it is estimated that 80 or more systems exist across the world, and despite its late acceptance, dance notation systems have existed since movement began to be formalized, such as in the English Morris dance in the 13th century and the French *Beauchamp-Feuillet* in the 16th century. Even so, contemporary artists, choreographers and students continue to find notation systems hard to master – very few can read notation and even fewer can write it. This is in contrast to music, where scores are widely read and written.

Movement is immediate; it comes alive through the lived experience of the body, and is codified over years of practice and training. Thus, it is not surprising that when computer systems became widely available there was immediate interest in using them to compose, edit and store scores for all of the common dance notation systems (Venable 1970, Badler 1975, Ryman 1980, Eshkol 1970). The most useful general-purpose tool to emerge was Lucy Venable's LabanWriter in 1987 – this allowed scores to be created and edited on the Mac computer. Such tools have continued to be developed and are still available today.[2] It is interesting that, although computer programs have been adopted fairly widely for these purposes, this has also focused attention on the drawbacks of the notation systems themselves. While dance is created by the act of movement and structured through choreography, it is an ephemeral art form, and for dance scholars a significant issue for dance is the process of archiving and preserving choreographic works. Before the advent of computers, reconstruction of a dance was usually based on the available archives, e.g. notation scores, plus whatever notes and sketches a choreographer or choreologist may have provided. The computer-based notation editors developed for Labanotation, Benesh Notation and Eshkol-Wachman Notation provide a means to save and transmit scores in digital form for archives and for the future reconstruction of an artist's work.[3] The underlying difficulty is that human movement is supremely complex – and thus it is hard to find a simple yet effective representation for it. Even in music, which has an agreed notation form, emerging forms of musical composition in the 20th century have required continuous development of notational systems to support compositional concepts, structures, emerging practices and strategies. Based on experience with music, there have been proposals to make use of a computer-based extensible markup language such as XML. In analogy to MusicXML (which has resulted in over 170 notation programs),[4] DanceXML has been created, but it has not been widely recognized as a useful system due to the sheer complexity needed to represent movement. This is due in part to a large gap (or chasm) between capturing movement *data* and representing movement *meaning*. While movement *data* is more, and more readily, available in the burgeoning growth of consumer sensing technologies, from Microsoft Kinect, to Leap Motion, the iPhone, Fitbit, and the Apple iWatch, movement *meaning* is much more difficult to extract, recognize and tease out from all of this data. While statistical summaries of movement data abound in the newly emerging apps of the quantitative self, choreography has historically incorporated movement representation in more poetic, complex, nuanced, aesthetic and layered *qualitative* strategies to create meaning. From the perspective of computational representation of movement *meaning,* there continues to be a need to develop a semantic mapping between the low-level details of movement data and the higher-level semantic descriptors that can correlate movement data to movement meaning. Indeed, movement composition that supports choreographic thinking at a higher level has not been well supported with technology. Because choreography is the art of designing movement it was inevitable that computers move from *archiving* movement to *creating* movement.

A single moment in movement: notes on the keyframe

The creation of movement using a computer comes with numerous challenges for choreography, and some of these challenges can be articulated more clearly by looking at creative fields outside of choreography. Animators have been creating moving images of human beings for many years – first with the classical Walt Disney two-dimensional systems and later with three-dimensional models of the human body. Although procedural methods are used more frequently today, most animation is based on keyframes. A keyframe is a three-dimensional representation of the body that captures a single moment in movement, and is normally represented as a particular pose within a sequence of movement. If there is very little movement change, keyframes may be spaced widely apart in time. If the movement is changing quickly in direction, speed or acceleration then the keyframes are placed much more closely together in time. The computer software calculates (or interpolates) the frames between any two adjacent keyframes. Applied to dance, these single-movement moments capture the key poses adopted by a dancer, and the algorithm to calculate the in-betweens can provide control over the *quality* of the movement, indicating emotion and other affect.

Historically, this use of animation keyframes to represent movement in dance has the great advantage that the keyframes can be read visually without the need to learn notation. While notation can also indicate the intention of the choreographer, the keyframe indicates the result. In our own historical development of choreographic software, and at a time when creating movement with the computer was in its infancy, we adopted keyframe-based solutions as a starting point in the development of DanceForms and LifeForms. Keyframe-based choreographic tools visualize the form of the body *in an image of the body itself.* A keyframe directly and mimetically represents a body shape *as a body shape*; it can be turned and manipulated in three dimensions. Crucially, the keyframe is not an abstract symbol of movement that needs to be translated. And therefore, as Merce Cunningham noted, the keyframe is easy to recognize and recreate in a studio. But keyframe-based choreographic tools also have disadvantages: the manipulation of a body shape requires individual rotation of multiple joints on the body model. This is time-consuming, requiring a choreographer to sit (not move) in front of a computer and does not correlate to the way we move or to the way choreographers create movement. So, historically, LifeForms/DanceForms moved towards a visual representation of the body in order to free choreographers from symbolic forms of notation and to enable more direct visualization of movement. At the same time, the mechanics of creating keyframes in the early 1990s were detail-oriented and temporally at odds with the choreographer's practice in the studio. Today, with the design of iDanceForms, we are moving towards multiple other ways to creatively generate and capture movement, and while keyframing remains the underlying structure of the LifeForms/DanceForms and iDanceForms platform, we are augmenting concepts of real-time movement sketching, camera keyframing and parameterized movement composition.

LifeForms/DanceForms software collaboration

While Merce Cunningham predicted at an early stage that computers could help to create new dances, most of the initial interest in software for dance in the 1970s was related to notation. However, computer-based dance tools that enable the visualization of movement in three dimensions open up new creative and archival processes for choreographers that complement other forms of recording and viewing. Beginning in the late 1980s our team at Simon Fraser University saw other possibilities in the increasing options for three-dimensional visualization techniques, and so we explored a different avenue that incorporated the visual and kinaesthetic relationship between movement and representation.

Working closely with choreographers and computer scientists, we set out to develop a choreographic software system inspired by choreographic practice in the studio. While working within the numerous constraints of the technology at hand, we premised our understanding of choreography on the experience of movement generated from the body. Because there already existed many animation systems, our primary goal was to focus on human movement in general and dance in particular.[5] The technological climate of the late 1980s and early 1990s introduced the emerging possibility of consumer-based 3D graphics. However, our first software experiments were developed on relatively high-end workstations especially designed for 3D modelling and animation. While these workstations were out of reach for the everyday consumer, our initial software designs found their way to the Macintosh desktop platform within a few years of our initial prototypes. We began with a set of poses (or keyframes) representing still shapes that could be recombined or edited by shifting relationships in time and space. The result was a choreographic authoring system implemented on an Evans and Sutherland workstation, initially called Compose (1989), then ported to a Silicon Graphics workstation and renamed LifeForms (1992), and now DanceForms (2005). In 1989, through some happy and perhaps even fateful coincidences, Merce Cunningham became a master user guiding the evolution of the system. Cunningham's influence included the integration of chance procedures and the use of error as a creative compositional strategy.

From the outset we recognized that there was a need for multiple viewpoints and levels of choreographic abstraction that included an ability to edit for the body itself, as well as creating spatial and temporal relationships that could be manipulated independently of one another. As we were able to make the software more generally available on the Mac and then on Windows, choreographers used the software as a basis for improvisation, or to sketch movement ideas in a general or abstract way. The DanceForms software is organized around the set of windows shown in Figure 15.1: the Stage, the Studio, the Score and the Performance Windows. The choreographer

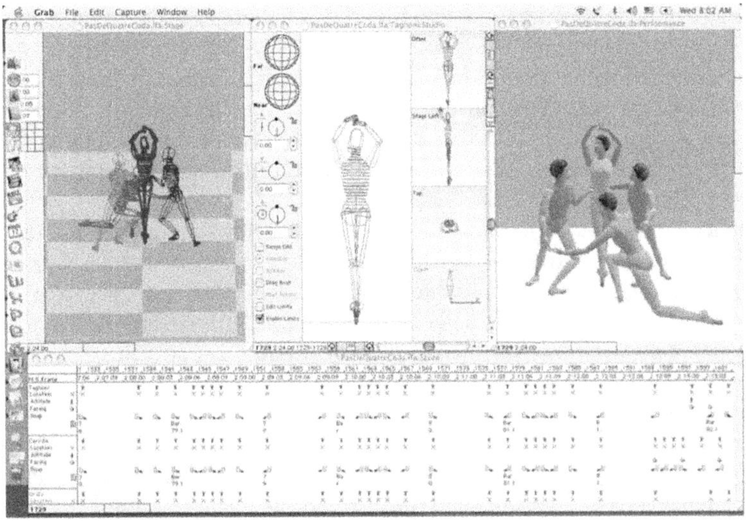

Figure 15.1 The DanceForms Stage (top left), Studio (top centre), Performance (top right) and Score (bottom) windows. Courtesy of © Thecla Schiphorst

creates body positions in the STUDIO by defining body keyshapes and can order and reorder sequences of body keyshapes by positioning them in the SCORE, much the same as music notes are sequenced from left to right in an orchestral score. The choreographer also orients the dancer(s) spatially in the STAGE. To view the dance in process, the choreographer views the PERFORMANCE. The PERFORMANCE brings all the elements together, displayed by a rendered virtual dancer or dancers. The choreographer can view the dance from any perspective and at varying tempos.[6]

Movement in and meaning out: other forms of movement input

While movement can be *created* in DanceForms through on-screen manipulation of the figure into keyframe poses with the computer providing the in-between movement, movement can also be *captured* through motion capture data files and then inputted into DanceForms. Motion capture has been used in entertainment, sports, and medical computer applications, in creating feature films and in the development of video games. In motion capture sessions for dance, images from multiple cameras are used to calculate three-dimensional positions to generate movement. This movement data can be mapped to a computer-generated 3D model so that the model performs the same actions as the original dancer.[7] The 3D models that result from motion capture can be mapped readily to the 3D keyframe poses created to animate a dance. Motion capture data brings with it challenges because motion capture frames are generated at a uniform rate of 30 or 60 frames/second and this makes the movement difficult to edit. DanceForms does have a feature that allows the user to think the motion capture sequence and remove all content except those frames that are "key". Thus movement can be inputted directly from the physical dancer and then manipulated in the software by the choreographer.[8]

DanceForms also includes movement libraries: pre-made movement keyframe sequences that can be imported onto the dancer/model. Currently the largest library is Rhonda Ryman's Ballet Moves II. This tutorial-based collection provides over a hundred keyframe-based animation clips made up of ballet steps and positions from each of the English and Russian schools of movement, as well as select samples from the Italian school. In addition, a ballet repertoire is provided that contains animations of the "Peasant Pas de Deux" from *Giselle*, the male and female "Bluebird Variations" from *Sleeping Beauty*, and the "Dance of the Four Cygnets" from *Swan Lake*, each accompanied by the appropriate midi sound file.[9] These various libraries of pre-constructed poses and movement sequences provide shortcuts for faster movement creation, which can be used in their original form or further manipulated by the choreographer.

DanceForms is used widely by choreographers and students of choreography, and offers several options for input and manipulation of movement. However, there is still far more potential that could be explored in digital tools for dance, particularly those that address the many challenges of the original DanceForms design. Some of these include incorporating more natural uses of movement data to create keyframes (such as mobile phones, tablets or other accelerometer-based sensor devices), exploring more aesthetic approaches to rendering the dancer's body, and incorporating higher-level compositional techniques that can generate various outcomes that the choreographer can select from.

Shaken not stirred: affordances of using mobile devices with digital dance tools

The evolution of computer systems, from laptops to smartphones and tablets, has integrated new sensors, which bring new opportunities for sensing movement in innovative ways. One such outcome is iDanceForms, developed as an app for the iPad and other mobile tablet devices. One of the goals of iDanceForms is to "use movement to create movement". This strategy *transforms* the concept of animation (manipulating joint angles with a mouse) towards the concept of physically *moving* the body through gesture or whole-body interaction *to create* movement phrases that can be used in composition.

The mobile paradigm incorporates sensors that can detect, activate, and respond to physical movement, including the accelerometer/gyroscope, the embedded camera, the microphone, and touchscreen surfaces that support the exploration and sensation of movement. While it may seem that tablet devices are merely a smaller and lighter form of laptop computer, the mobile tablet has distinct affordances, which include the ability to be carried without being "opened"; become a framing device (through the camera's eye); be stroked, pinched and caressed; and be moved, rotated, spun, shaken, and even "stirred". These are all embodied affordances that were not as easily included in the now normative laptop computer's design. Mobile technology *affords* movement in an entirely new way. And as we have been envisioning, this has the potential to support what we have seen to be the choreographer's historic vision: to find new ways to create dance in the studio, the stage, the street, and the world.

We have re-envisioned the act of keyframing in iDanceForms by taking advantage of the tablet's touch surface, which creates a more intuitive physical interaction for moving, exchanging and inserting keyshapes. In the usability workshops that we have conducted, dancers and choreographers were also able to improvise using the tablet's touch surface, exploring choreographic manipulation in ways that would not have been possible in the studio (see Figure 15.2).

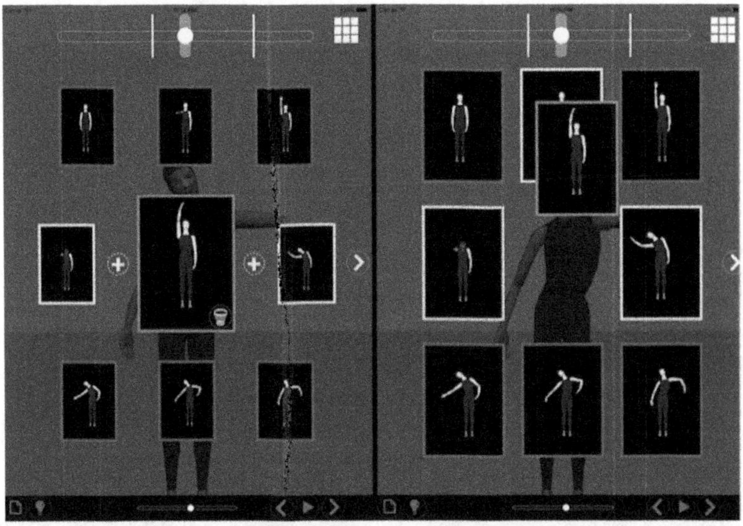

Figure 15.2 iDanceForms supports visual keyframing. It is easy to drag and drop, or exchange keyframes in a highly visual form. Courtesy of © Thecla Schiphorst

Since iDanceForms is portable and moveable, and since tablets and smartphones have an integral camera, it is natural to ask whether dance keyframes could be captured directly from the movement of a live dancer. In principle, this is quite feasible but in practice there are technical challenges. One of the most challenging problems of computer vision is to capture the 3D structure of a human body from a single 2D image. And while we are currently implementing camera keyframing utilizing the embedded 2D camera, we are also preparing for the integration of 3D depth cameras, which will enable 3D capture of both keyshapes and movement phrases. The current camera keyframing mode in iDanceForms provides the dancer/choreographer with a way to sketch movement phrases through the lived experience of movement itself. The camera is a simple, direct way to capture keyshapes and can be easily and quickly used in many environments (inside a studio or outdoors) without the need for a complete motion capture facility. Figure 15.3 illustrates our implementation of camera keyframing in iDanceForms. The choreographer simply takes a photograph of the dancer's movement. The photographic image is recognized by mapping it to a keyshape contained within an internal database, and if necessary interpolating between database keyshapes to match the image.

Some of our future work with iDanceForms includes extrapolating from the concepts found in music composition software. Music and sound composition software has been able to translate notes and tones to harmonics, polyphonics, timbral manipulation and phrasing, and has the capacity to

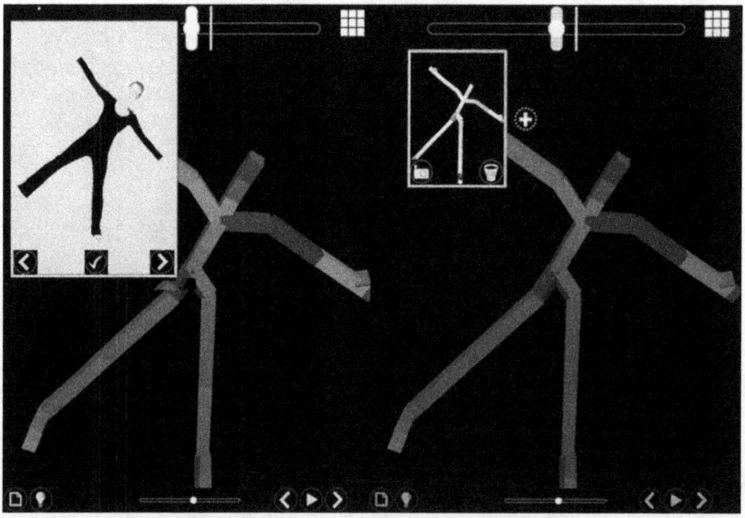

Figure 15.3 Camera keyframing enables the choreographer to create keyframes directly through the movement of the dancer. Courtesy of © Thecla Schiphorst

sample musical structures, decompose those structures, and translate them into multiple new sonic forms and formulas. Why not strategize for these *meaningful* translations in choreographic software? Indeed, our future ability to sketch, play, improvise looks set to increase as technological support tools for choreography continue to have great future promise.

Hand-held, mobile and urban choreography: iDanceForms and *MovingStories*

In 1979, Tom Calvert posed what can still be considered the "killer app" question: will computer tools be able to explore movement expression with the skilled nuance and emotional resolution that genuinely emerges from embodied movement experience?

> One can wonder about the accuracy of expression that will be possible. Can the soul of a dance be animated? Or is this too much to ask... this aspect could lead to interesting studies in body language and communication. Can the *effort* of a movement be computerized?... there is potential value in [computer systems] being developed in regard to dance... we must always remember that dancing is for people. (Calvert et al. 1979: 175)

Our current trajectory in the *MovingStories* research partnership seeks to explore digital tools for movement, meaning and interaction. Movement meaning emerges from movement expression, and movement expression is represented through the *qualitative* aspects of movement. Dance itself embodies and challenges the creation of a broad range of movement possibilities. *MovingStories* is an interdisciplinary, collaborative research partnership leading to the design and application of more richly articulated human movement knowledge within digital technology interaction.[10] This partnership focuses on designing digital tools for movement, meaning and interaction, investigated through the intersection of three unique domains: Laban Movement Analysis, digital interaction technologies, and social and cultural applications and outcomes (see Figure 15.4). *MovingStories* supports an innovative and interdisciplinary approach to redesigning digital technologies based on overlapping differentiated epistemological approaches to understanding movement. We are interested in how choreographic practice and its inextricable foundation in the experience of lived bodily movement can inform technology design, particularly in the design of movement interaction, including movement recognition, generation and composition.

The creative collaboration between science and art remains a vital locus where new ways to understand, represent and create human movement can be developed. The *MovingStories* project offers a key example of such research in the extension of DanceForms to the mobile application iDanceForms,

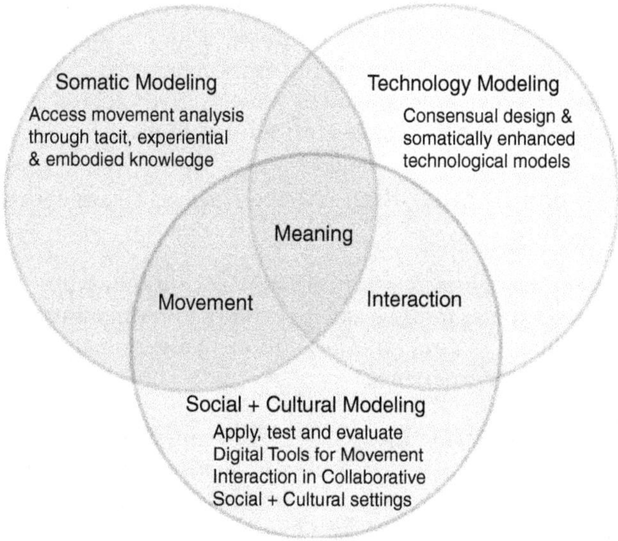

Figure 15.4 *MovingStories* intersects models of somatic experience, technology, and social and cultural applications. Courtesy of © Thecla Schiphorst

making creative use of new sensing technologies that focus on movement creation in more direct bodily-oriented ways.

In addition to digital cameras, smartphones and tablet computers include an accelerometer and gyroscope, which detect the movement of the device itself. To date, the accelerometer has been most frequently used to change the orientation of a browsing window for email or Internet access, but it is also capable of detecting movement through the measurement of the acceleration of the smartphone or tablet. This means that while discussing a movement sequence the choreographer and dancers can capture the dynamics of a live movement by recording the acceleration/deceleration signal. This cannot replace whole body motion capture but it provides a very direct way to demonstrate the quality (affect) of movement. These explorations with contemporary consumer movement sensors are part of *MovingStories,* which integrates movement and technology research.

Our *MovingStories* middleware (named m+m: movement+meaning middleware) enables a dancer or choreographer to use his or her own movement to manipulate the shape of an iDanceForms pose by linking the iPhone's accelerometer data as a movement controller to iDanceForms' keyframe editing (see Figure 15.5). We use this same mobile gestural recognition software in other applications in which *movement* of mobile phones can control visual, sonic or other effectors.

The ability to share iPhone or android smartphone accelerometer data has many potential application uses. *MovingStories* has also utilized the m+m

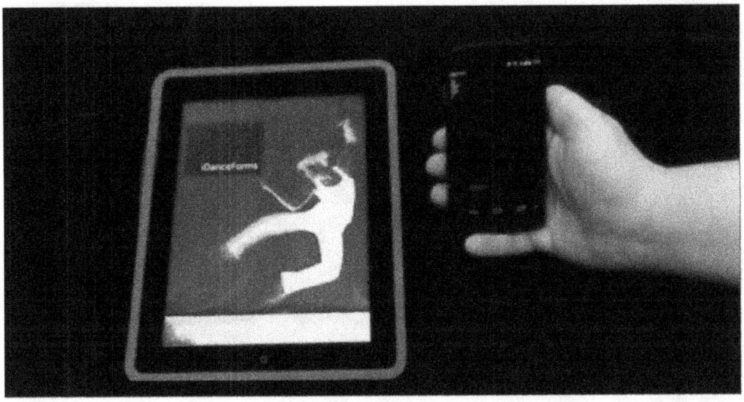

Figure 15.5 m+m middleware platform allows external sensors such as an iPhone accelerometer to provide more embodied ways to interact with the iDanceForms software. Courtesy of © Thecla Schiphorst

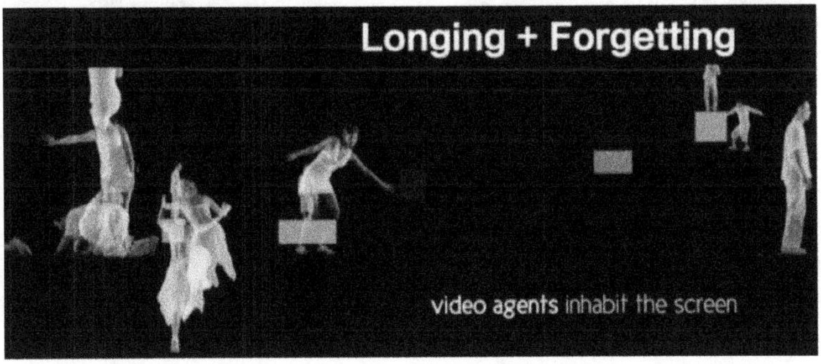

Figure 15.6 Video agents inhabit the urban screen, their movement generated in real time using Artificial Intelligence algorithms, and their paths controlled in real time by participants using mobile phones. Courtesy of © Thecla Schiphorst

to create an interactive urban screen artwork entitled *Longing + Forgetting* that uses participants' iPhone accelerometer data to interact with video agents that inhabit a large projected outdoor screen. This mobile interaction is based on the same algorithms that allow a choreographer to control movement in iDanceForms (see Figures 15.6, 15.7 and 15.8).

In *Longing + Forgetting*, interaction is controlled in a public urban space through the use of mobile phone gestures. Participants download an app and connect to a local network. Gesture recognition software parses accelerometer and gyroscope data from the smartphones. The software handles agent artificial intelligent video rendering and mobile interaction, also enabling an architectural façade map to be used to create agent motion paths.

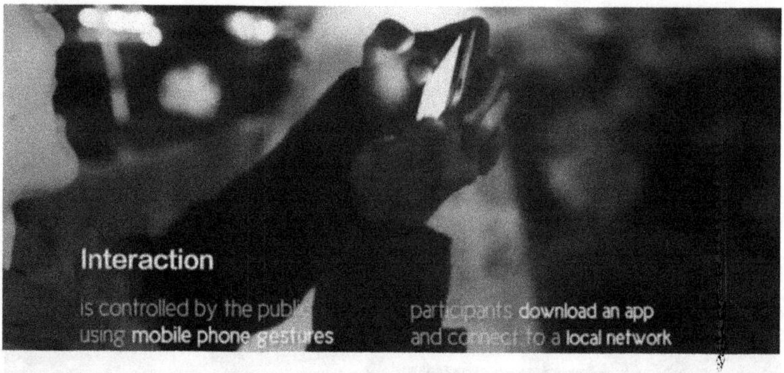

Figure 15.7 Interaction is controlled by the public using mobile phone gestures. Courtesy of © Thecla Schiphorst

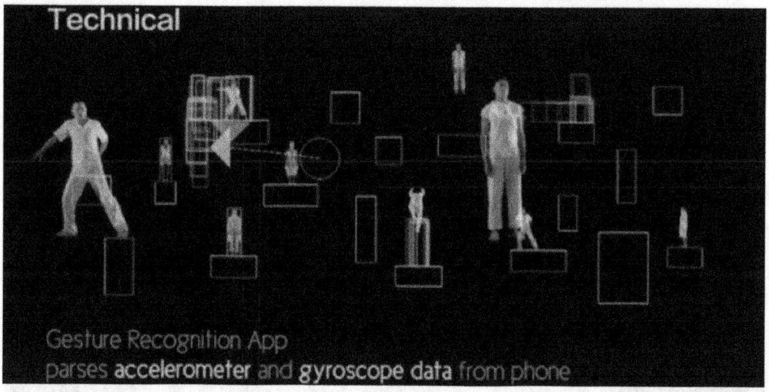

Figure 15.8 Gesture control is mapped to agent behaviours, linking smartphone movement data to the movement of the video agents, creating a collaborative choreography between video sprites and the public. Courtesy of © Thecla Schiphorst

MovingStories is exploring movement interaction from multiple perspectives including the representation, recognition, creation and generation of movement meaning. Our current research includes the:

- recognition and generation of movement qualities (exploring Effort/Shape properties in Laban Movement Analysis, or LMA) through applications of machine learning
- development of an open-source movement database that can be annotated and shared between multiple movement and dance software platforms
- continued development of iDanceForms including new sensor constructs such as camera keyframing, use of external sensors such as iPhones to

interact with iDanceForms movement, and the addition of higher-level movement compositional strategies
- recognition and generation of movement semantics such as personal movement signatures and movement styles that can be applied to movement in the database (see Figure 15.9)

MovingStories continues to innovate movement with computational models that can be applied within choreography and other movement domains. Our research group synthesizes interdisciplinary expertise in both movement and computational modelling. Our collaboration is based on this interdisciplinary exploration that explores expertise and fluency shared through process and practice.[11]

Discussion, conclusions and future work

This chapter has explored the intersection of dance and technology posing larger questions relating to the way technology has afforded, problematized or expanded choreographic thinking, and the way in which choreography, in its own right, can also challenge technological design processes. We have situated our discussion in the historical context of collaboration between two epistemologies of practice – technological and embodied – and provided examples from our own research in movement and technology.

The history of choreography has illustrated that dance embodies and challenges the creation of a broad range of movement possibilities. We have also explored how dance and choreographic thinking can create technological possibilities. Perhaps the key question is not, as the CNN reporter stated, how technology can come to the rescue of choreographers, but rather how dance knowledge can continue not only to innovate movement possibilities,

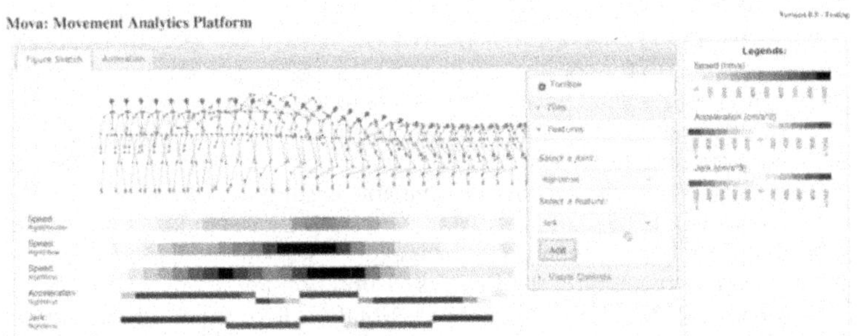

Figure 15.9 MOVA is a web-based environment that can display movement from our database, extract features based on movement annotation data and provide a movement analytics platform that can also generate and annotate movement style. Courtesy of © Thecla Schiphorst

but also to create possibilities for embodied and human developments in collaboration with technology. In Merce Cunningham's words, the interdisciplinary common ground between movement and technology can provide us the opportunities to "discover something that we do not already know".

We often think of virtuosity when we think of dance, but virtuosity comes in many forms: most certainly the physical and biomechanical virtuosity of a well-trained body, but also the conceptual and creative virtuosity of a well-trained mind that lies not only in the studio but in the nuance of everyday movement, and in the ability to see movement afresh through defamiliarization techniques such as "making strange" (Loke and Robertson 2007, Carlson, Schiphorst, and Shaw 2011). Virtuosity also lies in challenging interdisciplinary knowledge through the ability to create computational models that can reflect modes of movement experience, and in the social and cultural forms that emerge within the intersection of movement, technology and social meaning. After all, what is learned from a computer tool in creating dance can be generalized to other forms of human movement observation, creation, social communication, motion planning, understanding cultural difference and social negotiation, and even strategies for sustainable developments in human compassion and empathy.

Our future work in *MovingStories* includes the further articulation of a movement semantic framework that will make it easier to share movement data as meaningful descriptive and rich movement language. This could help us to comprehend the potency and intelligence of movement as a language for understanding and affect, and help to map movement expressivity to social and cultural knowledge, shaping methodological strategies for our ever-changing landscape of personal and collective knowledge.

Notes

1. The LifeForms/DanceForms software was initially published as Compose (1989), then LifeForms (1992), DanceForms (2005), and most recently iDanceForms (2014). We refer to the software by its two most widely known titles where we are discussing features and concepts that relate to the wider history of the software. We use only one title when referring to a specific manifestation of the software, and the latter part of this paper explores new features of the iDanceForms version.
2. Today LabanWriter continues to evolve and is supported by the dance department at the Ohio State University and personally by David Ralley (http://dance.osu.edu/research/dnb/laban-writer, accessed 19 September 2014). More recently, in an attempt to combine the benefits of notation and dance visualization, we have worked with the New York Dance Notation Bureau and others to develop LabanDancer, a partial prototype for translating Labanotation to animation (Wilke et al. 2005). Within its limitations the prototype is useful, but we found it difficult to address all aspects of the notation in animation.
3. Support for Labanotation is provided by the New York Dance Notation Bureau (www.dancenotation.org, accessed 19 September 2014), which has a repository of dance scores in Labanotation with significant holdings of videotapes,

photographs, programme s, and production information. Similar support for Benesh Notation is provided by the Benesh Institute within the London-based Royal Academy of Dance (www.rad.org.uk, accessed 19 September 2014).

4. See: http://www.musicxml.com/ (accessed 19 September 2014).
5. There are many general-purpose animation systems that could achieve the same results but the generality results in commands and user interfaces that are much more complicated. The most widely used general-purpose animation systems include Maya, MotionBuilder, and 3DS Max, all from AutoDesk, and iclone, blender and quite a few others.
6. DanceForms is widely used today in schools and colleges and by individual dancers and choreographers to teach or experiment with choreography.
7. This process may be contrasted to the older technique of rotoscoping, where the motion of an actor was filmed, then digitized (often by hand) for the frame-by-frame motion of an animated figure.
8. The classic motion capture studio, as used by game companies and motion picture studios, typically costs $50,000 to $500,000 or more, which puts it out of the reach of most dance companies. Such a studio would typically have multiple cameras (10–25) capturing images of the movement simultaneously. A powerful computer identifies the important landmarks on the body and computes the coordinates for a stick figure model of the dancer's body. The landmarks may be marked by colour (e.g. white dots) or they may transmit an infrared signal to the camera. Typical manufacturers include Motion Analysis Corp., Vicon Motion Systems Ltd, and OptiTrack from NaturalPoint, Inc. Recently there has been rapid development of much less expensive systems as consumer products for computer games, e.g. Nintendo Wii, Microsoft Kinect. The Kinect has a camera and depth sensor and uses proprietary software to provide full-body 3D motion capture and costs between $150 and $250. A number of developers (including *MovingStories*) are investigating how this can be directly applied to dance motion capture and choreography – it will be interesting to see what results.
9. In addition to Ballet Moves II, Credo Interactive also makes available to DanceForms users two other libraries which are not dance specific. Power Moves 1 and 2 provide a set of 750 keyframe-based animation and motion capture clips for many different movements including office moves, sports, walking, running, and pedestrian motions. In addition, there are also 21 general dance clips and 38 ballet clips. The MeGa MoCap V2 collection contains 500 motion capture clips. An added feature is that these clips are available in formats for a number of other animation systems as well as DanceForms. Several other libraries are also freely available on the web, e.g. the Carnegie Mellon University collection (www.mocap.cs.cmu.edu, accessed 19 March 2015).
10. The research is led by Dr Thecla Schiphorst and its institutional partners include Simon Fraser University (SFU), the University of Illinois at Urbana-Champaign (UIUC), the Laban Institute of Movement Studies in New York (LIMS), Emily Carr University of Art and Design (ECUAD), the University of British Columbia Theater Department (UBC), and the Vancouver-based enterprises Credo Interactive and H+ Technology.
11. *MovingStories* researchers include Thecla Schiphorst, Tom Calvert, Philippe Pasquier, Karen Bradley, Karen Studd, Guy Garnett, Donna Cox, Steve Dipoala, Maria Lantin, Lyn Bartram, Bernhard Reicke, Cheryl Prophet, Maria Lantin, Sarah Fdili Alaoui, Ulysses Bernardet, and numerous graduate students at SFU, ECUAD, UIUC and LIMS.

16

Dance without Bodies: Using Motion Capture and Particle Animation to Create NoBody Movement

Martine Époque and Denis Poulin

Introduction

Based on our background as movement artists and researchers working at the intersection of dance, film and video animation, our present query was to identify creative ways of cutting the corporeal figure of the dancer's body to call attention to the digital discretization of dance motion.[1] Portraying human movement has always fascinated artists from various disciplines. Among the notable works that achieve an expression of motion in the ambit of screendance, and which seem most appropriate when introducing this chapter, are *Nu descendant un escalier* (1911–18), a series of five paintings by the French artist Marcel Duchamp (inspired by a chronophotographic sequence produced by Eadweard Muybridge),[2] and *Pas de deux*, a dance film by the well-known animation filmmaker Norman McLaren.[3] In the concluding section of this short film, the movement of two dancers is broken up into traces that do not particularize the dancers as individual subjects. In addressing the trajectory and trace of movement, as opposed to the individual bodies of the moving figures, these two works constitute an important historical and artistic basis for our own explorations in the creative expression of movement forms.

Our attempt to explore the mediated expression of movement signatures goes back to 1978, when we created our first professional dance film, entitled *Beyond Curtains* (National Film Board of Canada). In this work, three dancers were characterized by the manner in which they moved, rather than by their physical appearance. In 1989, and bearing this very same query in mind, we began to explore how a computer tool might support the creative enterprise of studying movement and producing digital screendance. We created a dancing avatar with one of the few 3D animation softwares available at the time[4] and used it to interpret a floor dance choreographed by Époque entitled *Amiboïsme*. It is not until the development of motion capture technology (MoCap), however, and its introduction as a creative tool in the movement arts in the late 90s, that the capabilities of computer technology became a key tool for the representation and expression of movement itself, as opposed to the body that carries it.

We can identify three key properties of the technology that are of significance in the field of movement art research. First, MoCap and 3D animation software can engage with the sensitive content (movement) that makes this type of work an art form – thus making live movement a digital material for the artist to work with. From this point of view, our work often interrogates whether MoCap and 3D animation software will register and give back to the viewer the dancers' presence. Second, MoCap technologies can be used to highlight the visual expression of dance motion, providing a kind of contemporary remediation of the motion vision found in the work of Duchamp and McLaren. Thirdly, MoCap and 3D animation software are valuable as production and methodological tools in screendance, as they enable compositional processes that emphasize the acuity of dance motion.

The aim of this paper is to present a conceptual framework underpinning our practice, based on the paradigms of "dance without body", "dancers' motion signature", and "particle dance", all of which are key to the creation of *CODA*, a stereoscopic CGI film produced using MoCap and particle modelling software, and which is set to the finale of the Stravinsky *Rite of Spring* piano score. Our position here is not to tackle theoretical aspects of the work we have carried out. It is the position of the practitioner that mostly interests us, and the concepts and insights that emerge from practising movement. We conduct practice in the hope that our example can contribute to heartening exploration and experimentation with novel tools so as to enlarge the corpus of digital movement art and so as to convey other conceptions of the moving body, and other ways of disseminating expressive dance.

On the digital representation of human motion and body

If emphasizing human movement instead of the person's body in art is nothing new, speaking about "dance without body" is something that can appear confounding. How is dance possible when there is no physical presence? How, if after all it is the body that gives dance its expressive and sensory existence? On-stage choreographies or performances traditionally feature living dancers (live or pre-recorded), even though, within the remit of the digital arts, practitioners are increasingly conveying new modalities of embodiment that problematize this dominant position taken by the living anatomy of the performer, in order to introduce post-anatomical and metaphysical paradigms of the moving body.[5] It is within this context, and particularly through the experience of working with MoCap technologies, that we came to enunciate that digital dance is a "dance without body"[6] and that the process of mediation carries with it what we call the dancer's "motion signature".[7] These concepts and paradigms require some clarification. The point here is to identify what bodily ontology is involved in digital dance, and whether this dance expression actually has a need for the bodily image in order to communicate dance motion and emotion.

Film and video animation became perhaps the first contemporary audiovisual technologies to substitute a virtual dancing body for a physical one. In Chapter 2 of this book, Salazar Sutil traced the origins of contemporary motion capture apparatuses to the work of Fischer and Braune in the 1890s, and especially to the work of Nikolai Bernstein in 1920s Russia. But since these early developments, and since the expansion of a commercial industry of motion capture technology, MoCap has completely pushed the technological recording and capture of live movement beyond its existing boundaries by translating the formal visualization of a dancer's movement into abstract languages (mathematical and computational), thus removing this technological medium from the presence of a concrete body. The medium of motion capture is to movement art what audio recording is to sound art and music, in the sense that audio recording technology allowed people to listen to someone's voice without a bodily presence. Similarly, thanks to MoCap recording technology, we can look at someone's movements freed of one's body-presence. So, even if MoCap data derives from a living dancer in motion, what matters here is not the dancer as such but the digital transcription or mapping of this motion. Like a drop of blood, which can be used to detect the DNA of a person so as to determine identity without giving any visual details or referring to physical look, so dance data allows us to recognize the dancer who performed a MoCap session without delivering any visual image of his body.

But if MoCap makes possible the release of the dance from the dancer's body, it undoubtedly does not allow a release of the movement from the dancer himself. On the contrary, as humans and animals are indeed characterized by their corporeal morphology and by their way of moving, MoCap perfectly captures and preserves the dancer's kinetic dynamics so that it can fully preserve presence by the particularity of movement – not by the particularity of bodily morphology. So, along with Alex Vasilescu (2002), who introduced the concept of "human signature", we have put forward the notion of a "dancer's motion signature" (2006) to describe the representation of the dancer in a way that is as distinctive as a shadow outline or a fingerprint.[8] Digital artworks created by Paul Kaiser, Shelley Eshkar, and Marc Downie[9] offer fine examples of how one might portray the dancing body in a way that is neither physicalized in the final end-product, nor indeed photorealistic in a stylistic sense. In the words of OpenEnded Group:

> Though the work's imagery comes entirely from a computer simulation, it bears an unmistakable human trace – that of a dancer [...] abstracted

from his physical body via a process of optical motion capture that preserves his movement but not his likeness. (OpenEnded Group 2014)[10]

This description of OpenEnded Group and Bill T. Jones's landmark motion capture piece *Ghostcatching* is conclusive about the "unmistakable human trace" that MoCap records, stores and gives access to again and again. We may conclude, with Alexandre Lazaridès, that "the dancing body released from his apparent sarcophagus points to invisible, pure movement" (Lazaridès 2006: 58).[11] Since digital dance is pure motion, we might as well address it without a reconstructed bodily appearance in the final, screened outcome. However, what is the nature of this dance? Does such a digital dance demand a specialized praxis? Does MoCap imply some transformation of the practical side of dance and of the professional skills required? If yes, what are they?

From dance to "dance without body": a collaborative choreographic methodology

Physical human activity, whether functional, leisurely or sportive, involves the anatomical body, taken as a biomechanical apparatus. As a physical activity, dance also requires the instrumentalization of this machine-body. Moreover, the discipline of the dance sets a bodily technique in motion in terms of an assemblage of moving abilities acquired through years of training and natural talent. But dance emerges when movement itself takes a step beyond the body-machine and announces itself as an aesthetic medium capable of sharing the private inner experience the dancer is living. As we said before, MoCap cameras do not see the physical body: rather they locate the markers attached on the embodied subject. So no silhouette, no appearance, no shapes. In a word, nothing concretely physical is stored in a MoCap system's memory: only and exclusively the markers' positions in time and space, which are stocked under a veneer of digits.

This observation led to the complete transformation of our choreographic approach and the way we direct our dancers during MoCap sessions. Since dynamic characteristics are accurately registered by MoCap systems, and since the individual kinetic signature of each of the dancers can be thus materialized, we decided that the external representation of movement shapes, and the aesthetic concern for the beauty of these shapes, was less of a concern – especially since 3D animation software allows us to render and revamp the appearance of a dancing avatar. What is not so easy to render at the digital phase of the process, however, are the reconstructed, lifelike dynamics of movement. For this reason, during the capture session we only emphasized the intentions and dynamic qualities underlying a movement. Our dancers were allowed to perform a kind of impulsive improvisation, whose intention was to obtain and maintain a certain degree of authenticity in the issuance of corporeal signatures. Of course, this approach was somewhat disturbing for the dancers, who were accustomed to technical perfection in the delivery of a choreography. Before long, the performer had to accept that the delivery of the movement was not entirely within their control, but that part of the process of aesthetic interpretation would be handed over to a digital artist, and to a creative collaboration. Thus, collaboration and interpretation became two kernels from which our method was developed.[12]

Our discovery of fluids motion modelling led us in 2004 to the utilization of particle motion for the creation of a *dance without body* film. *Realflow* – the software we used for the purpose of this experiment – showed that it was possible to pour particles of water into an unmoving object, i.e. a glass. The question was: *would it be possible to pour particles not into a motionless object but into a dancing character?*

This idea of making use of particles to represent MoCap digitized dance would be totally original.[13] It was challenging without a doubt, and definitely exciting because it had never been done at that time. Sure enough, particles are *alive elements,* as they are animated in relation to diverse attributes given by an animator: weight, speed, direction, dynamics, mass, and other key parameters. Once they have been given these attributes, *particles move on their own and freely*. So the option of spreading particles in and around a digital character appeared to us a totally new paradigm for the animation of the moving body, certainly dissimilar when compared to the paradigm of keyframe animation. Would it be possible to use particle animation as a compositional tool to produce digital dance movement possessed with aesthetic power? If achievable, this could be developed into a totally new approach for 3D dance animation. And does the relationship between particles as building blocks of physical life and the basic theme of life and regeneration in *The Rite of Spring,* the musical masterpiece by Igor Stravinsky, not seem plausible enough? With these questions in mind, we embarked on a project entitled *NoBody dance: the Rite of Spring, a particle dance without body and 3D film based on MoCap*[14] whose aim was to test, validate and experiment with this idea whilst producing a prototype lasting 3–4 minutes.

NoBody dance: the prototype

Having obtained the necessary funding and following the installation of an appropriate optical MoCap system, the LARTech team at the University of Quebec and Montreal (UQAM)[15] set out in June 2005 to bring together a group of collaborators that included dance students, computer scientists, and 3D animation and modelling experts. Our next challenges were to learn how to operate the MoCap system and its software (EVaRT); how to create marker sets related to the captured movement samples; how to clean the data; and finally how to insert them into the chosen 3D animation software[16] so as to create digital actors.

In order to identify whether using particles in and around a dancing character was possible, Poulin made use of *Mona*, a digital character created in 2001 for *Tabula Rasa* and LIFEanimation.[17] This initial experiment was successful and so confirmed the feasibility of our project. Furthermore, we created digital bipeds specially designed to "hold" or "throw out" particles. We also tested particle generation in relation to different dance movements, to better understand the relationship between particle motion modelling and dance at a choreographic level. The creation of 3D particle dance required building and using lengthy and complex production pipelines, which involved going from the MoCap samples to the conversion of the digital data into particles. The process was fraught with difficulties, since it was not possible either to visualize (or even imagine) the visual output until all stages of the process were complete. Moreover, such a creative process required having to deal with various software applications and to make several technical tests with the help of 3D animation specialists, in order to get the right artistic answers we were looking for, and within the boundaries set by the given tools. So, unable to anticipate what images we would achieve, we learned to work within the limits this situation imposed on us. As we were unable to change the machine, we identified what it could do well and adapted to its own strengths and provocations, rather than looking for the unconditional realization of our creative ideas. We called this "working through answers": a behavioural methodology intended to fine-tune one's aesthetic aspiration in response to what digital tools can do.

One of the interesting advantages of this method is that it enhanced our creative practice and pushed us beyond our usual boundaries. More often than not, the pathway taken during a research project may well be ensured by an unassailable method and a step-by-step procedure delineated a priori. In situations like these, however, the quest for innovation can be limited by our own creative tendency to unconsciously go back to the same artistic choices and behavioural way of working. Being confronted by unexpected questions posed by the digital devices, we were subsequently pushed in unplanned ways, which in turn defined the process as a continuous confrontation with the haphazard and unexpected provocations of the machine – an agency of its own that could not possibly be planned beforehand, but which had to be worked out through specific answers. Before we could yield any concrete results, we spent two years testing our ideas, methods and technologies, so as to match particle motion to human movement and retain the best combinations. Nevertheless, due to the important number of occlusions happening with optical MoCap, we finally had to adapt the original choreography and cut several movements which were impossible to capture.

Another unexpected problem occurred with the MoCap suits. Like most optical MoCap systems suits, the suits used in this project are made of stretched material, for the most part covered by female Velcro. The markers have to be attached to the appropriate parts of the body with male Velcro. Unwittingly, during the MoCap capture of our duets, physical contact between the dancers caused huge problems because some of the markers were being transferred from the suit of one dancer onto the suit of another, making the marker sets ineffective. To properly capture these types of dance sequences, we had to conceive and create a special suit[18] on which markers were firmly fixed.

NoBody dance saw the light as a first prototype at the end of 2007.[19] The success of this prototype prompted us to attempt a longer version featuring the entire finale of Stravinsky's *Rite of Spring.* We were doubtful, however, as to whether the prototype had revealed the creative possibilities of using particles as a compositional tool. Would it be possible to extend this tool to produce not only dance motion but a comprehensive corporeal computing choreography made up of several dancers showing aesthetic and sensitive qualities like video and dance films do?

From *NoBody dance* to *CODA*

Work on the *NoBody dance: the Rite of Spring* film project began in 2010. The first challenge was to find how to build the framework of our film. Indeed, the conventional production process of an animated film involves the writing of a scenario and the drawing of a storyboard. However, as *NoBody dance* was going be a dance film without dialogue, the only sequential basis we could use as a canvas for the film was the choreographic content: action, motion, number of dancers, music, and the topic we had in mind for it. Thus, we decided to print a time code all along a video of Époque's *Rite of Spring* stage choreography. This chronometric measurement became our framework, from which we identified different sections of the music[20] and of the choreographic project. It was next used to draw figurative tables, which produced a first visual schema of the action as a whole. The *NoBody dance* project comprised ten sections, the last one being a coda.[21]

Working with this framework was useful for the preparation and execution of the MoCap sessions and for positioning of the existing data sequences. But when we started to physically work with dancers and to collect MoCap data, we rapidly understood that the making of the full *Rite* would be very difficult and not relevant for a small research-creation team like ours. So we decided to keep only the *CODA* part of the project, the duration of the whole being about eleven minutes.[22] We subsequently made other figurative and written tables of this section with more details of all components of the film, including the number of dancers in each scene and the kind of particles we were looking for to embody the dancers and evoke the scenery. Finally, we

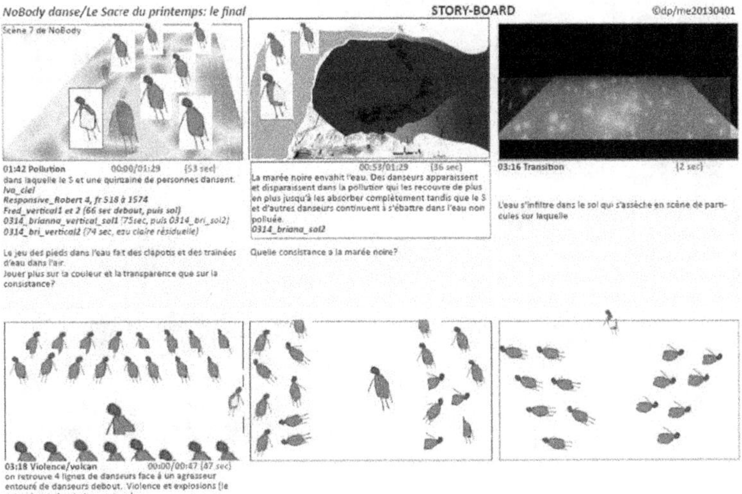

Figure 16.1 Storyboard (1) for *CODA*: finale of *NoBody dance*. Courtesy of © LARTech/Denis Poulin and Martine Époque

drew a visual graphic score that served as our storyboard, even though the device did not give a visual sense of the real appearance of the dance and environment that would appear in the final rendition of the film.

We were hoping that details of content as represented in these graphic scores would be suitable for helping our media artists understand what kind of images we were looking for. We also hoped that this structure would help convince producers of the merit of making this film, especially since it soon became clear that the technical process would exceed our financial possibilities. Unfortunately, the graphic score we produced did not convince producers in the first instance. As a matter of fact, if this visual reference clearly brought to light the structure of our film, it was not the kind of storyboard film producers are familiar with.

By the end of November 2012 we had produced enough dancing phrases to start the editing phase of the film. This process also involved making the layout of the film and carrying out a new type of choreographic work that was no longer realized at the embodied level but at the computational level. Using *Softimage* software, we created a layout by placing the dancers in virtual space, whilst synchronizing their coming in and out, as well as their dance sequences in relation to the music. This essential key operation required a very close collaboration between the choreographer, the technical director and the film director. Finally, making use of some images of this layout allowed us to make a more convincing storyboard. Even if this graphic score was still not made up of particles, it depicted the dance

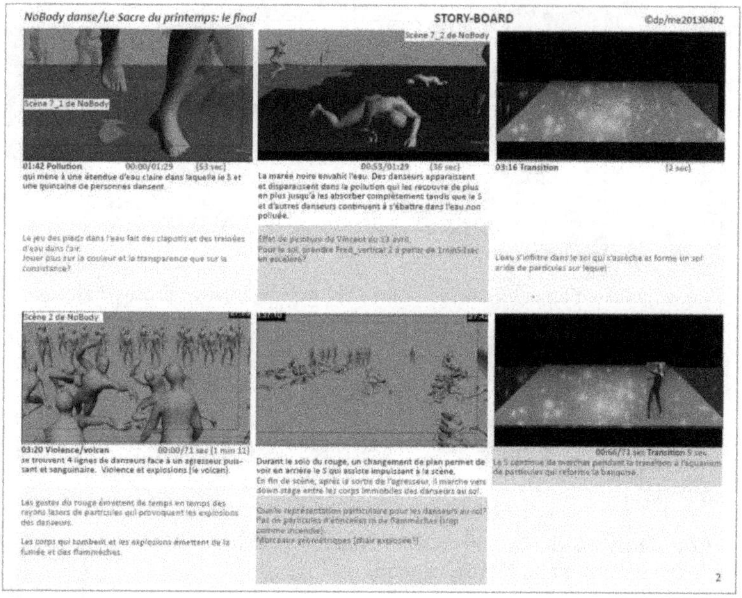

Figure 16.2 Storyboard (2) for *CODA*: finale of *NoBody dance*. Courtesy of © LARTech/Denis Poulin and Martine Époque

sequences in their environment under the form of humanoid avatars and realistic surroundings and in close relationship with the music duration.

This enhanced storyboard was the necessary reference for the data cleaning and for the particle motion design. As anyone familiar with the technology will know, optical MoCap data always requires plenty of cleaning, which can be a long, tedious and expensive process. In our case, this stage involved cutting out only the most glaringly obvious glitches in the chosen dance data, since rendition of the images to particle motion design would correct most of the smaller mistakes anyway. It is also from this point onward that all technical and artistic actions and decisions involved our whole production team: the particle maker, the choreographer, the technical director, and film director. Given various logistical problems, sustaining an intense and sensitive technical and artistic collaboration was not straightforward.[23] As Jean-Paul Fourmentraux points out:

> Today, several factors contribute to the extension of the domain of the creation to other actors than those formally established as artists. [...] Taken in an interdisciplinary movement which renews the modes of collective creation, actors whom we identified as technicians [...] can be

> otherwise brought to claim, at least to put on (to assume) an author's position. (Fourmentraux 2008)[24]

We noticed that such a process problematizes the place and responsibilities of each of the members of such a creation team. Since technological specialists become directly involved in the making of the work, can the computer choreographer and the film director be sole authors of the work? And as so much of the creative agency is handed over to the haphazard, sometimes even random, processes carried out by the computer, can the creative agency be said to be only human?

Doubtless, this kind of digital media production is mixing up art and science in such a profound way that it completely modifies the status of the members of the creative team and their relationship to the work itself. As part of a cross-traffic of skills and responsibilities exemplified by the production of the film *CODA*, artists became more and more involved in research whilst scientists became more and more involved in artistic creation.

Conclusion

The making of *CODA* gave us, and still gives us, the chance to fully live and appreciate – and sometimes find objections to – a typical creative process. Some of the conclusions drawn can be positive – others not so. A computer dance film made of MoCap material and particle animation is a long, difficult, and expensive way of producing artwork. The collaborative effort also emphasizes the need to create new tools, concepts and professional collaborations that can support the development of evolving digital art, so that we may expand the potential of human movement onto a range and medium of aesthetic communication that we are relatively new to (i.e. motion capture and particle animation).

By the same token, *CODA* shows how our brains are capable of programming and constructing a vision of movement with very few bits of visual information, which in turn demonstrates how much the presence of a dancer can be felt even without visually perceived embodiment. This is where the paradigms of "dance without body" and of "dancers' motion signatures" comes to the fore, particularly since this vision of movement without bodies, and without a visual cue, offers new conceptual, technological and non-visual approaches (perhaps more kinetic and proprioceptive) to the making of dance films and works of movement art. *CODA* finally validates that MoCap and particle animation represent a kind of revolutionary apparatus for computer choreography, as they help us reach and achieve great aesthetic power and a poetry of composed movement.

On the other hand, it is obvious from the expense and technical requirements of this technology that it is not a means of production accessible to everybody. A growing community of mixed skillsets, involving dance artists-teachers, computer technicians, and computer animators, both within universities and private studios or industry contexts, shows that the future of this kind of enterprise is a promising one. We believe that it is imperative to give dancers and choreographers these revolutionary tools to enable their exploration and experimentation, and to help them create works of digital computer art that express "dance without body".

Acknowledgements

The SSHRC and the FRQSC for their grant and Igor Stravinsky for his marvellous music | Hourshid and Mehrshid Afrakhteh for their outstanding interpretation of the music | René Chénier (ONF/NFB) | Marc Côté and Chantal Bédard (FAKE Digital Entertainment) for their support | Pierre-Luc Boily for his technical expertise (tools, animation, MoCap operation) and friendly help | Vincent Fortin for his marvellous particles and his osmosis with our thoughts | Frédérick Gravel for his collaboration on the movement, scenario, and MoCap of dance | Brianna Lombardo, Robert Meilleur and Severine Lombardo, dancers, for their kind collaboration on the MoCap of movements | the *CODA* team at FAKE Digital Entertainment for their remarkable involvement | Benoît Dame for the amazing soundtrack he created for *CODA* | Jean-Paul Vialard for the excellence of the sound mix | Manuel Marie for the quality of the recording of Stravinsky's music | Geoffrey Mitchell for the enhancement brought to the music with its 5.1 version | Gisèle Trudel, Hexagram-UQAM director, for having been of assistance | Hexagram-UQAM for its MoCap studio | The Dance Department and UQAM for their administrative collaboration | Indiegogo and those who contributed financially to *CODA* for their fine support.

Notes

1. This chapter is largely based on Poulin's PhD thesis, *Les espaces infochorégraphiques d'une danse sans corps* (UQAM 2013).
2. "*Nude Descending the Staircase* displays [Duchamp's] views of the importance of time in its relation to space. [...] Regardless of the distortions of the subject, since the painting takes a fourth-dimension of time into account it is in a way a more true to life rendering of reality than a static image [...]" (Bodish 2009: 538).
3. This short film produced by the National Film Board of Canada (1968) is a cinematic study of a ballet choreography by Ludmilla Chiriaeff, danced by Margaret Mercier and Vincent Warren, in which backlit and visual effects give movement a stunning stroboscopic-like appearance.
4. *Étude infochorégraphique*, by Denis Poulin, choreography by Martine Époque. Software: Swivel 3D.
5. In the Introduction of her book *Moving without a Body, Digital Philosophy and Choreographic Thoughts* (2013), Stamatia Portanova says that "many forms of physical expression have yielded to the same numerical abstraction, our dancing,

playing and even suffering bodies being increasingly supported by the development of digital technologies of all sorts. [...] Based on the same technological ground, diverse domains such as dance, game design and medical engineering are associated by an overarching idea, a fundamental underlying concept that is inspiring myriad innovative productions: the possibility to capture, store, and manipulate movement, abstracting it from the body and transforming it into numerical information, a data flow that can be used to activate further physical or mental, technical or creative process."

6. We enunciated this concept for the grant we requested from the Social Sciences and Humanities Research Council of Canada (SSHRC) in 2004.
7. We enunciated this concept for the grant we requested from the Fonds de Recherche du Québec – Société et Culture (FRQSC) in 2006.
8. For illustration, go to http://www.lartech.uqam.ca/collectionenglish.htm (accessed 20 March 2015).
9. Of particular interest are their works *Hand-drawn Spaces* (1998), *BIPED* (1999, choreographies by Merce Cunningham), *Ghostcatching* (1999) and *After Ghostcatching* (2010, choreographies by Bill T. Jones), in which they used MoCap and abstract drawing representation of the body. See: http://www.openendedgroup.com/artworks.html (accessed 15 February 2014).
10. See: http://openendedgroup.com/artworks/agc.html (accessed 15 February 2014). OpenEnded Group is made up of three digital artists – Marc Downie, Shelley Eshkar, and Paul Kaiser – whose pioneering approach to digital art frequently combines three signature elements: non-photorealistic 3D rendering; the incorporation of body movement by motion capture and other means; and the autonomy of artworks directed or assisted by artificial intelligence.
11. Original French citation: "Le corps dansant ressuscité de son sarcophage apparent est annonceur d'invisible, lui-même mouvement pur."
12. This collaborative choreographic methodology, in addition to the accuracy of the dancer's motion signatures, gave us the idea of producing a *Digital Collection of Dancer' Signatures* (LARTech, 2010). This interactive application makes it possible to observe, distinguish and recognize its thirteen dancers through the motion of four different abstract bipeds.
13. In 1971, Gunnar Johansson did some visually similar representations of human motion in his educational film *Motion Perception II: Three-Dimensional Motion Perception*, which presents his perception research. This film shows the motion of two dancers through a few light dots attached onto the main joints of a body. Johansson points out: "the motion of the living body was represented by a few bright spots describing the motions of the main joints. It is found that 10–12 such elements in adequate motion combinations in proximal stimulus evoke a compelling impression of human walking, running, dancing, etc" (1973: 201). We can also mention that Point Cloud movement visualization began to be used as of the early 80s. However, none of these techniques has anything to do with particles.
14. To the SSHRC.
15. A Motion Analysis system with 20 infrared cameras.
16. *Motion Builder*, *Maya*. In the last three years, we have added *Softimage* and *Unity*.
17. *Tabula Rasa* is a floor work of 40 minutes with 22 real dancers and a virtual one by Martine Époque (2001). *LifeAnimation* was a dance creation software developed by LARTech with the Montreal film animation company D 'n' A (2002). As this company closed in 2004, the software production stopped.

18. Conceived and made by Denis Lavoie, director of the *Carré Vert* Suits Company in Montreal.
19. From February 2008, this experimental computer-dance HD Blu-ray work lasting 8′ 45″ was part of *Beyond Images*, an interactive touring exhibition produced by the Sciences and Nature Museum of Sherbrooke (Quebec). Its host sites were: the Sherbrooke Museum of Nature and Science, QC – 17 February to 20 April 2008; la Cité des Télécoms, Pleumeur-Bodou, France – 2 February2009 to 31 December 2010; the McAuliffe-Shepard Discovery Center, Concord, New Hampshire, USA – 7 July to September 5 2011; THEMUSEUM, Kitchener, ON – 15 September 2011 to 22 January 2012.
20. Piano version by Michael Tilson Thomas and Ralph Grierson. Duration: 32′ 58″.
21. In music and in choreography, a coda is the final part, in which some of the main sections are revisited as a conclusion to the work.
22. TwinMuse, our pianists, registered this last section of Stravinsky's music in 9′ 04″.
23. For this we used a Google drive account and, from time to time, we exchanged hard data and up-to-date mounted sequences.
24. Translation. The original French text is: "Aujourd'hui, plusieurs facteurs concourent à l'extension du domaine de la création à d'autres acteurs que ceux conventionnellement désignés comme artistes. [...] Pris dans un mouvement interdisciplinaire qui renouvelle les modes de création collective, des acteurs que l'on identifiait comme des techniciens [...] peuvent être amenés sinon à revendiquer, du moins à endosser, une position d'auteur."

Conclusion: Reflections, Interventions, and the Dramaturgy of Interactivity

Mark Coniglio

Keynote speech given at the Corporeal Computing Conference, University of Surrey, 2 September 2013

Recently I was invited to give another keynote speech at the kick-off of the Intermediality Performance and Research Group at DeMontfort University, Leicester, UK. At the conclusion of that talk, I offered three mandates to my colleagues in their role as pedagogues of intermedial performance. One of those was to place a rigorous and forceful dramaturgical eye upon their students. Today, I want to follow my own advice and take a stab at turning that critical eye upon our own practice.

I'll start with a quote from my wonderful dramaturg, Peter von Salis, who told me on our first day working together: "If you don't want to kill your dramaturg, he's not doing his job." We can check in at the end of this talk to see if I've taken it to the point where you are wishing me some degree of bodily harm.

What is the dramaturgy of tracking?

Now, while I'm sure most of you know the term, if you don't work regularly in the theatre, then you might not have a good sense of what the word dramaturgy means. An exhaustive definition would take much longer than we have today, but let me offer one definition I like. Dramaturg Norman Frisch said, "Dramaturgy is the dialogic relationship between subject matter and its theatrical framing" (De Hulst and Hoogenboom 1994). In other words, dramaturgy makes sure the subject and its frame are having a conversation. The subject and the frame are like two great actors who carry with them an equally great dose of ego. Individually, each wants to highlight his own strongest attribute and let himself shine. Yet, when actors are truly great, they will, at just the right moment, choose to subvert their ego and yield gracefully to the other. When that give and take happens, the stage can become electric. Similarly, when the subject and its frame are having that dramaturgical conversation – when the push and pull between them is carefully and precisely balanced – we are rewarded with a truly great work of art.

That's quite a worthy goal, no? And, of course, we should seek to achieve no less with our ultra-responsive-Kinect-and-Arduino-powered-intelligent-agent-based-media-intensive-low-latency-real-time-interactive-hyper-instruments, right?

OK. Hopefully that adjective-laden phrase had the proper comic effect. If you found it amusing, it is likely because you've seen something far too similar before. And when you saw it, you immediately felt the palpable imbalance between the artwork and the technology that was purportedly there to support it.

The thing is, this imbalance is nearly unavoidable. Combining "old" art and "new" media instantly creates a dramaturgical problem that was succinctly articulated by my longtime colleague Dawn Stoppiello: "When you add technology to a piece, you have no choice but to spell it with a capital *T*."[1]

Unavoidable content

When one chooses to include advanced sensing technology in a live performance, a fundamental and unavoidable dramaturgical shift occurs in the content, the frame, and the conversation between them.

For instance, consider a dance work in which sensors are used to allow the performer to interactively control some aspect of the media. The content shifts because every movement-sensing system requires a specific physicality before it can function properly. So, at the very least, choreographic choices are pushed in a particular direction dictated by the sensory interface. Add to this the de facto interactive video scenography that so often accompanies such works, and you find yourself going down a very specific road of content indeed.

But, more importantly, the frame also shifts. This isn't just a performer on stage; it's a technologically extended body whose influence reaches into domains beyond its physical sphere of influence. To extend the body in this way is "amazing" – full of magic and mystery, two of the dirtiest shortcuts to pure spectacle. The fact that creators often trumpet the technological aspects of their work in the press or in programme notes simply throws fuel on the fire.

Given this, our choice to use computer systems in our work requires us to acknowledge that *it* – the technology – is now one of the topics, and maybe even *the* topic, of the piece. The reason you can't get round this nasty intruder is that, unlike a 'cello, a theatrical light or a loudspeaker, sensing technology – at least the advanced kind employed by the leading practitioners in our field – is simply not yet part of the vernacular.

I hope you can agree that once we have crossed the red line and placed a computer between our performer and our audience, we can't erase that damned capital T. The fact that the BBC, the *New York Times*, and every major news outlet has an entire section devoted to "Technology" should be proof enough of our cultural fascination with/fetishization of technology. In that light, it seems clear that the mere appearance of advanced sensing technology in a work will unavoidably shift the content, the frame and the dialogic relationship between them.

Sidebar: instruments

I want to include a little sidebar here. It is important to note that tracking systems employed in interactive performance are often referred to as "instruments". I myself have done so many, many times. Certainly, all instruments are, by their nature, transmedial: whether it's a violin, a surgeon's scalpel, or a motion-tracking system working in tandem with computer visualization software, the purpose of an instrument is to transform the gestures of the actor from one medium into another. While making this correlation is fair enough, I now find the analogy a bit of a dodgy scheme. I get a sinking feeling that it is a way to skirt dramaturgical responsibility by denying that technology is, at some level, the subject of your piece.

How is this a dodge? Because no one would question you if you said, I want to write a piece for solo violin. No one would feel the need to ply you with questions like, "why exactly do you want to write a piece for violin?" or "how does the violin influence the content of the piece?" Even if they did ask, and your answer was "because I like the sound of a violin", it probably wouldn't raise an eyebrow.

Referring to the latter, you could also say, "I like what I can do with a 24-channel Vicon motion-tracking system." But, unlike the violin, which has been with us since, give or take, the 15th century, your decision in the second case makes an unavoidable statement. Your chosen instrument of expression is not inert. It is, itself, content.

Protagonist/antagonist: reflection/intervention

So, you've already made a critical dramaturgical decision by choosing to include advanced sensing technology in your piece. At that moment, another crucial dramaturgical decision must be made: will the technology empower the performer, or antagonize her?

"Really?" you might ask. "Is that our only choice?"

Yes. I think so.

We've established that our friend the capital T is part of the content, which means its relationship to the protagonist, who is also part of the content, must be considered. This decision is polarized because, as any Aristotelian protagonist worth his salt will tell you, "whoever is not with me, is against me". There's no two ways about it.

In simple terms, the technological systems must either support the protagonist – serving as a tool of his expression – or they must work against him as a force of restraint or control. Whatever the non-technological dramatic relationships in your piece might be, you must decide whether the technology is a protagonist or an antagonist.

After reflecting on the canon of media dance works that I have both made and seen, I realized that the choice to make the technology a protagonist – a *Digital Reflection*, as I would choose to call it, is more common. But it has not led us to create artworks that are as compelling as one might hope, especially when considering their long-term impact. But, if our desire as creators and performers is to expand our vocabulary in unexpected ways, so that we might create work that is both memorable and profound, I believe that technology as antagonist – a *Digital Intervention* – offers us far more potential. An order of magnitude more in fact.

Let me explain why.

The Digital Reflection

In the Digital Reflection, sensory devices convert the actions of the performer into data. That data is usually manipulated or interpreted in some way by a computer algorithm or other digital system. Then, the computer presents us with a derived media result. Said more poetically, the media is a *reflection* of the body. It may be a sort of "fun house" mirror, where the image is less immediately recognizable because it is distorted in some way. But still, it is a representation of the body.

In my humble opinion, and specifically in the field of contemporary dance, this approach has got us nearly nowhere.

First, it has not led us to any significant shifts in choreographic practice. It is easy to cite previous fundamental shifts: Cunningham technique, the Judson Church movement, the influence of conceptual choreographers like Jérôme Bel or Xavier Le Roy. Each of these led a generation of dance makers into unexplored new frontiers of movement expression. But the technology we speak of today has simply not played a role in precipitating similar choreographic shifts. In fact, it might have done the opposite: when I peruse the offerings of compelling young choreographers in Berlin, I sense more of an anti-technology stance than anything else.

Instead, what we have generally seen over the past 20 years is a scenario where choreographers rely on an established choreographic practice and attempt to "splice in" the technological elements. The media component most regularly takes shape in the form of a quite spectacular visual scenography. I am unconvinced that these works, at least in relation to their media components, have approached us so unexpectedly – or moved us so deeply – that they remain sublimely emblazoned into our psyche.

Take a moment – right now – to make an inventory of the media-intensive performances that have left a truly profound mark on your mind or heart; search for the ones you can never forget. Now take that inventory again, but open it up to every performance you've seen. The discrepancy between the results is likely to be substantial, and will not be purely related to the fact that fewer media-intensive performances are made.

Please don't misunderstand me. There are works I truly admire from the *oeuvre*: the stunning visuals in Merce Cunningham's *Biped* (1999), created by Paul Kaiser and his team, or the wonderfully dynamic visual accompaniment to Gideon Obarzanek's *Glow* (2006), created by Frieder Weiss. These are fantastic and extremely artistic visualizations of human movement. But, I don't believe these truly beautiful visual accompaniments can leave an indelible mark on our psyche because we simply haven't *earned* them.

Here I want to share a quote by Bertolt Brecht (1930) from his essay "The Modern Theatre is Epic Theatre". He is speaking about the Wagnerian concept of the *Gesamtkunstwerk*, the total, integrated work of art, in contrast to his own notion of "epic theatre".

When the epic theatre's methods begin to penetrate the opera the first result is a radical separation of the elements. The great struggle for supremacy between words, music and production – which always brings up the question, "which is the pretext for what?" Is the music the pretext for the events on the stage, or are these the pretext for the music, etc? – can simply be bypassed by radically separating the elements.

So long as the expression "Gesamtkunstwerk" means that the integration is a muddle, so long as the arts are supposed to be "fused" together, the various elements will all be equally degraded, and each will act as a mere "feed" to the rest. The process of fusion extends to the spectator, who gets thrown into the melting pot, too, and becomes a passive (suffering) part of the total work of art. Witchcraft of this sort must of course be fought against. Whatever is intended to produce hypnosis, is likely to induce sordid intoxication, or creates fog, has got to be given up.

In another translation (Kattenbelt 2008), I found a different rendering of the phrase "sordid intoxication" that zeroes in directly on my point:

"unworthy ecstasy"

When we are presented with ecstasy without having earned it through conflict, we are unworthy, and its impact is severely reduced. Specifically, when the technology functions in the mode of the Digital Reflection, when the capital T attempts to fuse with the protagonist by protagonizing along with it, that big T does not give us much. What we have is essentially the love duet from *Romeo and Juliet*: there's no conflict to be found between the performer and media; there's just a lot of moon-eyed longing as they look into each other's eyes. And without conflict, we simply haven't earned all of that gorgeous imagery. Brecht is right in saying that any attempt to integrate or fuse the elements (in our case, art and technology) on the stage only gets us into more trouble. I'm sorry to say it, but such "unworthy ecstasies" are not going to generate any earthquakes in our practice as choreographers, nor will they lead our audiences to deeply consider their relationship to the world or to those around them. They are a great pleasure to watch to be sure, but great art is never about pleasure alone.

I think this lack of conflict would be less of a problem if the sensory "instrument" – the one I spoke about earlier – was so incredibly responsive that it would allow a performer to express him or herself with extreme variety and subtlety.

I often use the violin as an example of a great instrument, because it is so incredibly sensitive to human gesture. As my mentor Mort Subotnick once said, "When playing a single note, a great violinist can impose vibrato and other inflections with such precise and delicate control that you will be brought to tears."

Following that line of thinking, the impetus behind giving interactive control to the dancers in my own work has often been to take advantage of their great sensitivity as performers; they can *feel* the audience as they perform, and will respond in situ, imposing subtle changes in timing and dynamics that intensify the viewer's experience. How did they develop this ability? Through long hours spent practising with an instrument – their body – that is incredibly high-resolution and responds very dependably to the commands sent to it by their brain.

I can think of no digital gesture-sensing system that offers anything near the same level of resolution and responsiveness. No. Even the boldest attempts have resulted in quite low-resolution sensory systems whose responses are too often not repeatable with any useful degree of certainty. Performers can't use such a system to express themselves at the level they deserve because the instruments are simply too crude.

Faced with these inadequacies, some creators have sought to add "artificial intelligence" to their instruments and/or interactive systems. Their reasoning seemed to be, "if we can't make the instrument sensitive enough, then we'll make it smarter". But these instruments are not the least bit smart. And when I say "not smart", I mean that they lack the ability to understand context – the thing we admire so much in a great performer. The most important thing that an artistic partner or co-protagonist can offer is the ability to surprise. Surprise is not randomness, the belief so often held by younger students I have taught. Randomness is noise. Instead, surprise means that one has responded in a manner whose unpredictability is in contrast to a context. While the claim of "artificial intelligence" has been made in our field, I have yet to see a work in which I can clearly perceive any meaningful intelligence in action on the stage.

The Digital Intervention

The Digital Intervention inverts the relationship found in the Digital Reflection. It falls more in line with Brecht's edict of separation, and offers more opportunity for powerful expression and the production of a *worthy* ecstasy.

In the Digital Intervention, one invokes a system to which the performers must adapt with extreme rigour. If the system is designed properly, then it forces inhuman choices upon the creators and/or performers that are either impossible to realize, or at least very nearly so. As a sensor, human beings are simply *fantastic*. They are very high-resolution – as both an input and output device. And, most importantly, they are very smart indeed; they understand context at the highest level we know.

So what happens when you present a human being with an impossible problem, and you command them to solve it? In the day-to-day world, I suppose many would quickly give up. But dancers – and probably all talented performers – are quite a special breed. If you present a dancer with an impossible problem, I promise you they will find a solution that surprises you.

Certainly, this notion of intervention is not new. The chance procedures of Cage and Cunningham, or the hyper-rigorous twelve-tone compositions of Milton Babbitt, are also interventions. The system determines everything, and what falls outside of the rules is where individual creativity can be applied. But specific to our topic today, I think Cunningham's use of the animation software LifeForms in the latter part of his career provides a wonderful example of what I'm talking about.

As explained by my colleague Thecla Schiphorst in her 1986 Master's thesis, "Cunningham has often used the method of tossing three coins to determine the outcome of a hexagram, and it is this method he employed to create movement in LifeForms..." So, the arrangement of the body and limbs of animated, virtual dancers was determined by chance. Sometimes this would result in movements or arrangements of the limbs that were impossible to realize in the physical world. But Cunningham didn't edit this score to accommodate the physical world. Instead, the dancers were asked to realize exactly what they saw. They were confronted with the impossible.

How a creative interpreter handled this was made clear in the performance *Story of a Dancer*, created by Jérôme Bel in collaboration with Cédric Andrieux. In this work, Andrieux – who danced for Cunningham for many years – described a moment when Cunningham instructed him to replicate a LifeForms animation in which, among other movements, the body levitated in mid-air for several seconds. In the piece, Andrieux recounts his memory of that moment, telling the audience "I was humiliated". As a dancer who had been trained to realize the desire of the choreographer completely and without hesitation, his inability to achieve the stated goal filled him with shame. Yet, that shame also prevented him from giving up. He struggled, eventually finding a way to come as close as he could. It did not look like what he saw on the screen. It looked like something else. The impossibility of the problem led to a novel and unexpected solution. It led to something surprising.

loopdiver: Digital Intervention in practice

In Troika Ranch's work *loopdiver,* Dawn Stoppiello and I had our first encounter with a Digital Intervention. We started with a simple premise: use looped choreography to make a dance. Since loops are fundamental to all computer algorithms, this seemed an almost obvious choice for two creators focused on digital dance. We had no idea that this decision would result in a seminal and long-lasting shift in our working process.

To begin our research, I created a special looping tool in my software Isadora that could impose looping structures on any recorded media. Generally speaking, the musical loops with which we are most familiar are "fixed": the starting and ending points are always the same. The looping tool offered new possibilities, allowing one to specify not only how many times the loop would repeat, but also how the "loop points" would change over time. For instance, you could instruct the tool to "make the starting point 0.05 seconds later" or "make the loop 0.1 seconds shorter" after each repetition. Finally, the tool allowed one to sequence hundreds of these instructions. The tool allowed one to compose not with musical or movement phrases, but with loops.

After innumerable experiments, looping videos of Dawn's choreography and recordings of my music, we came to the following plan:

- Create a five-minute long dance, and a five-minute long music to accompany it.
- Make a video recording of this "base material".
- Use the looping tool to impose complex looping structures on the video recording.
- Render the result.

While using the tool to compose *loopdiver,* Dawn and I discovered it was impossible to work alone. Every change she made in the loop structure to accommodate the choreography fundamentally affected the musical result; every change I made to serve the musical composition had the same effect on her choreography. We had always been close collaborators, but never had we worked as intimately as we did on this project. The system simply demanded that we compose everything together.

The rendered video of this looped dance and music composition was about 45 minutes long and contained nearly 3,000 edits. But it was the next step that was most critical, and which seems the most mad in retrospect: we gave the rendered video to our six dancers and said, "the looped video is your score: memorize what you see, and perform the movement exactly as you see it in the video."

What was the first word out of their mouths? "Impossible!" But, just as Cédric Andrieux had done for Merce Cunningham, they did not relent. They began working arduously to accomplish the goal that we had put before them. After a long period of intense rehearsal, they had not only memorized the material, but were moving in a way that mimicked the score very, very closely indeed.

The resulting piece was surprising in several ways. First was the simple fact that the dancers actually were able to memorize the entire score. (Dawn and I were never sure about this until some weeks into rehearsal.) Second, the choreography and music were totally different from anything we had made previously; the dictates of the system caused us to create in a completely new way. Third, and most importantly, was the feeling that we experienced upon watching the piece. We later described the dancers as being locked in a "prison of repetition". It was difficult to watch them move in this inhuman way, and the desire to see them break free was both immediate and strong. If there was any ecstasy in this piece, it was most definitely earned.

loopdiver changed us as artists because we embraced the Digital Intervention. We adopted a rigorous process of composition that was, down at its core, technological in nature. The system itself fundamentally *intervened* in the creation and realization of this piece. Dawn and I were in a state of constant negotiation because our métiers were locked to each other by the loop structure. The dancers had to invent new systems – each one quite personal – to allow them to memorize a vast amount of material that was, on first inspection, impossible to memorize. And, burdened by the laws of physics, they had to struggle to stay true to the edits in the video that were simply not possible for a real body to realize.

For us, *loopdiver* was a game changer. And it has led me to believe that the Digital Intervention offers powerful answers to the problems I have identified with the Digital Reflection.

First, it immediately generated conflict because the technology had become a quite brutal antagonist. Just as it had for Cédric Andrieux when faced with Cunningham's impossible LifeForms score, the digitally inspired system we imposed upon ourselves in *loopdiver* forced all of us to come face-to-face with a series of impossible instructions. As a choreographer and a composer, Dawn's and my commitment to rigorously following the rules of the system led us to repeatedly make numerous novel and surprising choices. It was the same with the dancers. Their solutions to what they saw in the score forced them to move in ways that they had not previously imagined. For everyone concerned, the limitation imposed upon us became the central point of conflict. It made things harder for us, not easier. And upon finishing the piece, it became clear that this conflict was visible on the stage. It was part of the dramaturgical conversation.

Second, the "sensory system" for this piece – the dancers – was high bandwidth and very, very sensitive. The dancers took on precisely the same role that the sensors and computer play in the Digital Reflection. They looked at the incoming data – the score – and followed the rules of the system to determine what output they would produce. Take the fact that they brought their personalities and entire life history to the table, and you've easily got the most sophisticated "instrument" on the planet – one that is better than a violin, and most certainly superior to any computer-based interactive system that has yet been devised.

Finally, and related to the previous point: they were extremely smart. As I described a moment ago, Dawn and I were constantly surprised at the inventive ways the dancers found to strictly follow the rules of the system while simultaneously finding ways to move that expressed their personality. But, even better than that, they did something no computer has ever done to me:

Sometimes, they tried to cheat.

Not only did they try to cheat, they did so in incredibly clever ways. In the end, our dramaturg, Peter von Salis, often had to take on the role of the "enforcer", because it wasn't just the dancers who did this. Dawn and I also tried to sneak past the system more than once. These attempts to circumvent the system were extremely significant, because they demonstrate how badly the ego wanted to be released from the constraints imposed upon it.

It was interesting to show *loopdiver* to fans of Troika Ranch's previous works. "Where's the media? Where's the interaction?" they would ask. The thing is, *loopdiver* was by far the most mediated piece we've ever made. Every step of the choreography had passed through a system that was digital by nature. And the conflict we encountered in trying to live up to that system was visible from start to finish. For me, it was a much deeper media piece than any that Dawn and I had made before.

Solutions

Well, as you can see, I am a champion of the Digital Intervention. I am also discouraged by what we have achieved so far with works that rely on the model of the Digital Reflection. But I wouldn't come here today and be so cranky without at least offering a potential solution.

But first, let me say this: I would ask you all to consider how the notion of a Digital Intervention could enhance your creative practice. Again, you don't have to take my word for it. Merce Cunningham and John Cage were two of the highest practitioners of the art of intervention; and we know the inventive and wonderful results they achieved. We even heard today from Sally Jane Norman about Oskar Schlemmer's "malice of the object" (see Chapter 1, this volume) and the fact that he was creating choreographic instructions that were essentially a computer program – more interventions, I suspect. Digital systems are the perfect source material for such an endeavour, and the creativity of human beings provides us with the pinnacle of sensitivity, an understanding of context and the ability to respond creatively within the strict limits of a system. For me, the Digital Intervention is rich territory that I intend to continue exploring.

As for the Digital Reflection, I think the essential problem is that the systems that have come so far are just not sensitive enough. I hope that any of you who are developing sensory systems would aspire to become the Stradivarius of Digital Sensing, taking the violin as your model. Everything you do should be measured against this centuries-old instrument. Don't try to make these systems smart, because it's not necessary. (The violin is, in and of itself, quite "dumb".) Instead, your systems must be *extremely* sensitive to human gesture; they must provide a rich variation in the output that is directly tied to variations at the input, and they must ensure that the output for a given input should be both reliable and repeatable.

The latter point is critical because it rewards practice. In Malcolm Gladwell's book *Outliers* (2011), he posits that if one spends 10,000 hours with an instrument, he or she will become a master of it. It is probably impractical to ask a dancer, who has already invested 10,000 hours or more becoming a virtuoso of his or her body to invest a second 10,000 to master a sensory system whose long-term future is unsure at best. But, somehow, I think we need to develop new instruments so compelling and robust that someone would at least be *tempted* do to so.

Beyond the model of the violin, there's a further step I think we can take. Here I would mention again my colleague Thecla Schiphorst and the *Moving Stories* project at Simon Fraser University. The project website describes *Moving Stories* as an "interdisciplinary, collaborative research partnership leading to the design and application of more richly articulated human movement knowledge within digital technology interaction".[2] There is a key quote from their site I want to share with you. It takes low-level measurements – like position and acceleration – to task: "What has been underexplored is how these properties can be used to recognize higher-context qualities of human movement for creating a semantics of expressive motion."

Hear, hear, I say! As far as I can see, the holy grail of digital sensing technology, beyond pure sensitivity, is to get these systems to have some sense of context – to be able to sense the quality of a movement, and provide an appropriate interactive response. If a computer system could sense movement within the context of recent movements, or, better yet, within the context of all the other visual and aural materials on stage, I believe we would be thrust into an entirely new position of exploration. Why? Because we'd be genuinely surprised by the responses that the interactive system would give us. Think about the times when you've been powerfully surprised in your life: you're thrown off balance, so you immediately become hyper-aware of your situation and start trying to adapt. In other words, surprises are excellent interventions. Like the impossible instruction, they leave one no choice but to think creatively in the moment.

What I'm proposing here is not easy. In fact, it is also an "impossible problem" – one that I hope will inspire surprising creativity in you. But of this I'm sure: if the Digital Reflection is to have a long-lasting impact, if it is to inspire new ways of composing and performing, if it is to fundamentally shift how a generation of artists view their creative practice, the digital sensory systems you are inventing need to do one very simple thing: they need to want – and to be able – to *cheat.* On the day when we see a sensing system trying to get around the rules that we've imposed upon it, I think we can all agree that we will be very surprised indeed. The sensors will intervene at the same time as they reflect. And that will mark the beginning of a new and important era in digital performance practice.

Notes

1. Dawn Stoppiello in personal conversation with the author (2009).
2. The *Moving Stories* website is available at http://www.movingstories.ca (accessed 18 July 2014).

Bibliography

Acocella, J. (2009) Walking on the Moon. Michael Jackson in Motion. *New Yorker*, 27 July, p. 77.

Adams, P. (1998) Bruce Nauman and the Object of Anxiety. *October*, pp. 96–113. Available at: http://browse.reticular.info/text/collected/october/Bruce%20Nauman%20and%20the%20Object%20of%20Anxiety.pdf (accessed 3 May 2014).

Adorno, T. W. (1997 [1970]) *Aesthetic Theory* (trans. R. Hullot-Kentor). London: Athlone Press.

Agamben, G., and C. Casarino (2000) Notes on Gesture. In G. Agamben, *Means without End* (trans. V. Binetti). Minneapolis and London: University of Minnesota Press.

Alderson, E. (1991) *Proceedings of the ADUCC (Association for Dance in Universities and Colleges in Canada) Conference*, Concordia University, Montreal, October.

Appadurai, A. (1996) *Modernity at Large: Cultural Dimensions of Globalization*. Minneapolis: University of Minnesota Press.

Araujo, A. (2007) Introduction: A Pattern Constellation. *Haecceity Papers, University of Sydney*, Vol. 3, Issue 1, pp. 9–22. Available at: http://sydney.edu.au/sup/journals/haecceity/pdfs/3/Intro_3_1.pdf (accessed 25 March 2015).

Armitage, T. (2009) Toiling in the Data-mines: What Data Exploration Feels Like. Available at: http://berglondon.com/blog/2009/10/23/toiling-in-the-data-mineswhat-data-exploration-feels-like/ (accessed 3 September 2014).

Artaud, A. (1958) *The Theatre and Its Double* (trans. Mary Caroline Richards). New York: Grove Press.

Ashton, K. (2009) That "Internet of Things" Thing. *RFID Journal*. Available at: http://www.rfidjournal.com/articles/view?4986 (accessed 4 February 2014).

Austin, J. L. (1976) *How To Do Things with Words*. Oxford: Oxford University Press.

Azari, F. (1970 [1919]) Futurist Aerial Theatre: Flight as an Artistic Expression of States of Mind. The Drama Review 15 (1), pp. 128–30.

Baecker, R. M. (1974) Genesys – Interactive Computer-mediated Animation. In J. Halas (ed.) *Computer Animation*. New York: Hastings House, pp. 97–115.

Bailes, K. E. (1977) Alexei Gastev and the Soviet Controversy over Taylorism 1918–24. *Soviet Studies*, 29 (3), pp. 373–94.

Balla G., B. Cappa, F. Depero, G. Dottori, Fillia, F. T. Marinetti, E. Prampolini, M. Somenzi, and Tato (2009 [1929]) Manifesto of Aeropainting. In L. Rainey, C. Poggi, and L. Wittman (eds.) *Futurism: An Anthology*. New Haven and London: Yale University Press.

Barad, K. (2003) Posthumanist Performativity: Toward an Understanding of How Matter Comes to Matter. *Signs: Journal of Women in Culture and Society* 28 (3), pp. 801–31.

Baudry, J-L. (1978) *L'Effet cinema*. Paris: Albatros.

Beacham, R. (2006a) *Adolphe Appia: Künstler und Visionär des Modernen Theaters*. Berlin: Alexander Verlag.

Beacham, R. (2006b) Bearers of the Flame: Music, Dance, Design, and Lighting, Real and Virtual – the Enlightened and still Luminous Legacies of Hellerau and Dartington. *Performance Research*, 11 (4), pp. 81–94.

Benjamin, W. (2003 [1939]) The Work of Art in the Age of Technological Reproducibility. In H. Eiland, and M. W. Jennings (eds) *Selected Writings Volume 4 (1938–40)*. Cambridge, MA: Belknap Press of Harvard University Press, pp. 251–83.

Bennett, J. (2010) *Vibrant Matter: A Political Ecology of Things*. Durham: Duke University Press.

Bergson, H. (1922) *Creative Evolution* (trans. Arthur Mitchell). New York: University Press of America.

Bermúdez Pascual, B. (2013) (Capturing) Intention: The Life of an Interdisciplinary Research Project. *International Journal of Performance Arts & Digital Media*, 9 (1), pp. 61–81.

Bernstein, N. (1927) Kymozyklographion, ein neuer Apparat für Bewegungsstudium. In *Pflüger's Archiv für die gesamte Physiologie des Menschen und der Tiere*, 217 (1), pp. 782–92.

Bernstein, N. (1983) The Techniques of the Study of Movement. In H. T. A. Whiting (ed.) *Human Motor Actions: Bernstein Reassessed*. Amsterdam: Elsevier, pp. 77–120.

Bernstein, N. (1999 [1929]) Movements (trans. Mark Latash). *Motor Control*, 3 (2), pp.105–34.

Bethel, C. L., and R. R. Murphy (2008) Survey of Non-facial/Non-verbal Affective Expressions for Appearance-Constrained Robots. *IEEE Transactions on Systems, Man, and Cybernetics, Part C – Applications and Reviews*, 38 (1), pp. 83–92.

Bevilacqua, Frédéric (2007) Momentary Notes on Capturing Gestures. In S. deLahunta (ed.) *Capturing Intention*. Amsterdam: Emio Greco | PC and Amsterdam School of the Arts, pp. 26–31.

Bevnet (2008) SoBe Lizards Take Manhattan by Storm: Unleashing SoBe Life Water Thrillicious Movement in Times Square. Available at: http://www.bevnet.com/news/2008/02-27-2008-SoBe.asp (accessed 10 September 2014).

Birringer, J. (1998) *Media and Performance: Along the Border*. Baltimore: Johns Hopkins University Press.

Birringer, J. (2008) After Choreography. *Performance Research*, 13 (1), pp. 118–22.

Birringer, J. (2009) *Performance, Technology and Science*. New York: PAJ Publications.

Birringer, J. (2013) Bauhaus, Constructivism, Performance. *PAJ: A Journal of Performance and Art*, 104, pp. 39–52.

Birringer, J., and M. Danjoux (2013) The Sound of Movement Wearables: Performing *UKIYO*. *Leonardo*, 46 (3), pp. 232–40.

Blanchette, J-F. (2011) A Material History of Bits. *Journal of the American Society for Information Science and Technology*, 62 (6), pp.1042–57.

Blom, L. A., and L. T. Chaplin (1982) *The Intimate Act of Choreography*. Pittsburgh: University of Pittsburgh Press.

Blom, L. A., and L. T. Chaplin (1988) *The Moment of Movement: Dance Improvisation*. Pittsburgh: University of Pittsburgh Press.

Boden, M. A. (1998) Creativity and Artificial Intelligence. *Artificial Intelligence*, 103 (1), pp. 347–56.

Bodish, E. (2009) Cubism and the Fourth Dimension. *TIME*, 6 (3), pp. 527–40.

Boehner, K., P. Sengers, and S. Warner (2008) Interfaces with the Ineffable: Meeting Aesthetic Experience on Its Own Terms. *ACM Transactions on Computer-Human Interaction*, 15 (3), pp. 1–29.

Bolt, B. (2004) *Art Beyond Representation: The Performative Power of the Image*. London: I.B. Tauris.

Bourdieu, P. (1992) *Language and Symbolic Power*. Cambridge: Polity Press.

Bowlt, J. E. (1996) Ippolit Sokolov and the Gymnastics of Labor. *Experiment*, 2 (Fall), pp. 411–22.

Bowlt, J. E. (2013) Solomon Nikritin: The Old and the New. In C. Lodder, M. Kokkori, and M. Mileeva (eds) *Utopian Reality: Reconstructing Culture in Revolutionary Russia and Beyond*. Leiden: Brill, pp. 219–38.

Bowman, C. (2003) Heidegger, the Uncanny, and Jacques Tourneur's Horror Films. In S. J. Schneider, and D. Shaw (eds) *Dark Thoughts: Philosophic Reflections on Cinematic Horror*. New York: Scarecrow Press, pp. 65–83.

Bradford, J., and P. Cote-Laurence (1995) An Application of Artificial Intelligence to the Choreography of Dance. *Computers and the Humanities*, 29 (4), pp. 233–40.

Brand, M., and A. Hertzmann (2000) Style Machines. Proceedings of the Conference on Computer Graphics and Interactive Techniques, Annual Conference Series, ACM SIGGRAPH, pp. 183–92.

Brandstetter, G., and B. Wiens (eds) (2010) *Theater ohne Fluchtpunkt: Das Erbe Adolphe Appias – Szenographie und Choreographie im zeitgenössschen Theater/Theatre without Vanishing Points: The Legacy of Adolphe Appia – Scenography and Choreography in Contemporary Theatre*. Berlin: Alexander Verlag.

Brannigan, E. (2011) *DanceFilm: Choreography and the Moving Image*. Oxford: Oxford University Press.

Braune, W., and O. Fischer (1987) *The Human Gait*. Berlin: Springer-Verlag.

Brecht, B. (1964 [1930]) The Modern Theatre Is the Epic Theatre: Notes to the Opera *Aufstieg und Fall der Stadt Mahagonny*. In J. Willett (ed./trans.) *Brecht on Theatre: The Development of an Aesthetic*. London: Methuen, pp. 33–42.

Buchanan, R. (1996) Wicked Problems in Design Thinking. In R. Buchanan., and V. Margolin (eds) *The Idea of Design*. Cambridge, MA: MIT Press, pp. 3–20.

Budd, C., P. Huang, M. Klaudiny, and A. Hilton (2013) Global Non-rigid Alignment of Surface Sequences. *International Journal of Computer Vision*, 102 (1–3): 256–70.

Burke, C. (2007) Inspiring Spaces: Creating Creative Classrooms. *Curriculum Briefing*, 5 (2), pp. 35–9.

Butler, J. (1997) *The Psychic Life of Power: Theories in Subjection*. Stanford: Stanford University Press.

Cadoz, C., and M. M. Wanderley (2000) Gesture-Music. *Trends in Gestural Control of Music*, 12, p. 101.

Calvert, G. A., and T. Thesen (2004) Multisensory Integration: Methodological Approaches and Emerging Principles in the Human Brain. *Journal of Physiology-Paris*, 98 (1), pp. 191–205.

Calvert, T., J. Landis, and J. Chapman (1979) Notation of Dance with Computer Assistance. In D. T. Taplin (ed.) *New Directions in Dance*. Toronto: Pergamon Press.

Calvert, T., C. Welman, S. Gaudet, T. Schiphorst, and C. Lee (1991) Composition of Multiple Figure Sequences for Dance and Animation. *The Visual Computer*, 7 (2), pp. 114–21.

Calvert, T., L. Wilke, R. Ryman, and I. Fox (2005) Applications of Computers to Dance. *IEEE Computer Graphics and Applications*, 25 (2), pp. 6–12.

Camurri, A., and G. Volpe (eds) (2004) *Gesture-Based Communication in Human-Computer Interaction*. Berlin: Springer.

Camurri, A., S. Hashimoto, and M. Ricchetti, M. (2000) Eyesweb: Toward Gesture and Affect Recognition in Interactive Dance and Music Systems. *Computer Music Journal*, 24 (1), pp. 57–69.

Candy, L. (2007) Constraints and Creativity in the Digital Arts. *Leonardo* 40 (4) pp. 366–7.

Canetti, E. (1981) *Massa e Potere*. Milano: Adelphi.

Carifio, J., and R. Perla (2008) Resolving the 50-year Debate around Using and Misusing Likert Scales. *Medical Education*, 42 (12), pp. 1150–2.

Carlson, K., T. Schiphorst, and P. Pasquier (2011) Generating Movement Catalysts for Computer-Aided Choreography. Proceedings of the Second International Conference on Computational Creativity, pp. 123–28.

Carlson, K., T. Schiphorst, and C. Shaw (2011) A Visualization Tool for Contemporary Dance Analysis. *Computational Aesthetics in Graphics, Visualization, and Imaging*, pp. 113–20.

Carrol, E., and C. Latulipe (2009) The Creativity Support Index. *CHI '09 Extended Abstracts on Human Factors in Computing Systems*, New York, ACM, pp. 4009–14.

Carter, A., and R. Fensham (eds) (2011) Dancing Naturally: Nature, Neo-classicism and Modernity in the Early Twentieth Century. Basingstoke: Palgrave Macmillan.

Carter, P. (2004) *Material Thinking: The Theory and Practice of Creative Research*. Carlton: Melbourne University Publishing.

Châtelet, G. (2000) *Figuring Space: Philosophy, Mathematics, and Physics* (trans. R. Shaw, and M. Zagha). Dordrecht: Kluwer Academic Publishers.

Chen, J. (2007). Flow in Games (and Everything Else). *Communications of the ACM*, 50 (4), pp. 31–4.

Chernyshevsky, N. (1989) *What Is To Be Done?* (trans. Michael R. Katz.) Ithaca, NY: Cornell University Press.

Cloud-Dance (2012) http://www.cloud-dance-festival.org.uk/reviews/place-prize-semifinal-3.html (accessed 12 September 2013).

Cocker, E. (2013) Live Notation: Reflections on a Kairotic Practice. *Performance Research: A Journal of the Performing Arts*, 18 (5), pp. 69–76.

Coniglio, M. (2006) Material vs Content in Digitally Mediated Performance. In S. Broadhurst, and J. Machon (eds) *Performance and Technology: Practices of Virtual Embodiment and Interactivity*. Basingstoke: Palgrave Macmillan, pp. 78–84.

Coyne, R. (2010) *The Tuning of Place: Sociable Spaces and Pervasive Digital Media*. Cambridge, MA: MIT Press.

Cramer, F. (2013) Post-digital, a Term That Sucks But Is Useful. Available at: http://post-digital.projects.cavi.dk/?p=295 (accessed 4 October 2013).

Cranny-Francis, A. (2013) *Technology and Touch: The Biopolitics of Emerging Technologies*. Basingstoke: Palgrave Macmillan.

Cropley, A. J. (2000) Defining and Measuring Creativity: Are Creativity Tests Worth Using? *Roeper Review*, 23 (2), pp. 72–9.

Crowther, P. (1993) *Art and Embodiment: From Aesthetics to Self-Consciousness*. Oxford: Oxford University Press.

CSAILS (2011) Seraph: Dancing with Robots. Performance available at: http://www.csail.mit.edu/node/1640 (accessed 21 January 2014).

Csikszentmihalyi, M. (1997) *Flow and the Psychology of Discovery and Invention*. New York: Harper-Perennial.

Cunningham, M. (1955) The Impermanent Art. *7 Arts*, 3, pp. 69–77.

Cunningham, M. (1968) *Changes: Notes on Choreography*. New York: Something Else Press.

Cuykendall, S., E. Soutar-Rau, K. Cochrane, J. Freiberg, and T. Schiphorst (2015) Simply Spinning: Extending Current Design Frameworks for Kinesthetic Empathy. TEI '15 Proceedings of the Ninth International Conference on Tangible, Embedded, and Embodied Interaction, New York, ACM, pp. 305–12.

Cvejić, B., and A. Vujanović (2012) *Public Sphere by Performance*. Berlin: b_books.

Cybernetic Serendipity, Computer-programmed Choreography (1967) Archive of ICA. 1968 performance available at: http://cyberneticserendipity.net/ (accessed 3 November 2014).

Czompo, A. I. (1974) Dance: An Illiterate Art? *Journal of Health, Physical Education, Recreation*, 45 (2), pp. 73–6.

Dahl, S., and R. Bresin (2001) Is the Player More Influenced by the Auditory than the Tactile Feedback from the Instrument? Proceedings of the COST G-6 Conference on Digital Audio Effects (DAFX-01), December 6–8. Limerick, Ireland.

Dance with Drones (2014) Performance created by Daito Manabe. Performed at Spiral Hall, March. Available at: http://www.daito.ws/en/work/elevenplay_drones.html (accessed 1 August 2014).

Dautenhahn, K. (2007) Socially Intelligent Robots: Dimensions of Human-Robot Interaction. *Philosophical Transactions of the Royal Society*, 362: 679–704.

Davidson, D. (2003) What Is Creativity? Interview with University of California, Davis Dance Professor Della Davidson. Available at: http://www.youtube.com/watch?v=A7Bei0VSO3w (accessed 21 October 2014).

De Freitas, E., and N. Sinclair (2014) *Mathematics and the Body: Entanglements in the Classroom*. Cambridge: Cambridge University Press.

De Hulst, E. v., and M. Hoogenboom (1994) An Interview with Norman Frisch: Just Deal With It. *Theaterschrift*, 5 (6), pp. 150–80.

deLahunta, S. (2002) Software for Dancers: Coding Forms. *Performance Research Journal*, 7 (2), pp. 96–102.

deLahunta, S., and N. Z. Shaw (2008) Choreographic Resources Agents, Archives, Scores and Installations. *Performance Research*, 13 (1), pp. 131–3.

DeLanda, M. (2000) Deleuze, Diagrams, and the Genesis of Form. *Amerikastudien*, 45 (1), pp. 33–41.

Deleuze, G. (1992a) Postscript on the Societies of Control. *October*, 59, pp. 3–7.

Deleuze, G. (1992b) What Is a Dispositif? In T. Armstrong (ed.) *Michel Foucault: Philosopher*. New York: Routledge, pp. 159–68.

Deleuze, G. (1997 [1986]) *Cinema 1: The Movement-Image* (trans. H. Tomlinson, and B. Habberjam). Minneapolis: University of Minnesota Press.

Deleuze, G. (2001) *Difference and Repetition*. London and New York: Continuum.

Deleuze, G. (2004) *The Logic of Sense*. London and New York: Continuum.

Deleuze, G., and F. Guattari (2000) *Anti-Oedipus. Capitalism and Schizophrenia*. London and New York: Continuum.

Deleuze, G., and F. Guattari (2002 [1987]) *A Thousand Plateaus. Capitalism and Schizophrenia*. London and New York: Continuum.

Delsarte, F., and G. Stebbins (1887) *Delsarte System of Expression*. New York: Werner.

Demers, L. P. (2010) Machine Performers: Neither Agentic Nor Automatic. Paper presented at the 5th ACM/IEEE International Conference on Human-Robot Interaction, Osaka, March 2–5.

Diserens, C. (curator/ed.), in collaboration with C. R.Schlemmer (1999) *Oskar Schlemmer*. Marseille: Musées de Marseille, Réunion des Musées Nationaux.

Dissanayake, E. (1992) *Homo Aestheticus: Where Art Comes From and Why*. New York: Free Press, Macmillan Inc.

Dixon, S. (2007) *Digital Performance: A History of New Media in Theater, Dance, Performance Art, and Installation*. Leonardo Book Series. Cambridge, MA and London: MIT Press.

DIYdrones (2013) Project website available at: http://www.diydrones.com (accessed 1 November 2013).

Dolar, M. (2008) The Enjoying Machine. amber08: Art and Technology Festival, Istanbul: BIS, pp. 136–47.

Dontcheva, M., et al. (2003) Layered Acting for Character Animation. *ACM Transactions on Graphics*, 22 (3), pp. 409–16.

Époque, M., and D. Poulin (2008) NoBody Dance: An On-screen Choreography Introducing a "Dance without Body" Paradigm. TeDance: Perspectives on Technologically Expanded Dance (D. Tercio, Faculdade de Motricidade Humana, Université de Lisbonne), pp. 75–87.

Époque, M., and D. Poulin (2010) NoBody Danse: un Sacre du Printemps 3D en Infochorégraphie de Particules pour l'écran. In *Esthétique des Arts Médiatiques: Ensemble Ailleurs*. Ste-Foy: Presses de l'Université du Québec, pp. 181–94.

Époque, M., and D. Poulin (2013) La Présence du Danseur Sans Corps: Effet ou Réalité? In *Personnage Virtuel et Corps Performatif: Effets de Présence*. Ste Foy: Presses de l'Université du Québec, Collection Esthétique, pp. 233–52.

Époque, M., D. Poulin, and N. Marcy (2007) Le Projet NoBody Danse: une Recherché Creation Particulière. OLATS. Available at: http://www.olats.org/livresetudes/etudes/noBodyDanse.php (accessed 1 March 2014).

Evens, A. (2010) Digital Ontology and Example. In P. Gaffney (ed.) *The Force of the Virtual: Deleuze, Science, and Philosophy*. Minneapolis: University of Minnesota Press, pp. 69–202.

Fdili Alaoui, S., F. Bevilacqua, B. Bermudez, and C. Jacquemin (2013) Dance Interaction with Physical Model Visualization Based on Movement Qualities. *International Journal of Arts and Technology* (IJART), 6 (4), pp. 1–12.

Fensham, R. (2013) Choreographic Archives: Towards an Ontology of Movement Images. In R. Gade, and G. Borggren (eds) *Performing the Archive/Archiving Performance*. Museum Terculanum Press, pp. 146–62.

Flight Assembled Architecture (2012) Installation by Raffaello D'Andrea. Available at: http://raffaello.name/projects/flight-assembled-architecture/ (accessed 1 December 2012).

Flocon, A., and P. Soupault (1987) *Scénographies au Bauhaus, Dessau 1927–1930. Hommage à Oskar Schlemmer en Plusieurs Tableaux*. Paris: Garamont/Archimbaud.

Floridi, L. (2003) Informational Realism. In *Selected Papers from the Computers and Philosophy Conference* (CAP2003). Canberra: CAP2003, pp. 7–12.

Floridi, L. (2009) Against Digital Ontology. *Synthese*, 168 (1), pp. 151–78.

Flyfire (2010) SENSEable city project website available at: http://senseable.mit.edu/flyfire/ (accessed 1 December 2010).

Forsythe, W. (1999) *Improvisation Technologies. A Tool for the Analytical Dance Eye*. Interactive CD-ROM and book developed in collaboration with V. Kuchelmeister. Karlsruhe: Zentrum für Kunst und Medientechnologie.

Forsythe, W., M. Palazzi, and N. Zuniga Shaw (2000) One Flat Thing, Reproduced. Available at: http://synchronousobjects.osu.edu (accessed 14 July 2014).

Foster, S. L. (2011) *Choreographing Empathy: Kinesthesia in Performance*. London and New York: Routledge.

Fourmentraux, J. P. (2008) Œuvrer en commun. Dilemmes de la création interdisciplinaire négociée. *Négociations*, 2 (10), pp. 25–39.

Franko, M. (1989) Repeatability, Reconstruction and Beyond. *Theatre Journal*, 41 (1), pp. 56–74. Available at: http://www.jstor.org/stable/3207924 (accessed 25 March 2015).

Franko, M. (1995) *Dancing Modernism/Performing Politics*. Bloomington: Indiana University Press.

Freud, S. (2003 [1919]) *The Uncanny* (trans. D. McLintock). New York and London: Penguin.

Fülöp-Miller, R. (1927) *The Mind and Face Of Bolshevism*. London: Putnam and Sons.

Gallagher, S. (2005) *How the Body Shapes the Mind*. Oxford: Clarendon Press.

Galloway, A. (2011) Are Some Things Unrepresentable? *Theory, Culture & Society*, 28 (7–8), pp. 85–102.

Gastev, A. K. (undated) Как надо работать (сборник) (*How to Work: Selected Writings* by A. K Gastev). Available at: http://ruslit.traumlibrary.net/book/gastev-kak-nado-rabotat/gastev-kak-nado-rabotat.html#work006002 (accessed 10 June 2014).

Gauvreau, C. (2007) Entrevues: Chorégraphes de la Danse Sans Corps. *Journal L'UQAM*, 34 (7), p. 12.

Gerovitch, S. (2007) New Soviet Man Inside Machine: Human Engineering, Spacecraft Design, and the Construction of Communism. *Osiris*, 22 (1), pp. 135–57.

Ghedini, F., and M. Bergamasco (2010) Robotic Creatures: Anthropomorphism and Interaction in Contemporary Art. IEEE Ro-Man 2010 International Symposium in Robot and Human Interactive Communication, September 12–15, Viareggio, Italy.

Gladwell, M. (2011) *Outliers: The Story of Success*. New York: Back Bay Books.

GRASP (2012) Quadromotor Show at Saatchi and Saatchi New Director's Showcase, Cannes Lions Festival, 24 June. Available at: https://www.grasp.upenn.edu/success_story/meet_your_creator_quadrotor_show (accessed 1 December 2012).

Gray, J. (1984) Dance in Computer Technology: A Survey of Applications and Capabilities. *Interchange*, 15 (4), pp. 15–25.

Grotowski, J. (1968) *Towards a Poor Theatre*. New York: Simon and Shuster.

Guerin, A. (2010) Cue the Flying Robot Fairies! An Exploration of an Interdisciplinary *A Midsummer Night's Dream*. Available at: http://www.theatreanddance.txstate.edu/department/t-o-c/article-2.html (accessed 10 October 2013).

Gunkel, D., and P. Taylor (2014) *Heidegger and the Media*. Cambridge, UK and Boston: Polity Press.

Hackney, P. (1998) *Making Connections: Total Body Integration through Bartenieff Fundamentals*. Australia: Gordon and Breach Pub.

Hagendoorn, I. (2008) Emergent Patterns in Dance Improvisation and Choreography. In A. Minai, and Y. Bar-Yam (eds) *Unifying Themes in Complex Systems IV. Proceedings of the Fourth International Conference on Complex Systems*. Berlin and Heidelberg: Springer, pp. 183–95.

Haley, R. I. (1968) Benefit Segmentation: A Decision-Oriented Research Tool. *Journal of Marketing*, 32 (3), pp. 30–5.

Hall, S. (2004) *Culture, Media, Language: Working Papers in Cultural Studies 1972–79*. London: Routledge.

Hallnäs, L., L. Melin, and J. Redström (2002). Textile Displays: Using Textiles To Investigate Computational Technology as Design Material. *Proceedings of the Second Nordic Conference on Human-computer Interaction*, Aarhus, Denmark, pp. 157–66: ACM.

Halprin, D. (2002) *The Expressive Body in Life, Art, and Therapy: Working with Movement, Metaphor and Meaning*. London: Jessica Kingsley Publishers.

Hansen, L. A. (2013) Making Do & Making New: Performative Moves into Interaction Design. *International Journal of Performance Arts and Digital Media*, 9 (1), pp. 135–51.

Hansen, M. (2000) *Embodying Technesis: Technology Beyond Writing*. Ann Arbor: University of Michigan Press.

Hansen, M. (2006) *Bodies in Code: Interfaces with New Media*. New York: Routledge.

Hansen, M. (2015) *Feed-Forward. On the Future of 21st Century Media.* Chicago: University of Chicago Press.
Harman, G. (2002) *Tool-being: Heidegger and the Metaphysics of Objects.* Chicago: Open Court.
Harman, G. (2009) *Prince of Networks: Bruno Latour and Metaphysics.* Prahran, Victoria: Re.press.
Harman, G. (2013a) Materialism Is Not the Solution. Lecture presented at AIAS Guest Lecture in AIAS Auditorium, Aarhus.
Harman, G. (2013b) Seventy-Six Theses on Object-Oriented Philosophy. In *Bells and Whistles: More Speculative Realism.* Winchester, UK: Zero Books.
Harrison, C., J. Horstman, G. Hsieh, and S. E. Hudson (2012) Unlocking the Expressivity of Point Lights. Proceedings of the 30th Annual SIGCHI Conference on Human Factors in Computing Systems, CHI '12, Austin, Texas, May 5–10. New York, ACM, pp. 1683–92.
Heidegger, M. (1977 [1954]) *The Question Concerning Technology and Other Essays* (trans. W. Lovitt). New York and London: Harper Perennial.
Heidegger, M. (2007 [1962]) *Being and Time* (trans. J. Macquarrie, and E. Robinson). Malden, MA and Oxford: Blackwell Publishing.
Herbison-Evans, D. (1988) Dance, Video, Notation, Computers. *Leonardo*, 21 (1), pp. 45–50.
Hewitt, A. (2005) *Social Choreography: Ideology as Performance in Dance and Everyday Movement.* Durham: Duke University Press.
Holland, J. H. (1992) Genetic Algorithms. *Scientific American*, 267, pp. 62–72.
Hörtner, H. (2012) Spaxels, Pixels in Space: A Novel Mode of Spatial Display. Paper Presented at the 9th ICETE International Joint Conference on E-business and Telecommunications, Rome, July 24–7.
Humphrey, D. (2003) *The Art of Making Dances* (4th edition). Highstown, New Jersey: Princeton Book Company.
I-CARE-US (2012) http://projectoicareus.wordpress.com/ (accessed 28 October 2014).
Igarashi, T., et al. (2005) Spatial Keyframing for Performance-driven Animation. In Proc. ACM SIGGRAPH/Eurographics Symposium on Computer Animation, pp. 107–16.
Ingold, T. (2000) *The Perception of the Environment: Essays on Livelihood, Dwelling and Skill.* London and New York: Routledge.
Ingold, T. (2011) *Being Alive: Essays on Movement, Knowledge and Description.* London and New York: Routledge.
Jacob, M., A. Zook, and B. Magerko (2013) Viewpoints AI: Procedurally Representing and Reasoning about Gestures. Proceedings of DiGRA. Available at: http://adam.cc.gatech.edu/wpcontent/uploads/2013/10/Jacob.DIGRA13.Viewpoints.AI_.pdf (accessed 3 November 2014).
Jarry, A. (1965) *Selected Works of Alfred Jarry* (ed./trans. R. Shattuck, and S. Watson Taylor). New York: Grove Press.
Johansson, G. (1973) Visual Perception of Biological Motion and a Model for Its Analysis. *Percept. Psychophys.* 14, pp. 201–11.
Johnston, C. M. (1986) *The Creative Imperative.* Berkeley: Celestial Arts.
Johnstone, K. (1981) *IMPRO: Improvisation and the Theatre.* London: Methuen.
Ju, W., and L. Takayama (2009) Approachability: How People Interpret Automatic Door Movement as Gesture. *International Journal of Design*, 3 (2), pp. 1–10.
Kac, E. (1997) Foundation and Development of Robotic Art. *Art Journal*, 56 (3), pp. 60–7.

Kaldrack, I. (2011) Die Bauhaustänze Oskar Schlemmers. Bewegungs-Apparatur, dynamisierter Raum und mathematische Erkenntnisform. In *Maske und Kothurne, Internationale Beiträge zur Theater-, Film- und Medienwissenschaft*, 56 (1–2), pp. 123–40.

Kapijimpanga, S. (2005) Professional animator. Personal communication with Michael Neff.

Kaptein, M. C., et al. (2010) Powerful and Consistent Analysis of Likert-type Rating Scales. In *Proceedings of the SIGCHI Conference on Human Factors in Computing Systems*, CHI '10, New York, ACM, pp. 2391–4.

Kattenbelt, C. (2008) Intermediality in Theatre and Performance: Definitions, Perceptions and Medial Relationships. *Cultura, Lenguaje y Representación: La Intermedialidad*, 6, pp. 19–29.

Keay-Bright, W., and J. Gethin-Lewis (2011) Co-Creating Tools for Touch: Applying an Inspire-Create-Play-Appropriate Methodology for the Ideation of Therapeutic Technologies. Paper Presented at Include 2011, Royal College of Art, London, 18 April.

Keefe, D., D. Acevedo Feliz, T. Moscovich, D. H. Laidlaw, and J. J. LaViola Jr (2001) Cavepainting: A Fully Immersive 3D Artistic Medium and Interactive Experience. Proceedings of ACM Symposium on Interactive 3D Graphics. Available at: http://www.moscovich.net/tomer/papers/cavepainting.pdf (accessed 3 November 2014).

Keersmaeker, A. T. D., and B. Cvejic (2012) *A Choreographer's Score: Fase, Rosas danst Rosas, Elena's Aria, Bartok*. New Haven and London: Yale University Press.

Kirsh, D. (2013) Embodied Cognition and the Magical Future of Interaction Design. *ACM Transactions on Computer-Human Interaction*, 20 (1), pp. 1–30.

Knight, H. (2011) Eight Lessons Learned about Non-verbal Interactions through Robot Theater. Proceedings of the Third International Conference on Social Robotics, ICSR '11, pp. 42–51.

Kovar L., M. Gleicher, and F. Pighin (2002) Motion Graphs. *ACM Transactions on Graphics*, 21 (3), pp. 473–82.

Kozel, S. (2007) *Closer: Performance, Technologies, Phenomenology*. Cambridge, MA: MIT Press.

Krauthausen, K. (2010) Paul Valery and Geometry: Instrument, Writing Model, Practice. *Configurations*, 18 (3), pp. 231–49.

Kravchenko, A. (2010) Taylor and Gastev. *Expert*, 18 (May). Available at: http://expert.ru/expert/2010/18/teylor_i_gastev/ (accessed 20 June 2014).

Kumar, V. (2012) Quatrotors Perform the James Bond Theme. Presented at TED2012 Conference in Long Beach, California, 29 Februar. Available at: https://www.youtube.com/watch?v=_sUeGC-8dyk (accessed 1 December 2012).

Kwastek, K. (2013) *Aesthetics of Interaction in Digital Art*. Cambridge, MA: MIT Press.

Kwinter, S. (2008) *Far from Equilibrium. Essays on Technology and Design Culture*. Barcelona and New York: Actar.

Laban, R. (1947) *Effort*. London: Macdonald & Evans.

Laban, R. (1980 [1955]) *The Mastery of Movement* (4th edition). Plymouth: Macdonald & Evans.

Lamb, W. (1965) *Posture and Gesture: An Introduction to the Study of Physical Behaviour*. London: Duckworth.

Lansdown, J. (1978) The Computer in Choreography. *Computer* 11 (8), pp. 19–30.

Lapointe, F. (2005) Choreogenetics: The Generation of Choreographic Variants through Genetic Mutations and Selection. *The 2005 Workshops on Genetic and Evolutionary Computation*. ACM, pp. 366–9.

Lapointe, F., and M. Époque (2005) The Dancing Genome Project: Generation of a Human-Computer Choreography Using a Genetic Algorithm. *ACM International Conference on Multimedia*. ACM, pp. 555–8.

Lash, S. and Lury, C. (2007) *Global Culture Industry: The Mediation of Things*. Malden MA: Polity.

Laszlo, J., M. Neff, and K. Singh (2000) Interactive Control for Physically-based Animation. *Proceedings of SIGGRAPH*, pp. 201–8.

Laszlo, J., M. Neff, and K. Singh (2005) Predictive Feedback for Interactive Control of Physics-based Characters. Computer Graphics Forum (Eurographics), 24 (3), pp. 257–66.

Latour, B. (1992) Where Are the Missing Masses? The Sociology of a Few Mundane Artifacts. In Wiebe E. Bijker, and John Law (eds) *Shaping Technology/Building Society: Studies in Sociotechnical Change*. Cambridge, MA: MIT Press, pp. 225–58.

Latour, B. (2002) Gabriel Tarde and the End of the Social. In P. Joyce (ed.) *The Social in Question: New Bearings in History and the Social Sciences*. London: Routledge, pp. 117–32.

Latour, B., and E. Hermant (2006) Paris Ville Invisible. Available at: http://www.bruno-latour.fr/sites/default/files/downloads/viii_paris-city-gb.pdf. (accessed 14 October 2014).

Lazar, J., J. H. Feng, and H. Hocheiser (2010) *Research Methods in Human-Computer Interaction*. Chichester: Wiley.

Lazaridès, A. (2006) Merveilleuse Machine! *Cahiers de Théâtre JEU, Danser aujourd'hui*, 119 (2), pp. 57–60.

LeCoq, J. (2002) *The Moving Body: Teaching Creative Theatre* (trans. David Bradby). New York: Routledge.

Leder, D. (1990) *The Absent Body*, Chicago: University of Chicago Press.

Lee, W-G. (2008) Value, Performance and Narrative. *Journal of English Studies in Korea*, 15, pp. 179–211.

Lee, W-G. (2011) Culture and Finance: Cultural Logic of Financialization and the Performativity of Financial Crisis in Korea. *Discourse 201*, 14 (2), pp.105–35.

Lehman, A. L., and B. Richardson (curators/eds) (1986) *Oskar Schlemmer*. Baltimore: Baltimore Museum of Art.

Lenin, V. I. (1919) *The Soviets at Work: The International Position of the Russian Soviet Republic and the Fundamental Problems of the Socialist Revolution*. New York: Rand School of Social Sciences.

Lenoir, T. (1998). *Inscribing Science: Scientific Texts and the Materiality of Communication*. Stanford, CA: Stanford University Press.

Lepecki, A. (2007) Choreography as Apparatus of Capture. *TDR: The Drama Review*, 51 (2), pp. 119–23.

Leroi-Gourhan, A. (1993) *Gesture and Speech*. Cambridge, MA: MIT Press.

Lewis, D. (1971) Analog and Digital. *Noûs*, 5 (3), pp. 321–7.

Lillemose, J. (2008). Conceptualizing Materiality – Art from the Dematerialization of the Object to the Condition of Immateriality. Available at: http://umintermediai501.blogspot.co.uk/2008/01/conceptualizing-materiality-art-from.htmlries of Intermedia. N.p (accessed 23 July 2013).

Llinás, R. (2002) *I, of the Vortex: From Neurons to Self*. Cambridge, MA: MIT Press.

Loke, L., and T. Robertson (2007) Making Strange with the Falling Body in Interactive Technology Design. *Proceedings of Design and Semantics of Form and Movement* (DeSForM 2007), eds L. Feijs, S. Kyffin, and B. Young, Newcastle, UK.

Louppe, L. (ed.) (1994) *Traces of Dance: Drawings and Notations of Choreographers*. Paris: Editions dis voir.

Louppe, L. (2010) *Poetics of Contemporary Dance*. Alton, Hampshire: Dance Books.

Lu, D. (2012) Ontology of Robot Theatre. Paper presented at the ICRA workshop Robotics and Performing Arts: Reciprocal Influences. St Paul, Minnesota, May 2012.

Lury, C. (2014) *Inventive Methods: The Happening of the Social*. New York and London: Routledge.

Lyons, L., et al. (2012) Don't Forget about the Sweat: Effortful Embodied Interaction in Support of Learning. *Proceedings of the Sixth International Conference on Tangible, Embedded and Embodied Interaction*, TEI. New York, ACM, pp. 77–84.

Lyotard, J. (1984) *The Postmodern Condition: A Report on Knowledge*. Manchester: Manchester University Press.

MacKenzie, D. (2008) *An Engine, Not a Camera: How Financial Models Shape Markets*. Cambridge, MA: MIT Press.

Mäki-Patola, T. (2005) Musical Effects of Latency. *Suomen Musiikintutkijoiden*, 9, pp. 82–5.

Manning, E. (2009) *Relationscapes: Movement, Art, Philosophy*. Cambridge, MA: MIT Press.

Manning, E., and B. Massumi (2014) *Thought in the Act. Passages in the Ecology of Experience*. Minneapolis: University of Minnesota Press.

Manovich, L. (1995) The Labor of Perception. Available at: http://www.egs.edu/faculty/lev-manovich/articles/the-labor-of-perception/ (accessed 19 June 2014).

Marey, E. J., and E. Pritchard (1895) *Movement*. London: William Heinemann.

Marks, L. U. (2000) *The Skin of the Film: Intercultural Cinema, Embodiment, and the Senses*. Durham: Duke University Press.

Martin, K. (2014) 3D Simulations and the Two Sided Dress: Realizing Historic Fashion. Drexel Digital Museum Project. Available at: http://digimuse.westpha.drexel.edu/publicdrexel/index.php (accessed 5 March 2014).

Martin, T., and M. Neff (2012) Interactive Quadruped Animation. Motion in Games: *Proceedings of 5th International Conference, MIG 2012*, Rennes, France, pp. 208–19.

Massumi, B. (2002). *Parables for the Virtual: Movement, Affect, Sensation*. Durham: Duke University Press.

Massumi, B. (2011) *Semblance and Event. Activist Philosophy and the Occurrent Arts*. Cambridge, MA: MIT Press.

Maur, K. v. (1982) *Oskar Schlemmer der Maler, der Wandgestalter, der Plastiker, der Zeichner, der Graphiker, der Bühnengestalter, der Lehrer*. Munich: Prestel-Verlag.

Mazalek, A., S. Chandrasekharan, M. Nitsche, T. Welsh, P. Clifton, A. Quitmeye, F. Peer, F. Kirschner, and D. Athreya (2011) I'm in the Game: Embodied Puppet Interface Improves Avatar Control. *Proceedings of the Fifth International Conference on Tangible, Embedded, and Embodied interaction*, TEI. New York, ACM, pp. 129–36.

McKinney, J., M. Wallis, S. Popat, J. Bryden, and D. Hogg (2009) Embodied Conversations: Performance and the Design of a Robotic Dancing Partner. Undisciplined! Design Research Society Conference, Sheffield Hallam University.

McNeill, D. (1992) *Hand and Mind: What Gestures Reveal about Thought*. Chicago: University of Chicago Press.

McNeill, W. H. (1997) *Keeping Together in Time. Dance and Drill in Human History*. Cambridge: Harvard University Press.

Meredith, M. (2008) Intro. In T. Sakamoto, A, Ferrè, and M. Kubo (eds) *From Control to Design. Parametric/Algorithmic Architecture*. Barcelona and New York: Actar.

Merleau-Ponty, M. (2002 [1962]) *Phenomenology of Perception* (trans. Colin Smith). London: Routledge.

Metheny, E. (1968) *Meaning and Movement*. New York: McGraw Hill.

Michailidis, T., and J. Bullock (2011) Improving Performers' Musicality through Live Interaction with Haptic Feedback: A Case Study. *Proceedings SMC '11, 8th Sound and Music Computing Conference*, Padova, Italy.

Michaud, E. (1978) *Oskar Schlemmer. Théâtre et abstraction*. Lausanne: Editions L'Age d'Homme.

Misler, N. (1991) Designing Gestures in the Laboratory of Dance. In N. van Norman (ed.) *Theatre in Revolution: Russian Avant-Garde Stage Design 1913–1935*. London: Thames & Hudson, pp. 157–73.

Moholy-Nagy, L., et al. (1961) Die Bühne im Bauhaus. In W. Gropius (ed.) *The Theatre of the Bauhaus*. Middletown: Wesleyan University Press, pp. 17–48.

Mori, M. (2003 [1970]) The Uncanny Valley (trans. K. MacDorman, and T. Minato). *Energy*, 7 (4), pp. 33–5.

Morris, C. (ed.) (2006) *9 Evenings Reconsidered: Art, Theatre, and Engineering, 1966*. Cambridge, MA: MIT List Visual Arts Center.

Moten, F. (2003) *In the Break. The Aesthetics of the Black Radical Tradition*. Minneapolis: University of Minnesota Press.

Moynihan, D. S., and L. G. Odom (1984) Oskar Schlemmer's "Bauhaus Dances": Debra McCall's Reconstructions, *The Drama Review*, 28 (3), pp. 46–58.

Munster, A. (2006) *Materializing New Media: Embodiment in Information Aesthetics*. Hanover, NH: University Press of New England.

Murphy, R., D. Shell, B. Duncan, B. Fine, K. Pratt, A. Hopper, and T. Zourntos (2010) Paper presented at the workshop: What Do Collaborations with the Arts Have To Say about Human-Robot Interaction? Human-Robot Interaction Conference HRI2010, Osaka, Japan, March.

Nakazawa, M., and A. Paezold-Ruehl (2009) DANCING, Dance and Choreography: An Intelligent Nondeterministic Generator. Richard Tapia Celebration of Diversity in Computing Conference. ACM, pp. 30–4.

Nauman, B. (1966) Failing to Levitate in the Studio [Black and white photograph].

Neff, M., I. Albrecht, and H-P. Seidel (2007) Layered Performance Animation with Correlation Maps. *Computer Graphics Forum (Eurographics)*, 26 (3), pp. 675–84.

Negri, A. (2004) It's a Powerful Life: A Conversation on Contemporary Philosophy. *Cultural Critique*, 57, pp. 153–81.

Neumann, J. v. (1958) *The Computer and the Brain*. New Haven and London: Yale University Press.

Ng, K. (2002) Sensing and Mapping for Interactive Performance. *Organised Sound*, 7 (2), pp. 191–200.

Nissenbaum, H. (2001) How Computer Systems Embody Value. *Computer*, March, pp. 118–20. Available at: http://www.nyu.edu/projects/nissenbaum/papers/embodyvalues.pdf (accessed 3 November 2014).

Nissenbaum, H., M. Flanagan, and D. C. Howe (2008) Embodying Value in Technology: Theory and Practice. In J. van den Hoven, and J. Weckert (eds) *Information Technology and Moral Philosophy*. Cambridge: Cambridge University Press.

Noë, A. (2004) *Action in Perception*. Cambridge, MA: MIT Press.

Noland, C. (2009a) *Agency and Embodiment: Performing Gestures/Producing Culture*. Cambridge, MA: Harvard University Press.

Noland, C. (2009b) Coping and Choreography. *Proceedings of the Digital Arts and Culture Conference – After Media: Embodiment and Context*. UC Irvine, California,

December 12–15. Available at: http://www.escholarship.org/uc/item/0gq729xq (accessed 3 November 2014).

Noll, A. M. (1967) Choreography and Computers. *Dance Magazine*, 41 (1), pp. 43–5.

Norman, S-J. (2001) Corps/espaces interactifs. In C. Rousier (ed.) *Oskar Schlemmer, l'homme et la figure d'art*. Paris: Centre National de la Danse, pp. 152–64.

Norman, S-J. (2014) Grappling with Movement Models: Performing Arts and Slippery Contexts. *Proceedings of the 2014 International Workshop on Movement and Computing*, Paris, France, pp. 136–41.

Norman, S-J. (in press) Between Grammatisation and Live Movement Sampling. In M. Bleeker (ed.) *Transmission in Motion*. Cambridge, MA: MIT Press.

O'Hara, K., R. Harper, H. Mentis, A. Sellen, and A. Taylor (2013) On the Naturalness of Touchless: Putting the Interaction back into NUI. *ACM Transactions on Computer-Human Interactions*, 20 (1), pp. 1–25.

Ong, W. (1989) *Orality and Literacy. The Technologizing of the Word*. London and New York: Routledge.

Oore, S., D. Terzopoulos D., and G. Hinton (2002a) A Desktop Input Device and Interface for Interactive 3D Character Animation. Available at: http://www.graphicsinterface.org/proceedings/2002/219/paper219.pdf (accessed 3 November 2014).

Oore, S., D. Terzopoulos D., and G. Hinton (2002b) Local Physical Models for Interactive Character Animation. *Computer Graphics Forum*, 21 (3), pp. 337–46.

OpenEnded Group (2014). Available at: http://www.openendedgroup.com/about.html (accessed 1 February 2014).

Paiva A., I. Leite, and T. Ribeiro (forthcoming) Emotion Modeling for Social Robots. In R. A. Calvo, S. K. D'Mello, J. Gratch, and A. Kappas (eds) *Handbook of Affective Computing*. Oxford: Oxford University Press.

Panko, J. (2005) Professional animator. Personal communication with Michael Neff.

Parikka, J. (2010) Movement-Objects Track and Establish Connections Across Scales. Available at: http://mediacommons.futureofthebook.org/imr/2010/04/25/movement-objects (accessed 10 September 2014).

Pchelkina, L. (2013) The Biomechanics of Voice and Movement in Solomon Nikritin's Projection Theatre (1920s). In D. Zakharine, and N. Meise (eds) *Electrified Voices: Medial, Socio-Historical and Cultural Aspects of Voice Transfer*. Gottingen: V&R Unipress GmbH.

Phelan, P. (1995) Thirteen Ways of Looking at Choreographing Writing. In S. L. Foster (ed.) *Choreographing History*. Bloomington: Indiana University Press, pp. 200–10.

Place Prize (2013) Copter. Choreography and Performance by Nina Kov. Available at: http://www.theplaceprize.com/nina-kov (accessed 12 September 2013).

Plucker, J.A., and J. S. Renzulli (1999) Psychometric Approaches to the Study of Human Creativity. In R. J. Sternberg (ed.) *Handbook of Creativity*. Cambridge: Cambridge University Press, pp. 35–61.

Polydorou, D. (2011) *Immersion and Interaction: Creating Virtual 3D Worlds for Stage Performances*. PhD thesis, Brunel University.

Popper, F. (2007) *From Technological to Virtual Art*. Cambridge. MA: MIT Press.

Portanova, S. (2014) Infinity in a Step: On the Compression and Complexity of a Movement of Thought. *Inflexions*, 1.1. Available at: www.inflexions.org (accessed 2 January 2014).

Poster, M. (2002) High-Tech Frankenstein, or Heidegger Meets Stelarc. In J. Zylinska (ed.) *The Cyborg Experiments: The Extensions of the Body in the Media Age*. London and New York: Bloomsbury, pp. 15–32.

Poulin, D. (2013) Les Espaces infochorégraphiques d'une Danse sans Corps. Doctoral thesis, Études et pratiques des arts, Université du Québec à Montréal. Available at: http://www.archipel.uqam.ca/5979/ (accessed 28 October 2014).

Rabinbach, A. (1990). *The Human Motor: Energy, Fatigue, and the Origins of Modernity*. Oakland, CA: University of California Press.

Raunig, G. (2010) *A Thousand Machines* (trans. Aileen Derieg). Los Angeles: semiotext(e).

Reeves, N., E. Poncet, J. Nembrini, A. Martinoli, and A. Winfield (2005). [VOILES|SAILS], Self-Assembling Intelligent Lighter-than-air Structures. *Proceedings of 8th International Conference on Generative Art*, Milan, December.

Reilly, K. (2013) *Theatre, Performance and Analogue Technology: Historical Interfaces and Intermedialities*. Basingstoke: Palgrave Macmillan.

Reynolds, D., and M. Reason (2012) *Kinesthetic Empathy in Creative and Cultural Practices*. Bristol and Chicago: Intellect.

Rhizomatiks (2014). Available at: http://www.daito.ws/en/work/elevenplay_drones.html (accessed 16 March 2015).

Richardson, B. (1986) The Nimbus of Magic: An Album of Schlemmer's Stage Work. In A. L. Lehman, and B. Richardson (curators/eds) *Oskar Schlemmer*. Baltimore: Baltimore Museum of Art, pp. 161–94.

Rotman, B. (1993 [1987]) *Signifying Nothing: The Semiotics of Zero*. Stanford: Stanford University Press.

Rotman, B. (2000) *Mathematics as Sign: Writing, Imagining, Counting*. Stanford, CA: Stanford University Press.

Rotman, B. (2002) The Alphabetic Body. *Parallax*, 8 (1), pp. 92–104.

Rotman, B. (2008) *Becoming Beside Ourselves: The Alphabet, Ghosts, and Distributed Human Being*. Durham: Duke University Press.

Rotman, B. (2012) Topology, Algebra, Diagrams. *Theory, Culture & Society*, 29 (4/5), pp. 1–14.

Rotman, B., and N. Salazar Sutil (2009) Choreographing Category Theory: Ordinal 4. Performance presented as part of the Performing Topology colloquium, Lockwood Studio, Goldsmiths College, March. Available at: http://www.salazarsutil.net/#!gallery/cff9 (accessed 3 November 2014).

Runco, M.A., and S. O. Okuda Sakamoto (1999) Experimental Studies of Creativity. In R. J. Sternberg (ed.) *Handbook of Creativity*. Cambridge: Cambridge University Press, pp. 62–92.

Salazar Sutil, N. (2011) Binary Coded Performance. *Body, Space and Technology Journal*, 10 (2) Available at: http://people.brunel.ac.uk/bst/vol1002/nicolassalazarsutil/home.html (accessed 3 September 2014).

Salazar Sutil, N. (2015) *Motion and Representation: The Language of Human Movement*. Cambridge, MA: MIT Press.

Salmon, W. C. (2001) *Zeno's Paradoxes*. Indianapolis: Hackett.

Salter, C. (2009) Environments, Interactions and Beings: The Ecology of Performativity and Technics. In M. Chatzichristodoulou, J. Jeffries, and R. Zerihan (eds) *Interfaces of Performance*. Farnham, UK: Ashgate, pp. 27–42.

Sawchuk, A. A., E. Chew, R. Zimmermann, C. Papadopoulos, and C. Kyriakakis (2003) From Remote Media Immersion to Distributed Immersive Performance. *Proceedings of the 2003 ACM SIGMM Workshop on Experiential Telepresence*. Berkeley, CA: ACM.

Scheper, D. (1988) *Oskar Schlemmer, Das Triadische Ballett und die Bauhausbühne*. Berlin: Schriftenreihe der Akademie der Künste, Band 20.

Schiphorst, T. (1993 [1986]) A Case Study of Merce Cunningham's Use of the LifeForms Computer Choreographic System in the Making of Trackers. MA Thesis, B. G. S. Simon Fraser University. Available at: http://www.sfu.ca/~tschipho/publications/Schiphorst_M.A.Thesis.pdf (accessed 2 November 2014).

Schiphorst T. (1997) Body Noise: Subtexts of Computers and Dance. *Computer Graphics*, 31 (1), pp. 14–15.

Schiphorst, T., and S. Kozel (2002) Whisper: Wearable Body Architectures. Available at: http://www.fondation-langlois.org/html/e/page.php?NumPage=46 (accessed 2 June 2014).

Schiphorst, T., T. Calvert, C. Lee, C. Welman, and S. Gaudet (1990) Tools for Interaction with the Creative Process of Composition. CHI '90 Proceedings, Seattle, April.

Schlemmer, O. (1922) Playbill for opening night of *The Triadic Ballet*, Württembergisches Landestheater, Stuttgart, September 30, 1922. Reproduced in the exhibition *Oskar Schlemmer*, Marga Paz (curator), Museo Nacional Centro de Arte Reina Sofia, Madrid, 1996–7. Playbill and accompanying text by C. R.Schlemmer available at: www.museoreinasofia.es/sites/default/files/salas/informacion/206_oskar_schlemmer_ing.pdf (accessed 14 July 2014).

Schlemmer, O. (1925) Mensch und Kunstfigur. In W. Gropius (ed.) *The Theater of the Bauhaus* (trans. A. S. Wensinger). Middletown, Connecticut: Wesleyan University Press, pp. 16–46.

Schlemmer, O. (1929) Bühnenelemente. Lecture reproduced in D. Mahlow (curator/ed.) (1965) *Bild und Bühne, Bühnenbilder der Gegenwart und Retrospective*. Baden-Baden: Staatliche Kunsthalle, pp. 10–27.

Schlemmer, O. (1971 [1928]) *Man. Teaching Notes from the Bauhaus* (trans. J. Seligman/ed. H. Kuchling). London: Lund Humphries.

Schlemmer, O. (1972 [1958]) *The Letters and Diaries of Oskar Schlemmer*. Selected and edited by T. Schlemmer, trans. K.Winston. Middletown, Connecticut: Wesleyan University Press. (German edition *Oskar Schlemmer: Briefe und Tagebücher*, 1958, Munich: Albert Langen.)

Schlemmer, O. (1976 [1926]) The Mathematics of the Dance (trans. J. Stein). In H. Wingler (ed.) *The Bauhaus*. Cambridge, MA: MIT Press, pp. 118–9.

Schlemmer, O. (1978 [1928]) Neue Formen der Bühne. In E. Michaud, *Oskar Schlemmer. Théâtre et abstraction*. Lausanne: Editions L'Age d'Homme, pp. 1062–73.

Schön, D. A. (1983) *The Reflective Practitioner: How Professionals Think in Action*. New York: Basic Books.

Sederberg, T.W., and S. R. Parry (1986) Free-form Deformation of Solid Geometric Models. *SIGGRAPH Computer Graphics*, 20 (4), pp. 151–60.

Serra, R. (1968) *Hand Catching Lead* (16mm film transferred to DVD 3:02 min., b&w, silent). Museum of Modern Art, New York.

Serra, R. (1973) *Surprise Attack* (video, b&w, sound). Museum of Modern Art, New York.

Shapiro, A., Y. Cao, and P. Faloutsos (2006) Style Components. *Proceedings of Graphics Interface 2006*, pp. 33–9.

Sheets-Johnstone, M. (1999a) Emotion and Movement: A Beginning Empirical-Phenomenological Analysis of their Relationship. *Journal of Consciousness Studies*, 6 (11–12), pp. 259–77.

Sheets-Johnstone, M. (1999b) *The Primacy of Movement*. Amsterdam: John Benjamin.

Shinkle, E. (2008) Video Games, Emotions and the Six Senses. *Media, Culture and Society*, 30 (6), pp. 907–15.

Shneiderman, B. (2007) Creativity Support Tools: Accelerating Discovery and Innovation. *Communications of the ACM*, 50 (12), pp. 20–32.

Shusterman, R. (2008) *Body Consciousness: A Philosophy of Mindfulness and Somaesthetics*. New York: Cambridge Univesity Press.

Simmel, G. (1902) The Number of Members as Determining the Sociological Form of the Group: I and II. *American Journal of Sociology*, 8 (1), pp. 1–46.

Simmel. G. (2004) *The Philosophy of Money*. London: Routledge.

Simon, H. A. (1981) *The Sciences of the Artificial*. Cambridge, MA: MIT Press.

Sklar, D. (2008) Remembering Kinesthesia: An Inquiry into Embodied Cultural Knowledge. In C. Noland, and S. A. Ness (eds) *Migrations of Gesture*. Minneapolis: University of Minnesota Press, pp. 85–111.

Smirnov, A. (2013) *Sound in Z: Experiments in Sound and Electronic Music in Early 20th Century Russia*. London: Koenig Books.

Smith, W. (1956) Product Differentiation and Market Segmentation as Alternative Marketing Strategies. *Journal of Marketing*, 21 (1), pp. 3–8.

Sobchack, V. C. (1992) *The Address of the Eye: A Phenomenology of Film Experience*. Princeton, New Jersey: Princeton University Press.

Soga, A., B. Umino, and J. S. Longstaff (2006) Web3D Dance Composer: Automatic Composition of Ballet Sequences. In ACM SIGGRAPH Research Posters. ACM, p. 5.

Spaxels (2014). Ars Electronica FutureLab. Available at: http://www.aec.at/spaxels/ (accessed 1 August 2014).

Stark, T. (2012) Cinema in the Hands of the People: Chris Marker, the Medvedkin Group, and the Potential of Militant Film. *October*, 139, pp. 117–50.

Stern, D. N. (2010) *Forms of Vitality: Exploring Dynamic Experience in Psychology, the Arts, Psychotherapy, and Development*. Oxford: Oxford University Press.

Stern, N. (2013) *Interactive Art and Embodiment: The Implicit Body*. Canterbury: Gylphi Limited.

Sternberg, R. J. (2003) Contemporary Theories of Intelligence. *Handbook of Psychology Volume 2*, pp. 21–45.

Steyerl, H. (2012) *The Wretched of the Screen*. Berlin: e-flux, Inc/Sternberg Press.

Stites, R. (1989) *Revolutionary Dreams: Utopian Vision and Experimental Life in the Russian Revolution*. Oxford: Oxford University Press.

Stokes, P. D. (2005) *Creativity from Constraints: The Psychology of Breakthrough*. Berlin: Springer.

Sturman, D. J. (1998) Computer Puppetry. Computer Graphics and Applications. Available at: http://sse.tongji.edu.cn/ft/barnes/sturman_cga1998.pdf (accessed 4 November 2014).

Subyen, P., D. S. Maranan, K. Carlson, T. Schiphorst, and P. Pasquier (2011) Flow: Expressing Movement Quality Learning. The User in Flux Workshop at CHI '11, Vancouver, Canada.

Tarde, G. (1903) *The Laws of Imitation*. New York: Henry Holt and Company.

Tarde, G. (2012) *Monadology and Sociology*. Melbourne: re.press.

Tarver, J., and K. Bligh (1999) The Physical Art of the Performer. Workshop held at the Nightwood Theatre Studio, Toronto.

Taylor, F. W. (2005) *The Principles of Scientific Management*. Fairfield, IA: 1st World Publishing.

Terra, S. C. L., and A. Metoyer (2004) Performance Timing for Keyframe Animation. ACM SIGGRAPH/Eurographics Symposium on Computer Animation, pp. 253–8.

Thacker, E. (2009) On the Horror of Living Networks. Available at: http://www.networkpolitics.org/request-for-comments/dr-thackers-position-paper (accessed 10 September 2014).

Thorp, A., N. Owen, M. Neuhaus, and D. W. Dunstan (2011) Sedentary Behaviors and Subsequent Health Outcomes in Adults: A Systematic Review of Longitudinal Studies. *American Journal of Preventive Medicine*, 41 (2), pp. 207–15.

Thrift, N. (2005) Beyond Mediation: Three New Material Registers and Their Consequences. In D. Miller (ed.) *Materiality*. Durham, NC: Duke University Press, pp. 231–55.

Thrift, N. (2007) *Non-Representational Theory: Space, Politics, Affect*. London and New York: Routledge.

Toepfer, K. (1997) *Empire of Ecstasy: Nudity and Movement in German Body Culture, 1910–1935*. Berkeley: University of California Press.

Trimingham, M. (2004) Oskar Schlemmer's Research Practice at the Dessau Bauhaus. *Theatre Research International*, 29 (2), pp. 128–42.

Trimingham, M. (2011) *The Theatre of the Bauhaus. The Modern and Postmodern Stage of Oskar Schlemmer*. London/New York: Routledge.

Trotsky, L. (2005) *Literature and Revolution*. Chicago: Haymarket Books.

Troy, N. (1986) The Art of Reconciliation: Oskar Schlemmer's Work for the Theatre. In A. L. Lehman, and B. Richardson (curators/eds) *Oskar Schlemmer*. Baltimore: Baltimore Museum of Art, pp. 127–48.

Tsing, A. (2004) Inside the Economy of Appearances. In A. Amin, and N. Thrift (eds) *The Blackwell Cultural Economy Reader*. London: Blackwell.

txt (2009). Performance created by Fernando Galrito, Fernando Nabais and Stephan Jurgens. Available at: http://www.po-ex.net/taxonomia/materialidades/performativas/nabais-galrito-jurgens-ponto-txt (accessed 1 December 2012).

V2 (2012). Available at: http://v2.nl/archive/people/louis-philippe-demers (accessed 1 December 2012).

Varela, F. J., E. Thompson, and E. Rosch (1991) *The Embodied Mind*. Cambridge, MA: MIT Press.

Vasilescu, A. O. (2002) Human Motion Signatures: Analysis, Synthesis, Recognition. *Proceedings of the International Conference on Pattern Recognition*, Quebec City, Canada, August, Vol 3, pp. 456–60. Available at: http://alumni.media.mit.edu/~maov/motionsignatures/hms_icpr02_corrected.pdf. (accessed 1 March 2014).

Virilio, P. (1997) *Open Sky* (trans. Julie Rose). London: Verso.

Wardrip-Fruin, N. (2011) Digital Media Archaeology: Interpreting Computational Processes. In E. Huhtamo, and J. Parikka (eds) *Media Archaeology: Approaches, Applications, and Implications*. Berkeley, CA: University of California Press, pp. 302–22.

Wei, S. X. (2002) Resistance is Fertile: Gesture and Agency in the Field of Responsive Media. *Configurations*, 10 (3), pp. 439–72.

Weibel, P. (1996) The World as Interface. In T. Druckery (ed.) *Electronic Culture*. New York: Aperture, pp. 343–51.

Weibel, P. (2012) The Postmedia Condition. Available at: http://www.medialabmadrid.org/medialab/medialab.php?l=0&a=a&i=329 (accessed 17 August 2014).

Weiser, M., and J. Seely Brown (1996) The Coming Age of Calm Technology, Xerox PARC, October 5. Available at: http://homes.di.unimi.it/~boccignone/GiuseppeBoccignone_webpage/IUM2_files/weiser-calm.pdf (accessed 10 June 2014).

Whatley, S. (2012) The Poetics of Motion Capture and Visualisation Techniques: The Differences Between Watching Real and Virtual Bodies. In D. Reynolds, and M. Reason (eds) *Kinesthetic Empathy in Creative and Cultural Practices*. Bristol and Chicago: Intellect.

Wheeler, J. A. (1990) Information, Physics, Quantum: The Search for Links. In *Complexity, Entropy, and the Physics of Information: The Proceedings of the 1988 Workshop on Complexity, Entropy, and the Physics of Information*. Held in May/June, 1989, Santa Fe, New Mexico. Redwood City, CA: Addison-Wesley, pp. 309–36.

Whitehead, A. N. (1985) *Process and Reality*. New York: The Free Press.

WikiHow (2014) How to Do the Thriller. Available at: http://www.wikihow.com/Do-the-Thriller (accessed 9 September 2014).

Wilde, D., A. Cassinelli, and A. Zerroug (2012) Light Arrays. CHI '12 Extended Abstracts on Human Factors in Computing Systems, New York, ACM, pp. 987–90.

Wilke, L., T. Calvert, R. Ryman, and I. Fox (2005) From Dance Notation to Human Animation: The LabanDancer Project. *Computer Animation and Virtual Worlds*, 16 (3–4), pp. 201–11.

Winer, R. S. (2001) A Framework for Customer Relationship Management. *California Management Review*, 43 (4), pp. 85–105.

Wingler, H. M. (1976) *The Bauhaus* (ed. J. Stein/trans. W. Jabs, and B. Gilbert). 2nd adapted and enlarged edition in English (from German ed. *Das Bauhaus*, 1962, Cologne: Bramsche and M. Dumont Schauberg). Cambridge, MA: MIT Press.

Wittgenstein, L. (1980) *Culture and Value* (ed. C. H. von Wright/trans. Peter Winch). Oxford: Blackwell.

Wood, A. (2007) *Digital Encounters*. London: Routledge.

Young, K. (2011) Gestures, Intercorporeity, and the Fate of Phenomenology in Folklore. *Journal of American Folklore*, 124 (492), pp. 55–87.

Yu, T., and P. Johnson (2003) Tour Jeté, Pirouette: Dance Choreographing by Computers. *Genetic and Evolutionary Computation* – GECCO, pp. 156–7.

Zamyatin, Y. (1993) *We* (trans. Clarence Brown) London: Penguin Books.

Zimmerman, J., J. Forlizzi, and S. Evenson (2007) Research through Design as a Method for Interaction Design Research. In *HCI, CHI '07: Proceedings of the SIGCHI Conference on Human Factors in Computing Systems*, New York, ACM.

Index

Q

R

www.ingramcontent.com/pod-product-compliance
Lightning Source LLC
Chambersburg PA
CBHW060750150926
51501CB00004B/138
* 9 7 8 1 1 3 7 4 3 0 4 0 3 *